Research on Natural Bioactive Products-Based Pesticidal Agents

Research on Natural Bioactive Products-Based Pesticidal Agents

Topic Editors

Min Lv

Hui Xu

MDPI

Basel • Beijing • Wuhan • Barcelona • Belgrade • Novi Sad • Cluj • Manchester

Topic Editors

Min Lv
School of Marine Sciences
Ningbo University
Ningbo
China

Hui Xu
School of Marine Sciences
Ningbo University
Ningbo
China

Editorial Office
MDPI AG
Grosspeteranlage 5
4052 Basel, Switzerland

This is a reprint of the Topic, published open access by the journals *Agronomy* (ISSN 2073-4395), *Insects* (ISSN 2075-4450) and *Toxins* (ISSN 2072-6651), freely accessible at: https://www.mdpi.com/topics/pesticidal_agents.

For citation purposes, cite each article independently as indicated on the article page online and as indicated below:

Lastname, A.A.; Lastname, B.B. Article Title. *Journal Name* **Year**, *Volume Number*, Page Range.

ISBN 978-3-7258-7534-4 (Hbk)
ISBN 978-3-7258-7535-1 (PDF)
https://doi.org/10.3390/books978-3-7258-7535-1

Contents

Ruirui Zheng, Jinyu Zhao, Li Ma, Xingtao Qie, Xizhong Yan and Chi Hao
Behavioral, Electrophysiological, and Toxicological Responses of *Plutella xylostella* to Extracts from *Angelica pubescens*
Reprinted from: *Insects* **2023**, *14*, 613, https://doi.org/10.3390/insects14070613 1

Jianwei Xu, Min Lv, Shanshan Fang, Yanyan Wang, Houpeng Wen, Shaoyong Zhang and Hui Xu
Exploration of Synergistic Pesticidal Activities, Control Effects and Toxicology Study of a Monoterpene Essential Oil with Two Natural Alkaloids
Reprinted from: *Toxins* **2023**, *15*, 240, https://doi.org/10.3390/toxins15040240 17

Yongqing Tian and Zhixiang Zhang
Insecticidal Activities of *Sophora flavescens* Alt. towards Red Imported Fire Ants (*Solenopsis invicta* Buren)
Reprinted from: *Toxins* **2023**, *15*, 105, https://doi.org/10.3390/toxins15020105 33

Pan Li, Aijuan Zhao, Ru Li, Shibo Han, Na Li, Lusha Ji and Kang Lei
Identifying a Detoxifying Uridine Diphosphate Glucosyltransferase (UGT), MdUGT83K2, Which Can Glycosylate the Aryloxyphenoxypropionate Herbicide
Reprinted from: *Agronomy* **2023**, *13*, 306, https://doi.org/10.3390/agronomy13020306 45

Chao-Chao Wang, Ke Chen, Na Li, Xue-Kun Wang, Shi-Ben Wang, Pan Li, et al.
Discovery of 3-(1-Amino-2-phenoxyethylidene)-6-methyl-2*H*-pyran-2,4(3*H*)-dione Derivatives as Novel Herbicidal Leads
Reprinted from: *Agronomy* **2023**, *13*, 202, https://doi.org/10.3390/agronomy13010202 55

Tianze Li, Min Lv, Houpeng Wen, Yanyan Wang, Sunita Thapa, Shaoyong Zhang and Hui Xu
Synthesis of Piperine-Based Ester Derivatives with Diverse Aromatic Rings and Their Agricultural Bioactivities against *Tetranychus cinnabarinus* Boisduval, *Aphis citricola* Van der Goot, and *Eriosoma lanigerum* Hausmann
Reprinted from: *Insects* **2022**, *14*, 40, https://doi.org/10.3390/insects14010040 73

Weronika Kursa, Agnieszka Jamiołkowska, Jakub Wyrostek and Radosław Kowalski
Antifungal Effect of Plant Extracts on the Growth of the Cereal Pathogen *Fusarium* spp.—An In Vitro Study
Reprinted from: *Agronomy* **2022**, *12*, 3204, https://doi.org/10.3390/agronomy12123204 86

Meng Hao, Jianwei Xu, Houpeng Wen, Jiawei Du, Shaoyong Zhang, Min Lv and Hui Xu
Recent Advances on Biological Activities and Structural Modifications of Dehydroabietic Acid
Reprinted from: *Toxins* **2022**, *14*, 632, https://doi.org/10.3390/toxins14090632 104

Chunzhe Jin, Hui Han, Yongjian Xie, Baoling Li, Zhilin Zhang and Dayu Zhang
Toxicity, Behavioral Effects, and Chitin Structural Chemistry of *Reticulitermes flaviceps* Exposed to *Cymbopogon citratus* EO and Its Major Constituent Citral
Reprinted from: *Insects* **2022**, *13*, 812, https://doi.org/10.3390/insects13090812 133

Chaochao Wang, Ke Chen, Na Li, Shuyue Fu, Pan Li, Lusha Ji, et al.
Design, Synthesis, Mode of Action and Herbicidal Evaluation of Quinazolin-4(3*H*)-one Derivatives Based on Aryloxyphenoxypropionate Motif
Reprinted from: *Agronomy* **2022**, *12*, 1840, https://doi.org/10.3390/agronomy12081840 148

insects

MDPI

Article

Behavioral, Electrophysiological, and Toxicological Responses of *Plutella xylostella* to Extracts from *Angelica pubescens*

Ruirui Zheng, Jinyu Zhao, Li Ma, Xingtao Qie, Xizhong Yan * and Chi Hao *

College of Plant Protection, Shanxi Agricultural University, Jinzhong 030800, China;
zhengruirui2019@163.com (R.Z.); zhaojinyu312@163.com (J.Z.); mali890310@126.com (L.M.);
qxt123@nwafu.edu.cn (X.Q.)
* Correspondence: yanxizhong80@163.com (X.Y.); sxauhc@163.com (C.H.)

Simple Summary: *Plutella xylostella* is one of the most destructive insect pests affecting cruciferous vegetables. The use of various plant extracts for controlling this pest is increasing in popularity, as they are multi-bioactive, biodegradable, and ecologically safe. In this study, we evaluated the insecticidal oviposition deterrence and repellent activities of *Angelica pubescens* extract and its compounds against *P. xylostella.* The *A. pubescens* extract, caryophyllene oxide, and 3,4-dimethoxycinnamic acid, have been proven to have insecticidal and oviposition deterrent activities against *P. xylostella*. The *A. pubescens* extract and caryophyllene oxide also mediate *P. xylostella* repellence. Overall, our results highlighted the potential of *A. pubescens* extract, caryophyllene oxide, and 3,4-dimethoxycinnamic acid in controlling *P. xylostella* and in the development of effective eco−friendly "inhibit and kill" insecticide formulations.

Abstract: *Plutella xylostella* L. is a destructive pest affecting cruciferous vegetables, causing massive economic losses worldwide. Plant−based insecticides are considered promising insect control agents. The *Angelica pubescens* extract inhibited female oviposition, with an oviposition deterrence index (ODI) of 61.65% at 12.5 mg/mL. We aimed to identify the bioactive compounds in *A. pubescens* extract. The compounds from *A. pubescens* extract were analyzed using LC−MS techniques. The toxicity and behavioral responses of larvae and adults of *P. xylostella* to ten compounds were investigated. We found that the caryophyllene oxide and 3,4-dimethoxycinnamic acid inhibited female oviposition; the ODIs were 98.31% and 97.59% at 1.25 mg/mL, respectively. The *A. pubescens* extract, caryophyllene oxide, and 3,4-dimethoxycinnamic acid caused larval mortality, with LC_{50} values of 21.31, 4.56, and 5.52 mg/mL, respectively. The EAG response of females was higher than that of males under *A. pubescens* extract conditions, while the EAG response of males was higher than that of females in caryophyllene oxide and 3,4-dimethoxycinnamic acid conditions. The *A. pubescens* extract and caryophyllene oxide showed repellent activity against both female and male adults, while the 3,4-dimethoxycinnamic acid did not elicit any notable behavioral responses from *P. xylostella* adults. *A. pubescens* extract and caryophyllene oxide are potential insecticides, oviposition deterrents, and behavioral regulators against *P. xylostella*, and they could be potential candidates for the development of biological insecticides to control *P. xylostella*.

Keywords: *Plutella xylostella*; *Angelica pubescens*; extract/compounds; toxicity; repellent effect

1. Introduction

Plutella xylostella (Lepidoptera: Plutellidae) is one of the most damaging insect pests of cruciferous vegetables in global agricultural ecology [1]. *P. xylostella* adults have high fecundity and create a large larval population, and third and fourth-instar *P. xylostella* larvae have high feeding rates, causing direct damage to cruciferous vegetables and an estimated annual global economic loss of USD 4–5 billion [2]. Targeting the adult and larval stages of *P. xylostella* (i.e., using insect behavior-regulating stimuli to manipulate population structures and kill larvae) has been shown to be effective. Over the past few decades, due to excessive dependence on chemical pesticides to control *P. xylostella*, *P. xylostella* has developed a resistance to the traditional insecticides used to control it [3]. In addition, synthetic pesticides contain high levels of toxic residues, exert adverse effects on non−target organisms, and threaten human and environmental safety [4]. Thus, the development of new and effective "green" repellents and insecticides with different mechanisms of action could circumvent the current challenges in effectively controlling *P. xylostella* [5]. One benign alternative to synthetic insecticides is plant extracts and their derived compounds. Crude plant extracts could control *P. xylostella* in various ways, including via insecticidal [6] and repellent effects, growth and oviposition inhibition [7], and reducing their feeding [8]. Beyond their insecticidal properties, plant extracts have several advantages over traditional insecticides. They are generally considered safe for human and animal use, are biodegradable, and pose low toxicity risks for non-target organisms; these factors make them more environmentally acceptable than several synthetic insecticides.

Most moths rely on compounds from host plants to locate food resources and suitable oviposition sites. Compounds from non-host plants can cover and mask these host-plant-derived compounds, influencing orientation behaviors. Based on the literature, 42 different Apiaceae plant extracts have previously been demonstrated to possess biological insecticide activities, including *Foeniculum vulgare* essential oil's acute toxicity to *Myzus persicae* [9], *Trachyspermum ammi* essential oil's repellence activity to *Riptortus clavatus* [10], and *Angelica archangelica* extract, which has insecticidal, antifeedant, and growth inhibition activities against *Spodoptera littoralis* larvae [11]. The chemical compounds from Apiaceae plants, including sesquitterpenic lactones, furanocoumarins, monoterpene coumarins, polyacetylenes, volatiles pheylpropenes, and phethalides, have also been shown to possess a range of biological activities, including insecticidal, antifungal, and antioxidant properties [12,13]. Thus, Apiaceae plant extracts have considerable potential for application as biopesticides in the management of pests. *A. pubescens* belongs to the Apiaceae family, and its root is used in traditional Chinese medicine. Clinical applications have focused on its pharmacological activities, such as its analgesic, sedative, and hypnotic effects [14]. In previous studies, *A. pubescens* has been shown to have deworming [15] and insecticidal properties [16]. However, the oviposition deterrence and insecticidal activities of *A. pubescens* against *P. xylostella* and other Lepidopteran insects have not been reported.

The application of plant-extract-derived bioactive compounds as alternatives to synthetic insecticides has received considerable attention. *A. pubescens* extract contains coumarins, polyene−alkynes, phenolic acids, steroids, nucleoside elements, and other chemical compounds [15]. The main volatile compounds from the *A. pubescens* extract comprise polyene −alkynes and phenolic acids. For instance, the polyene−alkyne compounds alantolactone, isoalantolactone [17], and 1−hexene [18], and the phenolic acid compound caffeic acid, showed growth inhibiting and repellent effects against *Spodoptera litura* larvae and adults [19]. Thus, polyene−alkyne and phenolic acid compounds have potential in controlling *P. xylostella*. We studied the biological activities of *A. pubescens* extract and its constituents against *P. xylostella*, and conducted a preliminary study on its functional mechanisms.

In this study, for the first time, a hydroalcoholic extract was prepared from *A. pubescens*, a novel plant preparation with oviposition deterrence and insecticidal activities against *P. xylostella*. We used LC−MS techniques to identify and quantify the constituents of this *A. pubescens* extract. We aimed to test the following: (1) whether the constituents of *A. pubescens* extract have oviposition deterrence activity against *P. xylostella* adults and insecticidal activity against *P. xylostella* larvae (using the dual-choice oviposition deterrent bioassay and leaf dipping method); (2) the olfactory responses of *P. xylostella* to these bioactive compounds, investigated by electrophysiological tests, to assess the sensitivity of female and male antennae to the test compounds and a Y-tube olfactometer bioassay to evaluate the insects' behavioral response of the same compounds. Our findings regarding hypothesis (2) will help to further elucidate the relationship between the oviposition deterrent activity of these compounds and olfactory behavior of *P. xylostella*. This study provides preliminary data for elucidating the mechanisms of action of *A. pubescens* extract and its constituents to control *P. xylostella* adults and larvae.

2. Materials and Methods

2.1. Insects

Plutella xylostella larvae were collected from the vegetable fields of Shanxi Agricultural University Experimental Station in Taigu County, Jinzhong City, Shanxi Province, China (37°25′22″ N, 112°34′15″ E). Multi-generation larvae were successively reared in the laboratory of Shanxi Agricultural University, following the rearing protocol previously described by Yan et al. [20] and Guo et al. [21]. Larvae were fed on fresh cabbage leaves (*Brassica oleracea* (Gai Liang Qing Fen)). The moths were reared in a 30 × 40 × 60 cm cage (consisting of a plastic tube frame and fitted with insect-proof screens) and fed with a 15% honey solution. Cabbage leaves were placed in the cage to facilitate egg-laying by the moths. All pupae were placed in a 10 cm glass tube with a cotton plug to ensure access to fresh air. The insect house was maintained at a temperature of 25 ± 2 °C, relative humidity of 60 ± 10%, and photoperiod of 16:8 h (L:D). Experiments were performed on *P. xylostella* adults (24–48 h post-emergence) and on neonate uniform third instar larvae, which were starved for 4 h before experiments. The F10 generation was used for all bioassays.

2.2. Chemicals

The chemical compounds used for the bioassays were purchased from Aladdin (Shanghai, China) and Macklin (Shanghai, China) (Table S1). LC−MS grade methanol, acetonitrile, and formic acid were purchased from Merck KGaA (Germany) and Xiya Reagent (Shandong, China). Matrine (Hei Ye Ming) was purchased from Zhenge Biotechnology (Guangdong, China). 20% ethanol, 0.5% Triton−X100, and 0.5% DMSO (TDE solution) mixture was used as a solvent, and the compound test solutions and matrine were configured as 10, 5, 2.5, 1.25, and 0.63 mg/mL.

2.3. Extraction and LC−MS Analysis

Commercial *A. pubescens* root material was purchased from Renhe Pharmaceutical Co., Ltd. (Jiangxi, China) and ground to a powder using a grinding machine. As described by the extraction method of Domenico et al. [22], with slight modifications. Then, 300 g of the powder was dissolved in 600 mL of solvent (25% water, 25% 2-propanol, and 50% ethanol). This mixture was agitated overnight at room temperature (30 °C, September, Taigu, China) before sonication treatment for 30 min at 40 °C. The supernatant was filtered through filter paper with a vacuum suction filter. The sonication and filtration processes were repeated five times. The solvents were combined and evaporated under reduced pressure at 50 °C to obtain 100 mL extract solvents, which were stored in a refrigerator at 4 °C until further use.

A. pubescens extract was diluted in TDE solution to obtain 200, 100, 50, 25, and 12.5 mg/mL test solutions. Subsequently, the extract was vigorously mixed using a vortex machine for 5 min to dissolve it thoroughly.

LC−MS of the *A. pubescens* extract was performed by the Sci−Tech Innovation Quality Testing Company (Qingdao, China). The LC−MS analysis method is outlined in the Supplementary Materials (Method S1 and Table S2).

2.4. Dual-Choice Oviposition Assay

The oviposition deterrence assay method was performed as described by Yang et al. [23]. The assay was carried out using 5 sets of moths, 5 females and 15 males, which were introduced into a transparent plastic cylindrical lidded container (20 cm diameter × 8 cm height, with at least 40 pinholes for ventilation). The *A. pubescens* extract (200, 100, 50, 25, and 12.5 mg/mL) and its derived compounds of caryophyllene oxide and 3,4-dimethoxycinnamic acid (10, 5, 2.5, 1.25, and 0.63 mg/mL) were used as test solutions.

The TDE solution was used as a negative control. The 10 mg/mL matrine solution was used as the positive control. A cabbage leaf disc (6 cm in diameter) was placed into the test solution (5 mL) for 60 s, and the solvent on the leaf disc was allowed to dry naturally via evaporation. The leaf disc was placed at a diagonal end position in a clear transparent plastic cylindrical container. We placed a small piece of absorbent cotton (soaked with 15% honey) in the center of the clear transparent plastic cylindrical container. Three replicates were prepared for each test solution. The number of eggs on the cabbage leaf disc was counted after 24 h of treatment.

2.5. Insecticidal Assay

The insecticidal assay was performed using the leaf dip method, as reported by Ratnaweera et al. [24]. The *A. pubescens* extracts (200, 100, 50, 25, and 12.5 mg/mL) and derived compounds of caryophyllene oxide and 3,4-dimethoxycinnamic acid (10, 5, 2.5, 1.25, and 0.63 mg/mL) were used as test solutions. A cabbage leaf disc (6 cm in diameter) was placed into the test solution (5 mL) for 60 s, the solvent was evaporated naturally, and the disc was put into a Petri dish padded with moistened filter paper. TDE solution was used as a negative control and 10 mg/mL matrine solution was maintained as a positive control. Twenty third-instar larvae were placed in the Petri dish. Five replicates were performed for each test solution. The numbers of dead larvae were counted at 24, 48, and 72 h of treatment. The larvae were considered dead when they demonstrated paralysis, tipping, and/or immobility when touched with the bristles of a fillet-type brush.

2.6. Electrophysiological (EAG) Assay

We used the EAG technique as descried by Song et al., with some modifications [4], to test the antennal response of *P. xylostella* adults to *A. pubescens* extract (200, 100, 50, 25, and 12.5 mg/mL) and its constituents caryophyllene oxide and 3,4-dimethoxycinnamic acid (10, 5, 2.5, 1.25, 0.63 mg/mL). Ice was used to reduce the moth's mobility before excising the antenna at the base of the flagellum and cutting 2 mm from the antenna tip. The antenna was positioned between two sharpened electrode probes (reference and record electrodes) containing a conducting gel. The TDE solution was used as a control to measure the potential antenna response. We pipetted 20 μL of test solution onto a piece of 0.5 × 5 cm Whatman® filter paper (the concentrations were 1.6, 0.8, 0.4, 0.2, and 0.1 g/cm^2 and 0.08, 0.04, 0.02, 0.01, and 0.005 g/cm^2) and inserted it into a Pasteur pipet (15 cm in length). The Pasteur pipet device was connected to the stimulation controller via a silicone tube. When the baseline was stable, each antenna was calibrated with the control solution, followed by the test solutions at five concentrations from low to high, then the control again. The duration of each stimulation was 0.5 s, with 30 s intervals between successive stimulations. For the extract and each constituent sample, new sets of six virgin females and six virgin males (with two replications of each stimulus) were tested.

2.7. Y-Tube Olfactometer Assay

The behavioral response of *P. xylostella* adults to *A. pubescens* extract (200, 100, 50, 25, and 12.5 mg/mL) and its constituents caryophyllene oxide and 3,4-dimethoxycinnamic acid (10, 5, 2.5, 1.25, and 0.63 mg/mL) were tested using a Y-tube olfactometer (internal width of 24 mm × length of 20 cm (choice arm) and 29 mm × 20 cm (main arm) with a 30° angle between arms), as described by Song et al. [4]. For the treatment arm, 20 μL of test solution was applied to a piece of Whatman® filter paper (5 cm in length and 5 mm in width). For the control arm, filter paper containing 20 μL 15% honey water was provided. The *P. xylostella* adults were placed at the main arm.

Using an air sampler, atmospheric air was allowed to flow through an activated charcoal cartridge. Purified air was pushed into the Y-tube olfactometer at a rate of 300 mL/min per pump. Each arm was ventilated for 3 min to fill it with test gas prior to the experiment. During the test, a moth was considered to have made a choice when it crossed the midpoint of one of the upwind choice arms and stayed there for more than 30 s.

If a moth did not move after 5 min in the main arm, it was recorded as "no choice". To avoid position effect bias, the test solvents were switched between arms when 10 insects had made their choice. Behavioral tests were performed from 14:00 to 18:00, under red light and room-temperature conditions (25 ± 1 °C). Olfactometers were cleaned at the end of each day with bleach and water, and left to dry overnight. In total, 50 virgin males and 50 virgin females were tested for each test solvent.

2.8. Statistical Analysis

The oviposition deterrence index (ODI) of adults and the calibrated mortality rate of larvae were calculated using Abbott's formula [25]. The ODI, calibrated mortality, and EAG response data were subjected to one-way ANOVA, followed by Tukey's multiple comparison tests to determine significant differences ($p < 0.05$). The numbers of eggs in the treatment and control groups were subjected to independent-samples t-tests to determine significant differences. Probit analyses of larvae-calibrated mortality data were performed to determine the median lethal dose (LC_{50}) and corresponding 95% confidential intervals.

$$\text{Calibrated mortality} = \frac{\text{rate of mortality in the treatment} - \text{rate of mortality in the control}}{1 - \text{rate of mortality in the control}}$$

$$\text{Oviposition deterrence index} = \frac{\text{number of eggs in the treatment} - \text{number of eggs in the control}}{\text{number of eggs in the treatment} + \text{number of eggs in the control}}$$

EAG response values were calculated as $(b - a)/a \times 100$, where a is the average EAG amplitude value of the control stimulus and b is the average EAG amplitude value of treated stimulus.

For the behavioral results, the proportion of *P. xylostella* which chose the treatment and control was calculated. Chi-squared tests were used to determined significant differences. Any moths that did not make a definitive choice were not used in the statistical analysis.

3. Results

3.1. Bioassay Activity and LC−MS Analysis for A. pubescens Extract

In previous study that had tested 20 plant species, the *A. pubescens* extract demonstrated oviposition deterrent and insecticidal activities against *P. xylostella*, with an oviposition deterrence index of 96.88% at 100 mg/mL, and the larval mortality rate was 97.00% at 100 mg/mL (Table 1). The polyene−alkynes and phenolic acids were the main volatile constituents of *A. pubescens* extract. Thus, we selected ten compounds from the polyene−alkynes and phenolic acids, with contents of at least 0.01% in *A. pubescens* extract, for subsequent experiments. The retention time, composition content, and chemical structures are presented in Table 2 and Figure S1.

Table 1. Calibrated mortality of *P. xylostella* larvae to *A. pubescens* extract at five concentrations.

A. pubescens Extract Concentration	Calibrated Mortality (%)	Oviposition Deterrence Index (%)
200 mg/mL	100.00 ± 0.00 a	100.00 ± 0.00 a
100 mg/mL	97.00 ± 0.01 a	96.88 ± 0.03 ab
50 mg/mL	94.00 ± 0.04 a	69.80 ± 0.09 bc
25 mg/mL	40.00 ± 0.04 b	63.75 ± 0.08 c
12.5 mg/mL	33.00 ± 0.01 b	61.80 ± 0.08 c

Different letters represent significant differences at different concentrations (Tukey's test, $p < 0.05$).

3.2. Oviposition Deterrent Activity of 10 Compounds against Female Adults of P. xylostella

The results showed that the 10 compounds (at a concentration of 10 mg/mL) significantly ($p < 0.05$) affected the oviposition behavior of female *P. xylostella* adults. Caryophyllene oxide, nootkatone, and 4-coumaric acid showed significant oviposition deterrent activity against *P. xylostella* females, with ODI values of 100%, 100%, and 90.67% at 10 mg/mL, respectively, which were 1.3–1.5 times higher than that of the positive control matrine. The

ODI of 3,4-dimethoxycinnamic acid against *P. xylostella* females was 57.01% at 10 mg/mL, which was 0.8 times higher than that of the positive control matrine. The ODIs of other compounds against *P. xylostella* females did not exceed 50% at 10 mg/mL (Table 3).

Table 2. Chemical composition of 10 compounds (derived from the *A. pubescens* extract) used in subsequent tests against *P. xylostella*.

Sample	Retention Time (min)	Composition (%)	Sample	Retention Time (min)	Composition (%)
Artemisinin	6.337	0.163	Ethyl caffeate	9.284	0.079
4-Methylbenzylidene camphor	10.752	0.138	2-Hydroxycinnamic acid	1.207	0.074
Santonin	8.477	0.077	Caffeic acid	5.377	0.064
Caryophyllene oxide	10.959	0.072	3,4-Dimethoxycinnamic acid	11.527	0.014
Nootkatone	13.519	0.014	4-Coumaric acid	6.555	0.012

Table 3. Calibrated mortality and oviposition deterrence index of 10 compounds from *A. pubescens* extract (at a concentration of 10 mg/mL) against *P. xylostella*.

Sample	Calibrated Mortality (%)	Oviposition Deterrence Index (%)
Artemisinin	30.00 ± 0.02 e	33.66 ± 0.03 bcd
4-Methylbenzylidene camphor	67.00 ± 0.06 bc	24.45 ± 0.04 cd
Santonin	61.00 ± 0.06 cd	48.46 ± 0.06 bc
Caryophyllene oxide	85.00 ± 0.05 ab	100 ± 0.00 a
Nootkatone	26.00 ± 0.08 e	100.00 ± 0.00 a
Ethyl caffeate	34.00 ± 0.03 e	32.08 ± 0.11 bcd
2-Hydroxycinnamic acid	33.00 ± 0.07 e	38.03 ± 0.06 bcd
Caffeic acid	40.00 ± 0.10 de	38.92 ± 0.01 bcd
3,4-Dimethoxycinnamic acid	75.00 ± 0.02 bc	57.01 ± 0.02 b
4-Coumaric acid	40.00 ± 0.02 de	90.67 ± 0.09 a
Matrine	100.00 ± 0.00 a	69.16 ± 0.02 b

Different letters represent significant differences at different concentrations (Tukey's test, $p < 0.05$).

3.3. Insecticidal Activity of 10 Compounds against P. xylostella Larvae

The 72 h mortality results showed that the caryophyllene oxide and 3,4-dimethoxycinna mic acid (at 10 mg/mL) had a calibrated mortality above 75%, which was 0.75 times higher than that of the positive control matrine. In contrast, the mortality under other compounds did not exceed 55% (Table 3).

3.4. Oviposition Deterrent and Insecticidal Activities of A. pubescens Extract, Caryophyllene Oxide, and 3,4-Dimethoxycinnamic Acid against P. xylostella

Based on the results of screening assays, *A. pubescens* extract, caryophyllene oxide, and 3,4-dimethoxycinnamic acid exhibited oviposition deterrent activity against *P. xylostella* female adults and insecticidal activity against *P. xylostella* larvae. Thus, these compounds were used in subsequent experimental studies.

The LC_{50} values of the *A. pubescens* extract, caryophyllene oxide, and 3,4-dimethoxycin namic acid in the larval stages were 21.31, 4.56, and 5.52 mg/mL, respectively (Table 4). The cabbage leaves treated with 200 mg/mL *A. pubescens* extract had no eggs and the total number of eggs was reduced; as the concentration decreased, the number of eggs of *P. xylostella* increased. In addition, no eggs were found on the cabbage leaves treated with caryophyllene oxide, and the total numbers of eggs were reduced at concentrations of 10, 5, and 2.5 mg/mL. The 3,4-dimethoxycinnamic acid demonstrated oviposition deterrence activity against female *P. xylostella* at 1.25 mg/mL, which surpassed that at the higher concentrations (10 mg/mL, 5 mg/mL, and 2.5 mg/mL), but the total number of eggs was reduced at concentrations of 5, 2.5, and 1.25 mg/mL (Figure 1).

Table 4. Lethal effects of *A. pubescens* extract, caryophyllene oxide, and 3,4-dimethoxycinnamic acid against third-instar larvae of *P. xylostella*.

Sample	Regression Equation	LC_{50} in mg/mL 95% CL	Chi-Square (DF = 23)	*p* Value
A. Pubescens extract	y = −3.69 + 2.78x	21.31(16.03−26.71)	71.74	0.00
Caryophyllene oxide	y = −1.16 + 1.76x	4.56(3.55−6.26)	46.86	0.02
3,4-Dimethoxycinnamic acid	y = −0.83 + 1.12x	5.52(4.00−8.87)	32.24	0.10
Matrine	y = −1.66 + 3.05x	3.50(2.99−4.13)	43.32	0.01

Each value represents the mean of five replicates, and each experiment included 20 individuals; 95%CL means confidence interval at the 95% confidence level.

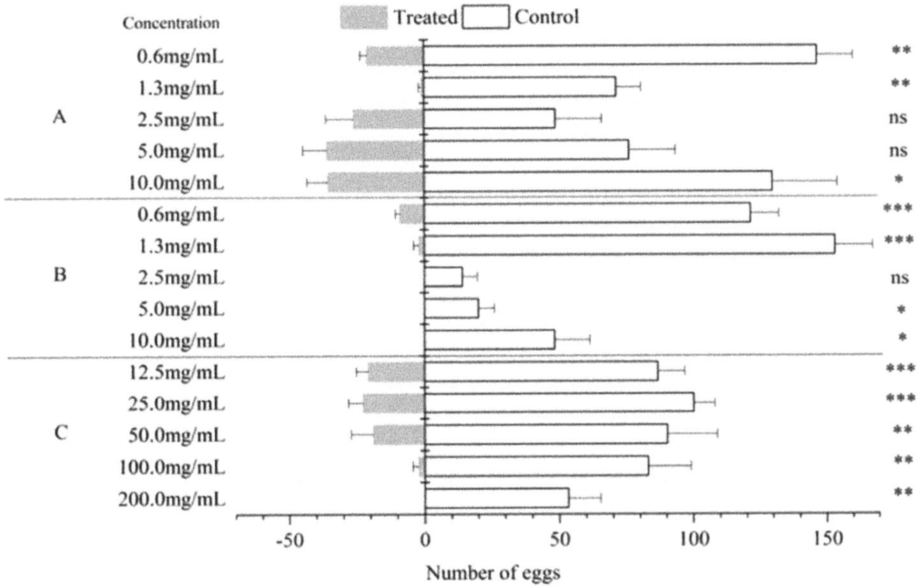

Figure 1. Number of *P. xylostella* eggs (mean ± SE) on leaves treated with five concentrations of 3,4-dimethoxycinnamic acid (**A**), caryophyllene oxide (**B**), *A. pubescens* extract (**C**), and with the control. Asterisks indicate significant differences: * = $p < 0.05$, ** = $p < 0.01$, *** = $p < 0.001$, ns = $p > 0.05$ (independent-samples *t*-test).

3.5. Differential EAG Responses to A. pubescens Extract and Individual Compounds

EAG responses were elicited by concentrations of 200, 100, 50, 25, and 12.5 mg/mL *A. pubescens* extract, concentrations of 10, 5, 2.5, 1.25, and 0.63 mg/mL caryophyllene oxide, and 3,4-dimethoxycinnamic acid treatment (Figure 2). In both sexes, the EAG responses were higher under different concentrations of caryophyllene oxide than these under 3,4-dimethoxycinnamic acid. In the *A. pubescens* extract treatment group, the EAG responses indicated that the female moths were slightly more affected than the male moths, with EAG values of 4.86 (female) and 1.13 mV (male) at 200 mg/mL, respectively. In the caryophyllene oxide treatment group, the EAG response of female adults was slightly lower than that of male adults at concentrations of 2.5, 1.25, and 0.63 mg/mL. The EAG responses of male adults treated with caryophyllene oxide at 0.63 mg/mL were higher than at the concentrations of 5, 2.5, and 1.25 mg/mL, which indicated that male receptors were more sensitive at lower concentrations. The EAG responses of *P. xylostella* treated with 3,4-dimethoxycinnamic acid showed that the female moths were less sensitive than the male moths, with EAG values of 1.00 (females) and 1.19 mV (males) at 10 mg/mL, respectively. The EAG responses of female moths treated with 3,4-dimethoxycinnamic acid did not differ significantly at 5, 2.5, 1.25, and 0.63 mg/mL, with values of 0.46, 0.59, 0.35, and 0.49 mV, respectively.

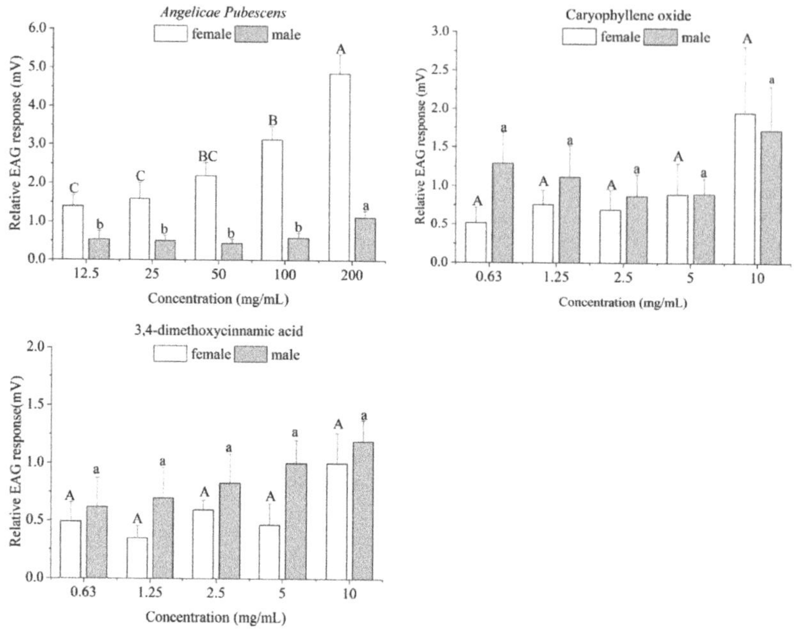

Figure 2. EAG responses (mean ± SE) of *P. xylostella* adults treated with five concentrations of *A. pubescens* extract, caryophyllene oxide, and 3,4-dimethoxycinnamic acid. Different letters represent significant differences at different concentrations (Tukey's test, $p < 0.05$).

3.6. Differential Behavioral Effects of A. pubescens Extract and Individual Compounds

The *A. pubescens* extract showed a significant repellent effect ($p < 0.05$) for both male and female adults of *P. xylostella* at a concentration of 25 mg/mL. There was a significant reduction in female *P. xylostella* who selected the *A. pubescens* extract arm ($p < 0.05$) at 100 mg/mL when compared with these which chose the control arm. Caryophyllene oxide at concentrations of 5 and 2.5 mg/mL showed a significant repellent effect for both sexes of *P. xylostella* ($p < 0.05$), and female *P. xylostella* preferred the control when the concentration was 0.63 mg/mL. Males also significantly preferred the control arm ($p < 0.05$) at 10 mg/mL. However, no significant differences in choice were observed for either sex when choosing between 3,4-dimethoxycinnamic acid at five concentrations or the control, except for the male selection control arm at 5 mg/mL (Figure 3).

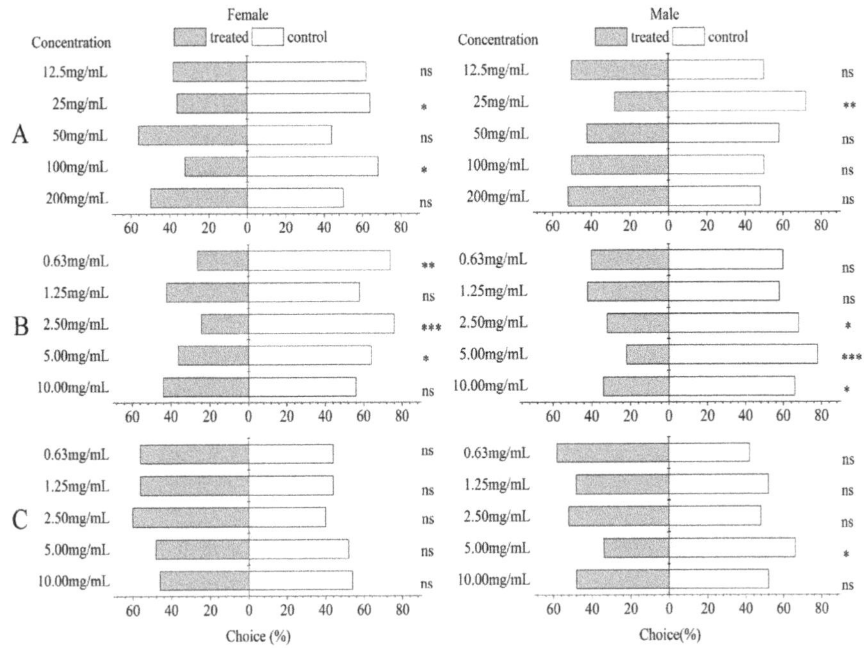

Figure 3. Behavioral responses of female and male *P. xylostella* adults to *A. pubescens* extract, caryophyllene oxide, and 3,4-dimethoxycinnamic acid in a Y-tube olfactometer. Both bars represent the percentages (percent choice n = 100) of moths that chose the treated (*A. pubescens* extract (**A**), caryophyllene oxide (**B**), and 3,4-dimethoxycinnamic acid (**C**) at five concentrations) or control arm. Asterisks indicate significant differences: * = $p < 0.05$, ** = $p < 0.01$, *** = $p < 0.001$, ns = $p > 0.05$ (Chi-squared (χ^2) test).

4. Discussion

The indiscriminate use of synthetic insecticides has caused a series of adverse effects on the environment and consumers. Novel pesticides to control *P. xylostella* are urgently needed to combat these environmental safety challenges. Previous studies have shown that plant extracts and derived bioactive compounds have a variety of biological activities to control pests, and they provide an alternative to synthetic insecticides due to their biodegradable nature and safety for non-target organisms [7,26,27].

Solvent-based (water, 2-propanol, and ethanol) sonication extraction methods have commonly been used to extract compounds from Chinese medicinal plants. This method is attractive because it avoids the use of other potentially harmful solvents, making the process more eco-friendly and safer for use in the field.

In our study, *A. pubescens* extract exhibited oviposition deterrence activity against *P. xylostella* adults, with an ODI of 61.50% at 12.5 mg/mL, and insecticidal activity against *P. xylostella* larvae, with an LC_{50} value of 21.31 mg/mL. Compared with other extracts, the ethanol/distilled water extraction of *Trillium govanianum* exhibited insecticidal activity against larvae of *P. xylostella*, with mortality not exceeding 50% at 72 h and an LC_{50} value of 11.040 mg/mL at 96 h [28]. The *Lantana camara* extracts at 8% and 10% showed ovicidal and oviposition deterrence activities, with a lower insecticidal activity against *P. xylostella* larvae [29]. The *n*-hexane fraction of *Zanthoxylum armatum* extract showed insecticidal activity against larvae of *P. xylostella* with an LC_{50} value of 29.88 mg/mL, which was similar to what was achieved in the present study [30]. Thus, *A. pubescens* extract can be used as a potential oviposition deterrent and insecticidal agent to control *P. xylostella*.

Active compounds derived from plants or plant products have important bioactive functions and potential applications in agricultural pest management [4,27,31]. They often have a similar mode of action to synthetic insecticides [32], and may be able to provide precursor compounds for synthetic pesticides. Caryophyllene oxide is a common polyene−alkyne that has been identified as one of the main constituents of various extracts with biological activities. For example, it is one of the main constituents of *Cyperus rotundus* essential oil, which showed contact toxicity and repellent activity against *Callosobruchus maculatus*, *Trogoderma granarium*, and *Oryzaephilus surinamensis* [33]. It is also a major constituent of *Lantana camara* extract, which had ovicidal activity and an oviposition deterrent effect on *P. xylostella* [29]. In addition, Huang et al. [31] reported that *Artemisia lavandulaefolia* extract and its constituent caryophyllene oxide had strong contact toxicity and fumigant activity against larvae and adults of *P. xylostella*, with LD_{50} values of 20.71 mg/mL and 1.06 mg/mL, respectively. In our study, caryophyllene oxide exhibited oviposition deterrence and insecticidal activities against *P. xylostella* adults and larvae, with an ODI of 98.31% at 1.25 mg/mL and an LC_{50} value of 4.59 mg/mL. The above findings suggested that caryophyllene oxide has multiple modes of action against various insects, including *P. xylostella*. Yulia [34] summarized the biological activity of β-caryophyllene and its oxidized form; these compounds can easily penetrate cell membranes [35], do not cause allergic reactions upon contact, even in persons with dermatitis [36], increased the cytochrome P450 activity of mice [37], and exhibited in vivo pain-alleviating and anti-inflammatory properties [38]. Thus, caryophyllene oxide could be a candidate in the development of new insecticides.

3,4-dimethoxycinnamic acid is a common phenolic acid, with medicinal value against tumor cells, methicillin-resistant *Staphylococcus aureus*, and *Plasmodium falciparum* [39]. In our study, 3,4-dimethoxycinnamic acid showed oviposition deterrence and insecticidal activities against *P. xylostella*, with an ODI of 97.59% at 1.25 mg/mL and an LC_{50} value of 5.52 mg/mL. 3,4-dimethoxycinnamic acid is functionally related to trans-cinnamic acid (NCBI−Pub chem), which showed larvicidal activity against *Aedes aegypti* [40] and nematicidal activities against *Meloidogyne incognita* and *Tylenchulus semipenetrans* [41]. In addition, cinnamic acid derivatives with a methylxanthine caffeine skeleton showed insecticidal potential against *Mythimna separata* [42]. Importantly, 3,4-dimethoxycinnamic acid is a key constituent of propolis, which is not cytotoxic to normal cell lines [39].

In addition, we observed that 3,4-dimethoxycinnamic acid showed oviposition deterrence activity against *P. xylostella*, with an ODI of 97.59% at 1.25 mg/mL. The biological effects decreased as the concentration increased, with an ODI of 57.01% at 10 mg/mL. Previous reports regarding the effects of these volatile compounds at different concentrations have reported varying results. For instance, (Z)-9-octadecenoic acid ethyl ester (OE) induced an oviposition preference and exhibited concentration-dependent bidirectional regulation. High concentrations of OE inhibited oviposition, while low doses promoted oviposition, with the Gr32a playing a role in oviposition stimulation [43]. Linalool (10 and 100 mg/L), and cis-3-hexen-1-ol (100 mg/L) exhibited repellent activities against *Cyrtorhinus lividipennis*. However, linalool (1 mg/L), and cis-3-hexen-1-ol (0.1 mg/L) appeared to attract *C. lividipennis* [44]. Additionally, the egg-laying preference of *Helicoverpa assulta* for heptanal disappeared when the concentration increased from 0.001 mol/L to 0.01 mol/L [45]. Nevertheless, our results showed that 3,4-dimethoxycinnamic acid has potential as an insecticide against *P. xylostella* at certain concentrations.

Numerous studies have shown that non-host plant extracts and derived compounds can cover or mask host plant compounds. This depends on the chemical communication involved in *P. xylostella* egg-laying and feeding. For instance, *Traiadica sebifera* extract and its fractions showed antifeedant activities against *P. xylostella* larvae [27]. *Ludwigia tomentosa* and *L. longifolia* extracts showed oviposition deterrence and insecticidal activities against *P. xylostella* [7]. Generally, insects utilize olfaction from their antenna to perceive and distinguish semiochemicals [46]. In our study, the EAG and behavioral responses of *P. xylostella* to *A. pubescens* extract and its compounds, caryophyllene oxide and 3,4-dimethoxycinnamic acid, were studied via EAG and Y-tube olfactometer experiments.

The EAG response is linked to the number of olfactory receptors in the insect's antennae [47]. In this study, both female and male adults of *P. xylostella* showed positive EAG responses to *A. pubescens* extract, caryophyllene oxide, and 3,4-dimethoxycinnamic acid, with responses ranging from 0.4 to 5.0 mV. For comparison, the EAG responses of *P. xylostella* to α-farnesene and β-myrcene ranged from 0.4 to 1.2 mV and from 0.1 to 1 mV, respectively [48]. Overall, the *P. xylostella* olfactory sensilla of both sexes were able to perceive *A. pubescens* extract, caryophyllene oxide, and 3,4-dimethoxycinnamic acid, which indicates that *A. pubescens* extract and its compounds are relevant in both female and male ecology. The EAG responses indicated a higher neuronal response to caryophyllene oxide than 3,4-dimethoxycinnamic acid in *P. xylostella* adults, which was correlated with increased receptor stimulation, and may have affected their behavioral responses.

In our study, the EAG response of female adults was more sensitive than that of male adults to *A. pubescens* extract. However, the male moths had slightly higher sensitivity than female moths in EAG response to different concentrations of caryophyllene oxide and 3,4-dimethoxycinnamic acid. Previous research has shown that this sexually dimorphic EAG response is linked to the volatile semiochemical detection cues of different numbers of olfactory receptors in male and female antennae, due to the different roles played by the same compound in the ecology of the different sexes [49]. The female insects use plant volatile compounds to find suitable oviposition positions [50]; the male insects use plant volatile compounds to find environments where females may be present [51]. For instance, female adults of *Cotesia glomerata* perceived more compounds than the males when undergoing EAG response assessments to benzyl cyanide [52]. The females also showed a higher EAG sensitivity to allyl isothiocyanate than males [48]. Thus, we hypothesized that *A. pubescens* extract may influence the oviposition behavior of *P. xylostella* females. In addition, previous studies have found that females of *P. xylostella* showed lower EAG responses than males to α-farnesene, β-myrcene, octanal, nonanal, and methyl jasmonate (at concentrations of 0.1) [48], while (Z)-3-hexen-1-ol elicited higher EAG responses in males of *Crioceris duodecimpunctata* than females [53]. In our study, the EAG responses of male moths to caryophyllene oxide and 3,4-dimethoxycinnamic acid were stronger than the responses of female moths; these two compounds may play an important role in interfering with the recognition of intraspecific sex pheromones by male moths.

Compounds that can trigger EAG activity in insect antennae do not necessarily lead to behavioral responses [26]. Therefore, olfactory assays were required to determine the behavioral responses of *P. xylostella* to *A. pubescens* extract and its compounds. In the Y-tube olfactory assays, female and male *P. xylostella* adults preferred the control arm over the treated arm with *A. pubescens* extract and caryophyllene oxide, which exhibited repellent effects and led to different behaviors at different concentrations. Previous research found that green leaf volatiles at different concentrations induced variable behavioral effects in European corn borer larvae [54]. Terpinene-4-ol and camphor showed concentration-dependent repellent activities against *P. xylostella* females [4], similarly to our results. The repellent response of *P. xylostella* adults to *A. pubescens* extract and caryophyllene oxide was due to the ability of their antennae to perceive these odorous substances. Some previous studies have reported that *Nilaparvata lugens* specimens were repelled by caryophyllene oxide, and *Nlug*OBP8 was highly expressed in multiple sensilla of *N. lugens* antenna. Antennae of ds*Nlug*OBP-injected BPHs completely lost the ability to sense caryophyllene oxide [55,56]. Nguyen et al. reported that caryophyllene oxide was ranked highly as a mosquito odorant-binding protein inhibitor (with larvicidal activity) based on its docking scores, and likely docks similar to permethrin [57]. Thus, we hypothesized that caryophyllene oxide has repellent activity against *P. xylostella* and may affect the odorant-binding protein of *P. xylostella*. Relevant research has reported that repellents prevent both female and male moths from achieving adequate orientation/location on the surfaces of treated leaves, thereby reducing the mating rates and egg-laying numbers [58]. The *A. pubescens* extract at a concentration of 25 mg/mL, caused a repellent effect for both female and male adults of *P. xylostella*, without influencing the overall egg-laying ability of *P. xylostella* female adults, which indicated that *A. pubescens* extract only had oviposition deterrence activity against *P. xylostella* female adults and did not affect mating. The caryophyllene oxide treatment group demonstrated repellent effects against both female and male adults of *P. xylostella* at 5 and 2.5 mg/mL; the oviposition deterrence results showed that the total egg numbers were decreased at 5 and 2.5 mg/mL, which indicated that caryophyllene oxide had repellent effects against both male and female adults, and also affected the mating rate and reduced the number of eggs laid.

In our study, the behavioral responses of female and male adults of *P. xylostella* to 3,4-dimethoxycinnamic acid were not significantly different from those of the control. Previous studies have reported that linalool and γ-terpinene cause EAG responses, but neither male nor female adults could distinguish between control and these compounds [50]. The EAG responses to these compounds were not always consistent with the behavioral responses. In contrast, 3,4-dimethoxycinnamic acid showed strong oviposition deterrence activity, with an ODI value of 97.59% at 1.25 mg/mL. Phytophagous insects have two stages of oviposition behavior: flying direction, landing, drumming on the leaf surface, and laying eggs. The first stage is mediated by visual and olfactory cues; the second stage is mediated by gustatory cues [23,59]. Rhodojaponin-III showed oviposition deterrence activity against *Spodoptera litura* without depending on antenna responses [60]. These results suggest that the oviposition deterrence activity of 3,4-dimethoxycinnamic acid is unrelated to antennae olfactory behavior, and it may involve different sensory neurons. Whether 3,4-dimethoxycinnamic acid influences other parts of the olfactory or gustatory systems requires further research.

The emergence of resistance in insect pest species is the main problem that must be taken into account for a new management strategy. Typically, phytophagous insects treated with natural and synthetic insecticides can use detoxification enzymes to metabolize xenobiotics [31]. Previous studies have reported that the Car E and GST activities of *P. xylostella* larvae significantly decreased when in contact with caryophyllene oxide [31]. Thus, long-term caryophyllene oxide use may cause insecticide resistance. However, caryophyllene oxide exhibited repellent activity on *P. xylostella*, reduced the *P. xylostella* population, reduced the use of chemical pesticides, and delayed the development of insecticide resistance. In addition, it is not currently widely available on the market; no reports on caryophyllene oxide resistance in the field have been published.

This study confirmed that *A. pubescens* extract and its compound derivatives, caryophyllene oxide and 3,4-dimethoxycinnamic acid, have repellent oviposition effects on *P. xylostella*, and directly affect their interactions with cruciferous vegetables. In acute toxicity experiments with the oral administration of *A. pubescens* extract, the LD_{50} of mice was determined to be 7.35 $\pm$ 0.62 g/kg [61]; however, its use has not been reported in clinical applications [15]. The chemical caryophyllene oxide, a volatile of tomato flower [62] and tobacco [45], had a positive effect on two pollinators, *Apis mellifera* and *Bombus terrestris*, which are efficient and reliable pollinators of greenhouse crops, and a parasitic wasp of *Campoletis chlorideae*, which is the main parasitoid of *H. assulta* larvae. These studies indicated that caryophyllene oxide is relatively safe for non-target organisms.

The EAG and behavioral experiments demonstrated that the oviposition-repellent behavior of caryophyllene oxide on *P. xylostella* was related to the olfactory nervous system. Odor molecules bind to OR proteins upon ligand binding, leading to signaling via G-proteins and intracellular second messengers to open membrane ion channels [63]. This depolarizes the sensory neuron to generate action potentials, which are conducted along its axon into the olfactory bulb of the brain and induce behavioral responses to molecules, rather than acting on neurotransmitters related to insecticide resistance. Thus, *A. pubescens* extract, caryophyllene oxide, and 3,4-dimethoxycinnamic acid can delay the development of insecticide resistance.

In addition, our results from EAG and Y-tube olfactometer tests provide evidence that *P. xylostella* responds to volatile compounds from *A. pubescens*. Previous studies employing these methods have found that *Ocimum basilicum* oil and its constituents showed repellent activity against *Musca domestica* [26]. Cao et al. reported that *Panax notoginseng*, *Angelica sinensis*, *Gastrodia elata*, and *Peucedanum praeruptorum*, and their constituents, showed attractive activity against *Stegobium paniceum* [64]. Song et al. reported that *Pelargonium hortorum* oil and its constituents showed repellent oviposition against *P. xylostella* [4]. Thus, the EAG and Y-tube olfactometer tests can be used effectively in screening new potential repellent extracts.

Several studies have found that multi-component phytochemical blends have greater repellent activity against various insect species than the corresponding single semiochemicals [65]. For instance, the b-caryophyllene/hexanal blend, at a close natural ratio of 2:1, showed greater attraction to *Holotrichia parallela* than the compounds did individually [66]. Behavioral effects depending on different chemical contents in mixtures are critical for non-host avoidance, as observed for the eight volatiles in *Rosmarinus officinalis* at ratios of 13:2:13:8:1:24:6:17 having stronger repellent effects against *Ectropis obliqua* than the constituents alone in equal doses [50]. The five compounds of *Pelargonium hortorum* substances at ratios of 1:5:3:4:3 significantly repelled egg-laying of female DBM, considerably more than demonstrated by the independent compounds [4]. Therefore, in order to test the potential synergistic effect of the compounds found in extract, it is best to use the natural ratios of these compounds derived from plant extracts, which warrants further research.

5. Conclusions

Overall, this study demonstrated the oviposition deterrence and insecticidal activity potential of *A. pubescens* extract, caryophyllene oxide, and 3,4-dimethoxycinnamic acid against adult and larval stages of *P. xylostella*. Furthermore, the EAG and behavioral responses of *P. xylostella* adults to *A. pubescens* extract and caryophyllene oxide indicated that these compounds can be used to control moth populations by repelling both male and female adults of *P. xylostella*, thereby deterring oviposition in female adults. 3,4-dimethoxycinnamic acid demonstrated strong oviposition deterrence activity against female adults of *P. xylostella*, but it was not associated with the antennal olfactory responses. Our findings showed that *A. pubescens* extract and its constituents have potential applications in the development of natural insecticides and controlling moth populations. Further research is needed to determine the optimal proportions of these different components for insecticide applications and to examine the toxicological implications of these compounds.

Supplementary Materials: The following supporting information can be downloaded at: https://www.mdpi.com/article/10.3390/insects14070613/s1, Table S1: List of tested compounds for *A. pubescens* extract; Method S1: LC–MS analysis; Table S2: LC solvent system conditions; Figure S1: Structures of 10 chemical constituents isolated from *A. pubescens* extract.

Author Contributions: Conceptualization, R.Z. and X.Y.; Data curation, R.Z., J.Z. and L.M.; Formal analysis, R.Z., J.Z. and L.M.; Investigation, R.Z. and J.Z.; Methodology, R.Z., J.Z. and X.Q.; Validation, R.Z., J.Z., L.M. and X.Q.; Visualization, X.Q. and X.Y.; Writing—original draft, R.Z.; Writing—review and editing, R.Z., J.Z., L.M., X.Q. and X.Y. Funding Acquisition, L.M., X.Y. and C.H. All authors have read and agreed to the published version of the manuscript.

Funding: This study was supported by the Natural Science Foundation of Shanxi Province (grant number: 20210302123387), the Leading Industrial & Engineering Research Project of Shanxi Agricultural University (grant number: CYYL23−25), and the Cultivation and Innovation Project of Plant Protection College of Shanxi Agricultural University (grant number: ZBXY223B−13).

Data Availability Statement: The authors confirm that all data are available in this paper.

Acknowledgments: We thank the Shanxi Agricultural University of Hui Cao for providing materials used for experiments. We thank Xue Feng, Penghua Bai, and Jingjing Dong for their assistance in writing the manuscript. Additionally, we thank Shaobo Lian and Rong Wang for their assistance in performing the experiments.

Conflicts of Interest: The authors declare no conflict of interest.

References

1. Zhu, L.H.; Li, Z.Y.; Zhang, S.F.; Xu, B.Y.; Zhang, Y.J.; Zalucki, M.P.; Wu, Q.J.; Yin, X.H. Population dynamics of the diamondback moth, *Plutella xylostella* (L.), in northern China: The effects of migration, cropping patterns and climate. *Pest Manag. Sci.* **2018**, *74*, 1845–1853. [CrossRef]
2. Zalucki, M.P.; Shabbir, A.; Silva, R.; Adamson, D.; Liu, S.S.; Furlong, M.J. Estimating the Economic Cost of One of the World's Major Insect Pests, *Plutella xylostella* (Lepidoptera: Plutellidae): Just How Long Is a Piece of String? *J. Econ. Èntomol.* **2012**, *105*, 1115–1129. [CrossRef] [PubMed]
3. Shen, X.; Cao, L.; Chen, J.; Ma, L.; Wang, J.; Hoffmann, A.A.; Wei, S. A comprehensive assessment of insecticide resistance mutations in source and immigrant populations of the diamondback moth *Plutella xylostella* (L.). *Pest Manag. Sci.* **2022**, *79*, 569–583. [CrossRef] [PubMed]
4. Song, C.; Ma, L.; Zhao, J.; Xue, Z.; Yan, X.; Hao, C. Electrophysiological and Behavioral Responses of *Plutella xylostella* (Lepidoptera: Plutellidae) to Volatiles from a Non-host Plant, Geranium, *Pelargonium × hortorum* (Geraniaceae). *J. Agric. Food Chem.* **2022**, *70*, 5982–5992. [CrossRef] [PubMed]
5. Isman, M.B. Botanical Insecticides in the Twenty-First Century-Fulfilling Their Promise? *Annu Rev Entomol.* **2019**, *65*, 233–249. [CrossRef]
6. Cheng, X.; He, H.; Dong, F.; Xu, C.C.; Zhang, H.; Liu, Z.; Lv, X.; Wu, Y.; Jiang, X.; Qin, X. Synthesis of Halopyrazole Matrine Derivatives and Their Insecticidal and Fungicidal Activities. *Molecules* **2022**, *27*, 4974. [CrossRef]
7. Ferreira, E.A.; Faca, E.C.; de Souza, S.A.; Fioratti, C.A.G.; Mauad, J.R.C.; Cardoso, C.A.L.; Mauad, M.; Mussury, R.M. Antifeeding and Oviposition Deterrent Effect of *Ludwigia* spp. (Onagraceae) against *Plutella xylostella* (Lepidoptera: Plutellidae). *Plants* **2022**, *11*, 2656. [CrossRef]
8. Li, Y.-F.; Sun, H.; Xi, N.; Zhang, Y. Effects of Cantharidin and Norcantharidin on Larval Feeding and Adult Oviposition Preferences of the Diamondback Moth (Lepidoptera: Plutellidae). *J. Econ. Èntomol.* **2019**, *112*, 1634–1637. [CrossRef]
9. Pavela, R. Essential oils from Foeniculum vulgare Miller as a safe environmental insecticide against the aphid *Myzus persicae* Sulzer. *Environ. Sci. Pollut. Res.* **2018**, *25*, 10904–10910. [CrossRef]
10. Sung-Chan, L.; Seon-Mi, S.; Min-Jung, H.; Jun-Hyeong, K.; Il, N.; Ji-Hong, P.; Il-Kwon, P. Behavioral and Electrophysiological Effects of Ajowan (*Trachyspermum ammi* Sprague) (Apiales: Apiaceae) Essential Oil and Its Constituents on Nymphal and Adult Bean Bugs, *Riptortus clavatus* (Thunberg) (Hemiptera: Alydidae). *Insects* **2020**, *11*, 104. [CrossRef]
11. Pavela, R.; Vrchotová, N. Insecticidal effect of furanocoumarins from fruits of *Angelica archangelica* L. against larvae *Spodoptera littoralis* Boisd. *Ind. Crops Prod.* **2013**, *43*, 33–39. [CrossRef]
12. Evergetis, E.; Michaelakis, A.; Haroutounian, S.A. Essential Oils of Umbelliferae (Apiaceae) Family Taxa as Emerging Potent Agents for Mosquito Control; Integrated Pest Management and Pest Control. In *Current and Future Tactics*; InTech Publisher: London, UK, 2012. [CrossRef]
13. Christensen, L.P.; Brandt, K. Bioactive polyacetylenes in food plants of the Apiaceae family: Occurrence, bioactivity and analysis. *J. Pharm. Biomed. Anal.* **2006**, *41*, 683–693. [CrossRef] [PubMed]
14. Zaugg, J.; Eickmeier, E.; Rueda, D.C.; Hering, S.; Hamburger, M. HPLC-based activity profiling of Angelica pubescens roots for new positive GABAA receptor modulators in *Xenopus oocytes*. *Fitoterapia* **2011**, *82*, 434–440. [CrossRef] [PubMed]
15. Lu, Y.; Wu, H.; Yu, X.; Zhang, X.; Luo, H.; Tang, L.; Wang, Z. Traditional Chinese Medicine of Angelicae Pubescentis Radix: A Review of Phytochemistry, Pharmacology and Pharmacokinetics. *Front. Pharmacol.* **2020**, *11*, 335. [CrossRef] [PubMed]
16. Tabanca, N.; Gao, Z.; Demirci, B.; Techen, N.; Wedge, D.E.; Ali, A.; Sampson, B.J.; Werle, C.; Bernier, U.R.; Khan, I.A.; et al. Molecular and Phytochemical Investigation of *Angelica dahurica* and *Angelica pubescentis* Essential Oils and Their Biological Activity against *Aedes aegypti*, *Stephanitis pyrioides*, and *Colletotrichum* Species. *J. Agric. Food Chem.* **2014**, *62*, 8848–8857. [CrossRef]
17. Kaur, M.; Kumar, R.; Upendrabhai, D.P.; Singh, I.P.; Kaur, S. Impact of sesquiterpenes from *Inula racemose* (Asteraceae) on growth, development and nutrition of *Spodoptera litura* (Lepidoptera: Noctuidae). *Pest Manag. Sci.* **2017**, *73*, 1031–1038. [CrossRef]
18. Yuan, G.; Zhao, L.; Du, Y.; Yu, H.; Shi, X.; Chen, W.; Chen, G. Repellence or attraction: Secondary metabolites in pepper mediate attraction and defense against *Spodoptera litura*. *Pest Manag. Sci.* **2022**, *78*, 4859–4870. [CrossRef]
19. Punia, A.; Singh, V.; Thakur, A.; Chauhan, N.S. Impact of caffeic acid on growth, development and biochemical physiology of insect pest, *Spodoptera litura* (Fabricius). *Heliyon* **2023**, *9*, e14593. [CrossRef]
20. Yan, X.-Z.; Deng, C.-P.; Xie, J.-X.; Wu, L.-J.; Sun, X.-J.; Hao, C. Distribution patterns and morphology of sensilla on the antennae of *Plutella xylostella* (L.)—A scanning and transmission electron microscopic study. *Micron* **2017**, *103*, 1–11. [CrossRef]
21. Guo, Z.; Kang, S.; Wu, Q.; Zhang, Y. Technical specifications for the mass-rearing of the diamondback moth, *Plutella xylostella* (L.). *Chin. J. Appl. Entomol.* **2015**, *52*, 492–497.

22. Domenico, R.; Patrizio, P.; Barbara, P.; Filomena, M. Antifungal activity of pomegranate peel extract against fusarium wilt of tomato. *Eur. J. Plant Pathol.* **2017**, *147*, 229–238. [CrossRef]
23. Yang, K.; Gong, X.-L.; Li, G.-C.; Huang, L.-Q.; Ning, C.; Wang, C.-Z. A gustatory receptor tuned to the steroid plant hormone brassinolide in *Plutella xylostella* (Lepidoptera: Plutellidae). *eLife* **2020**, *9*, e64114. [CrossRef] [PubMed]
24. Ratnaweera, P.B.; Jayasundara, J.M.N.M.; Herath, H.H.M.S.D.; Williams, D.E.; Rajapaksha, S.U.; Nishantha, K.M.D.W.P.; de Silva, E.D.; Andersen, R.J. Antifeedant, contact toxicity and oviposition deterrent effects of phyllostine acetate and phyllostine isolated from the endophytic fungus *Diaporthe miriciae* against *Plutella xylostella* larvae. *Pest Manag. Sci.* **2020**, *76*, 1541–1548. [CrossRef] [PubMed]
25. Sakuma, M. Probit analysis of preference data. *Appl. Èntomol. Zool.* **1998**, *33*, 339–347. [CrossRef]
26. Senthoorraja, R.; Subaharan, K.; Manjunath, S.; Pragadheesh, V.S.; Bakthavatsalam, N.; Mohan, M.G.; Senthil-Nathan, S.; Basavarajappa, S. Electrophysiological, behavioural and biochemical effect of *Ocimum basilicum* oil and its constituents methyl chavicol and linalool on *Musca domestica* L. *Environ. Sci. Pollut. Res.* **2021**, *28*, 50565–50578. [CrossRef]
27. Dolma, S.K.; Reddy, S.G.E. Characterization of *Triadica sebifera* (L.) Small Extracts, Antifeedant Activities of Extracts, Fractions, Seed Oil and Isolated Compounds against *Plutella xylostella* (L.) and Their Effect on Detoxification Enzymes. *Molecules* **2022**, *27*, 6239. [CrossRef]
28. Dolma, S.K.; Suresh, P.S.; Singh, P.P.; Sharma, U.; Reddy, S.G. Insecticidal activity of the extract, fractions, and pure steroidal saponins of *Trillium govanianum* Wall. ex D. Don for the control of diamondback moth (*Plutella xylostella* L.) and aphid (*Aphis craccivora* Koch). *Pest Manag. Sci.* **2020**, *77*, 956–962. [CrossRef]
29. Thanavendan, G.; Kennedy, J.S. Biochemical characterization and insecticidal activity of different solvent crude extracts of *Lantana camara* L. on diamondback moth, *Plutella xylostella* (Linn.). In Proceedings of the International Conference on Agriculture & Forestry, Colombo, Sri Lanka, 10–12 June 2015; Volume 1, pp. 117–138. [CrossRef]
30. Kumar, V.; Reddy, S.G.E.; Chauhan, U.; Kumar, N.; Singh, B. Chemical composition and larvicidal activity of *Zanthoxylum armatum* against diamondback moth, *Plutella xylostella*. *Nat. Prod. Res.* **2015**, *30*, 689–692. [CrossRef]
31. Huang, X.; Huang, Y.; Yang, C.; Liu, T.; Liu, X.; Yuan, H. Isolation and Insecticidal Activity of Essential Oil from *Artemisia lavandulaefolia* DC. against *Plutella xylostella*. *Toxins* **2021**, *13*, 842. [CrossRef]
32. Pour, S.A.; Shahriari, M.; Zibaee, A.; Mojarab-Mahboubkar, M.; Sahebzadeh, N.; Hoda, H. Toxicity, antifeedant and physiological effects of trans-anethole against *Hyphantria cunea* Drury (Lep: Arctiidae). *Pestic. Biochem. Physiol.* **2022**, *185*, 105135. [CrossRef]
33. Sajedeh, J.; Nooshin, Z.S.; Leila, R.; Antoni, S. Chemical composition and insecticidal efficacy of *Cyperus rotundus* essential oil against three stored product pests. *Int. Biodeter. Biodegr.* **2018**, *133*, 93–98. [CrossRef]
34. Gyrdymova, Y.V.; Rubtsova, S.A. Caryophyllene and caryophyllene oxide: A variety of chemical transformations and biological activities. *Chem. Pap.* **2022**, *76*, 1–39. [CrossRef]
35. Sarpietro, M.G.; Di Sotto, A.; Accolla, M.L.; Castelli, F. Interaction of β-caryophyllene and β-caryophyllene oxide with phospholipid bilayers: Differential scanning calorimetry study. *Thermochim. Acta* **2015**, *600*, 28–34. [CrossRef]
36. Matura, M.; Skold, M.; Borje, A.; Andersen, K.E.; Bruze, M.; Frosch, P.; Goossens, A.; Johansen, J.D.; Svedman, C.; White, I.R.; et al. Selected oxidized fragrance terpenes are common contact allergens. *Contact Dermat.* **2005**, *52*, 320–328. [CrossRef] [PubMed]
37. Lněniková, K.; Svobodová, H.; Skálová, L.; Ambro, M.; Matouková, P. The Impact of Sesquiterpenes β-caryophyllene Oxide and trans nerolidol on Xenobiotic-metabolizing Enzymes in Mice in vivo. *Xenobiotica* **2017**, *48*, 1089–1097. [CrossRef]
38. Chavan, M.; Wakte, P.; Shinde, D. Analgesic and anti-inflammatory activity of Caryophyllene oxide from *Annona squamosa* L. bark. *Phytomedicine* **2010**, *17*, 149–151. [CrossRef] [PubMed]
39. Ribeiro, V.P.; Arruda, C.; Aldana-Mejia, J.A.; Bastos, J.K.; Tripathi, S.K.; Khan, S.I.; Khan, I.A.; Ali, Z. Phytochemical, Antiplasmodial, Cytotoxic and Antimicrobial Evaluation of a Southeast Brazilian Brown Propolis Produced by *Apis mellifera* Bees. *Chem. Biodivers.* **2021**, *18*, e2100288. [CrossRef] [PubMed]
40. Cansian, R.L.; Staudt, A.; Bernardi, J.L.; Puton, B.M.S.; Oliveira, D.; de Oliveira, J.V.; Gomes, A.C.C.; Andrade, B.C.O.P.; Leal, I.C.R.; Simas, N.K.; et al. Toxicity and larvicidal activity on *Aedes aegypti* of citronella essential oil submitted to enzymatic esterification. *Braz. J. Biol.* **2023**, *83*, e244647. [CrossRef]
41. Kusakabe, A.; Wang, C.; Xu, Y.-M.; Molnár, I.; Stock, S.P. Selective Toxicity of Secondary Metabolites from the Entomopathogenic Bacterium *Photorhabdus luminescens sonorensis* against Selected Plant Parasitic Nematodes of the Tylenchina Suborder. *Microbiol. Spectr.* **2022**, *10*, e0257721. [CrossRef]
42. Zhang, S.; Liu, H.; Yang, N.; Xiong, L.; Wang, B. Synthesis and evaluation of novel xanthine-acrylamides and xanthine-acrylates as insecticidal agents. *Pest Manag. Sci.* **2022**, *78*, 2086–2095. [CrossRef]
43. Zhang, L.; Sun, H.; Grosse-Wilde, E.; Zhang, L.; Hansson, B.S.; Dweck, H.K. Cross-generation pheromonal communication drives Drosophila oviposition site choice. *Curr. Biol.* **2023**, *33*, 2095–2103. [CrossRef] [PubMed]
44. Jiang, N.N.; Mao, G.F.; Li, T.; Mo, J.C. The olfactory behavior response of *Cytorhinus lividipennis* to single component of rice volatiles. *Ecol. Environ. Sci.* **2018**, *27*, 262–267. [CrossRef]
45. Wang, C.; Li, G.; Miao, C.; Zhao, M.; Wang, B.; Guo, X. Nonanal modulates oviposition preference in female *Helicoverpa assulta* (Lepidoptera: Noctuidae) via the activation of peripheral neurons. *Pest Manag. Sci.* **2020**, *76*, 3159–3167. [CrossRef]
46. Wang, M.-M.; He, M.; Wang, H.; Ma, Y.-F.; Dewer, Y.; Zhang, F.; He, P. A candidate aldehyde oxidase in the antennae of the diamondback moth, *Plutella xylostella* (L.), is potentially involved in the degradation of pheromones, plant-derived volatiles and the detoxification of xenobiotics. *Pestic. Biochem. Physiol.* **2020**, *171*, 104726. [CrossRef] [PubMed]

47. Bruce, T.J.; Wadhams, L.J.; Woodcock, C.M. Insect host location: A volatile situation. *Trends Plant Sci.* **2005**, *10*, 269–274. [CrossRef]
48. Houjun, T.; Lin, S.; Chen, Y.; Chen, Y.; Zhao, J.; Gu, X.; Wei, H. Electroantennogram Responses to Plant Volatiles Associated with Fenvalerate Resistance in the Diamondback Moth, *Plutella xylostella* (Lepidoptera: Plutellidae). *J. Econ. Èntomol.* **2018**, *111*, 1354–1360. [CrossRef] [PubMed]
49. Germinara, G.; De Cristofaro, A.; Rotundo, G. Antennal olfactory responses to individual cereal volatiles in *Theocolax elegans* (Westwood) (Hymenoptera: Pteromalidae). *J. Stored Prod. Res.* **2009**, *45*, 195–200. [CrossRef]
50. Zhang, Z.; Bian, L.; Sun, X.; Luo, Z.; Xin, Z.; Luo, F.; Chen, Z. Electrophysiological and behavioural responses of the tea geometrid *Ectropis obliqua* (Lepidoptera: Geometridae) to volatiles from a non-host plant, rosemary, *Rosmarinus officinalis* (Lamiaceae). *Pest Manag. Sci.* **2014**, *71*, 96–104. [CrossRef]
51. Yang, Z.; Bengtsson, M.; Witzgall, P. Host plant volatiles synergize response to sex pheromone in codling moth, Cydia pomonella. *J. Chem. Ecol.* **2004**, *30*, 619–629. [CrossRef]
52. Blažytė-Čereškienė, L.; Aleknavičius, D.; Apšegaitė, V.; Būda, V. Response of Parasitic Wasp *Cotesia glomerata* L. (Hymenoptera: Braconidae) to Cabbage Plants of Two Varieties: Olfactory Spectra of Males and Females. *J. Econ. Èntomol.* **2022**, *115*, 1464–1471. [CrossRef]
53. Pistillo, O.M.; D'isita, I.; Germinara, G.S. Olfactory Response of the Spotted Asparagus Beetle, *Crioceris duodecimpunctata* (L.) to Host Plant Volatiles. *J. Chem. Ecol.* **2022**, *48*, 41–50. [CrossRef] [PubMed]
54. Piesik, D.; Rochat, D.; Delaney, K.J.; Marion-Poll, F. Orientation of European corn borer first instar larvae to synthetic green leaf volatiles. *J. Appl. Èntomol.* **2013**, *137*, 234–240. [CrossRef]
55. He, P.; Chen, G.; Li, S.; Wang, J.; Ma, Y.; Pan, Y.; He, M. Evolution and functional analysis of odorant-binding proteins in three rice planthoppers: *Nilaparvata lugens*, *Sogatella furcifera*, and *Laodelphax striatellus*. *Pest Manag. Sci.* **2018**, *75*, 1606–1620. [CrossRef]
56. Duan, S.-G.; Lv, C.-L.; Liu, J.-H.; Yi, S.-C.; Yang, R.-N.; Liu, A.; Wang, M.-Q. NlugOBP8 in *Nilaparvata lugens* Involved in the Perception of Two Terpenoid Compounds from Rice Plant. *J. Agric. Food Chem.* **2022**, *70*, 16323–16334. [CrossRef]
57. Hung, N.H.; Quan, P.M.; Dai, D.N.; Satyal, P.; Huong, L.T.; Giang, L.D.; Hung, L.T.; Setzer, W.N. Environmentally-Friendly Pesticidal Activities of *Callicarpa* and *Karomia* Essential Oils from Vietnam and Their Microemulsions. *Chem. Biodivers.* **2023**, *20*, e202200210. [CrossRef] [PubMed]
58. Haselton, A.T.; Acevedo, A.; Kuruvilla, J.; Werner, E.; Kiernan, J.; Dhar, P. Repellency of α-pinene against the house fly, *Musca domestica*. *Phytochemistry* **2015**, *117*, 469–475. [CrossRef]
59. Ozaki, K.; Ryuda, M.; Yamada, A.; Utoguchi, A.; Ishimoto, H.; Calas, D.; Marion-Poll, F.; Tanimura, T.; Yoshikawa, H. A gustatory receptor involved in host plant recognition for oviposition of a swallowtail butterfly. *Nat. Commun.* **2011**, *2*, 542. [CrossRef]
60. Yi, X.; Liu, J.; Wang, P.; Hu, M.; Zhong, G. Contacting is essential for oviposition deterrence of Rhodojaponin-III in *Spodoptera litura*. *Arch. Insect Biochem. Physiol.* **2014**, *86*, 122–136. [CrossRef]
61. Zhi-Wen, D.; Yu-Min, Z.; Hai-Shan, L.I.; Jia-Lun, W.; Ling, Z.; Wei, S. The Research on Acute and Long Time Toxicity of Angelicaroot Capsule. *J. Shenyang Med. Coll.* **2002**, *4*, 4–9.
62. Liu, J.; Zhang, J.; Shen, J.; Zhao, H.; Ma, W.; Jiang, Y. Differences in EAG Response and Behavioral Choices between Honey Bee and Bumble Bee to Tomato Flower Volatiles. *Insects* **2022**, *13*, 987. [CrossRef]
63. Sato, K.; Pellegrino, M.; Nakagawa, T.; Nakagawa, T.; Vosshall, L.B.; Touhara, K. Insect olfactory receptors are heteromeric ligand-gated ion channels. *Nature* **2008**, *452*, 1002–1006. [CrossRef] [PubMed]
64. Cao, Y.; Pistillo, O.M.; Lou, Y.; D'Isita, I.; Maggi, F.; Hu, Q.; Germinara, G.S.; Li, C. Electrophysiological and behavioural responses of *Stegobium paniceum* to volatile compounds from Chinese medicinal plant materials. *Pest Manag. Sci.* **2022**, *78*, 3697–3703. [CrossRef] [PubMed]
65. Hieu, T.T.; Choi, W.S.; Kim, S.-I.; Wang, M.; Ahn, Y.-J. Enhanced repellency of binary mixtures of *Calophyllum inophyllumnut* oil fatty acids or their esters and three terpenoids to *Stomoxys calcitrans*. *Pest Manag. Sci.* **2014**, *71*, 1213–1218. [CrossRef] [PubMed]
66. Zhang, M.; Cui, Z.; Zhang, N.; Xie, G.; Wang, W.; Chen, L. Electrophysiological and Behavioral Responses of *Holotrichia parallela* to Volatiles from Peanut. *Insects* **2021**, *12*, 158. [CrossRef]

toxins

MDPI

Article

Exploration of Synergistic Pesticidal Activities, Control Effects and Toxicology Study of a Monoterpene Essential Oil with Two Natural Alkaloids

Jianwei Xu [1], Min Lv [1,*], Shanshan Fang [1], Yanyan Wang [1], Houpeng Wen [1], Shaoyong Zhang [2] and Hui Xu [1,*]

[1] College of Plant Protection, Northwest A&F University, Xianyang 712100, China
[2] Key Laboratory of Vector Biology and Pathogen Control of Zhejiang Province, College of Life Science, Huzhou University, Huzhou 313000, China
* Correspondence: author: lvmin@nwsuaf.edu.cn (M.L.); orgxuhui@nwsuaf.edu.cn (H.X.)

Abstract: With the increasing development of pest resistances, it is not easy to achieve satisfactory control effects by using only one agrochemical. Additionally, although the alkaloid matrine (**MT**) isolated from *Sophora flavescens* is now utilized as a botanical pesticide in China, in fact, its pesticidal activities are much lower in magnitude than those of commercially agrochemicals. To improve its pesticidal activities, here, the joint pesticidal effects of **MT** with another alkaloid oxymatrine (**OMT**) (isolated from *S. flavescens*) and the monoterpene essential oil 1,8-cineole (**CN**) (isolated from the eucalyptus leaves) were investigated in the laboratory and greenhouse conditions. Moreover, their toxicological properties were also studied. Against *Plutella xylostella,* when the mass ratio of **MT** and **OMT** was 8/2, good larvicidal activity was obtained; against *Tetranychus urticae,* when the mass ratio of **MT** and **OMT** was 3/7, good acaricidal activity was obtained. Especially when **MT** and **OMT** were combined with **CN**, the significant synergistic effects were observed: against *P. xylostella,* the co-toxicity coefficient (CTC) of **MT/OMT** (8/2)/**CN** was 213; against *T. urticae,* the CTC of **MT/OMT** (3/7)/**CN** was 252. Moreover, the activity changes over time of two detoxification enzymes, carboxylesterase (CarE) and glutathione *S*-transferase (GST) of *P. xylostella* treated with **MT/OMT** (8/2)/**CN**, were observed. In addition, by scanning electron microscope (SEM), the toxicological study suggested that the acaricidal activity of **MT/OMT** (3/7)/**CN** may be related to the damage of the cuticle layer crest of *T. urticae.*

Keywords: matrine; 1,8-cineole; ternary complex; pesticidal activity; detoxification enzymes; SEM

Key Contribution: Joint pesticidal activities of **MT–OMT–CN** were related to the mass ratio of ternary complex and controlling pests. These results will lay the foundation for the future study of different combinations of **MT–OMT–CN** and the use of **CN** as a synergist with other bioactive plant natural products as pesticidal agents in crop protection.

1. Introduction

Plutella xylostella Linnaeus (diamondback moth; Lepidoptera: Plutellidae) as the major migratory insect pest mainly endangered cruciferous vegetables and caused economic loss in the world of up to US$ 4–5 billion annually [1,2]. *P. xylostella* has strong reproductive ability and strong stress resistance. Currently, it has been proved that *P. xylostella* has resistance to a variety of commercial insecticides, which makes it extremely difficult to control [3]. *Tetranychus urticae* Koch (Acari: Tetranychidae) as the major miscellaneous feeding mite had the characteristics of the wide host and high reproduction rate [4,5]. In recent years, with the continuous expansion of vegetable cultivation area, the damage of *T. urticae* becomes more and more serious. The role of chemical insecticides in the management of these pests was indisputable; however, the extensive and irrational use of chemical agrochemicals has resulted in increasing resistances and environmental problems [6–8]. Cantharidin and

(S)-(-)-palasonin showed different toxicity to *P. xylostella*, with median lethal concentrations, and quinolizidine alkaloids and neem-based products exhibited moderate acaricidal activities against *T. urticae* [9–11]. Therefore, the research and development of potential pesticide alternatives from bioactive plant natural products has received much attention in recent years [12–16].

Matrine (**MT**, Figure 1) and oxymatrine (**OMT**, Figure 1) were two main alkaloids isolated from the plant *Sophora flavescens* (Kushen). Matrine was one of the promising botanical pesticides registered in China due to its good insecticidal, antibacterial, and other agricultural activities [17–20]. Oxymatrine showed a wide range of biological properties, such as pesticidal [21,22], antiviral [23], anti-inflammatory [24], and anti-tumor effects [25]. The monoterpene 1,8-cineole (**CN**, Figure 1, eucalyptol), an essential oil isolated from the eucalyptus leaves, exhibited a variety of interesting activities, such as antimicrobial [26,27], anti-inflammatory [28], and insecticidal properties [29,30]. The plant essential oils used as botanical pesticides for pest control will be a trend in the foreseeable future [31–38]. Due to their synergistic effects, a mixture of two or three bioactive compounds could play the vital role for the management of pests [39–46]. The unique advantages of plant essential oils can greatly reduce the dosage of pesticides and effectively improve their efficacy level [47,48]. Due to the problems of poor efficiency of matrine and oxymatrine, the wide application of matrine and oxymatrine as the pesticide is restricted. Therefore, in this paper, the pesticidal activities of the different complexes of 1,8-cineole with matrine and oxymatrine were investigated against *P. xylostella* and *T. urticae*. Meanwhile, detoxification enzymes in insects usually play an important role in the process of metabolizing pesticides and producing drug resistance [49]. The change in detoxification activity level may be a response to environmental stress [50]. Carboxylesterase (CarE) and glutathione *S*-transferase (GST) are two important metabolic detoxification enzymes in insects. Moreover, the epidermis, a common component of insect body walls, is an extracellular matrix released by dermal cells and plays a crucial role in maintaining insect body form, reducing water loss and resisting microbial infection and predation [51]. So, the enzyme activity changes over time of CarE and GST of treated *P. xylostella*, and the effects of compounds on the cuticles of *T. cinnabarinus* were also investigated.

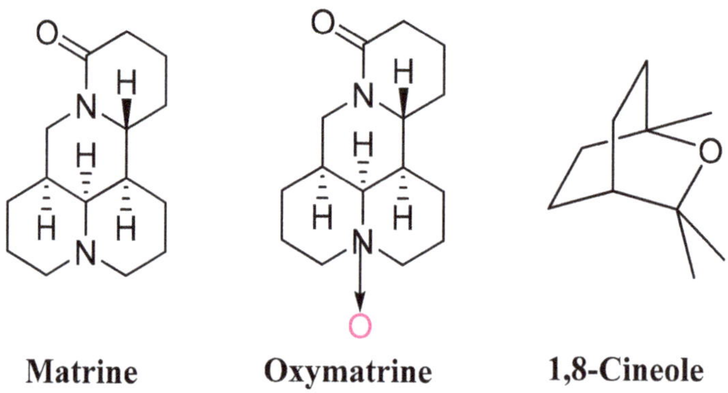

Figure 1. Chemical structures of matrine, oxymatrine and 1,8-cineole.

2. Materials and Methods

2.1. Insects and Chemicals

Tetranychus urticae (female adults) were reared for many generations in our laboratory with cowpea seedlings as the host plants to establish a stable population (temperature: 26 ± 1 °C; RH: 60–80%; photoperiod: light/dark = 14/10 h). *Plutella xylostella* (3rd instar larvae) were obtained from a laboratory population reared for many generations in the School Key Laboratory of Applied Entomology, Northwest A&F University. The cabbage net seedlings as the host plants were continuously planted in the greenhouse of our laboratory (temperature: 25 ± 2 °C; RH: 70 ± 10%; photoperiod: light/dark = 16/8 h). The host plants did not contact any pesticides. **MT** and **OMT** (98% purity) were purchased from Baoji Haoxiang Biotechnology Co. Ltd. (Baoji, China). **CN** (99% purity) was purchased from Aladdin Chemistry Co., Ltd. (Shanghai, China). β-Cypermethrin (96.6% purity) was bought from Hubei Supur Chemical Co., Ltd. (Hubei, China). Spirodiclofen (98.2% purity) was purchased from Shanghai Yuanye Biotechnology Co., Ltd. (Shanghai, China).

2.2. Biological Assay

2.2.1. Insecticidal Activity of Matrine (**MT**), Oxymatrine (**OMT**) and Their Binary Complexes against *P. xylostella* by the Leaf-Loading Poison Method

The solutions of matrine (**MT**), oxymatrine (**OMT**) and their binary complexes were prepared in acetone at 10 mg/mL (toosendanin as the positive control and acetone as CK) (Table 1). For each compound, 45 robust third instar larvae of *P. xylostella* were selected out (15 insects per group). The corresponding solution (1 μL) was evenly spread on a cabbage leaf disc (surface area: 0.25 cm^2). One piece of the above discs was added and eaten up by each *P. xylostella*, which was raised in each well of 12-well culture plates during 48 h (temperature: 25 ± 2 °C; RH: 70 ± 10%; photoperiod: light/dark = 16/8 h). Their corrected mortality rate (CMR) values (%) = $(T - C) \times 100/(100\% - C)$; C is the mortality rate of CK, and T is the mortality rate of the treated *P. xylostella* [52].

Table 1. Insecticidal activity of matrine (**MT**), oxymatrine (**OMT**) and their binary mixtures against *P. xylostella* at 10 μg/larva.

Compound	Corrected Mortality Rate (Mean ± SE, %)	
	24 h	**48 h**
MT	9.1 ± 2.2	20.9 ± 2.2 b [b]
MT/OMT (9/1) [a]	11.3 ± 0	16.2 ± 0 bc
MT/OMT (8/2)	13.6 ± 2.2	27.9 ± 2.2 a
MT/OMT (7/3)	11.3 ± 3.8	20.9 ± 2.2 b
MT/OMT (6/4)	2.3 ± 2.2	18.6 ± 2.2 bc
MT/OMT (5/5)	4.5 ± 0	20.9 ± 2.2 b
MT/OMT (4/6)	9.1 ± 2.2	18.6 ± 2.2 bc
MT/OMT (3/7)	4.5 ± 0	13.9 ± 2.2 c
MT/OMT (2/8)	2.3 ± 2.2	18.6 ± 2.2 bc
MT/OMT (1/9)	4.5 ± 0	16.2 ± 0 bc
OMT	2.3 ± 2.2	16.2 ± 0 bc
toosendanin	11.3 ± 0	32.5 ± 2.2 a

[a] Mass ratio; [b] Multiple range test using Duncan's test ($p < 0.05$). The same letters denote treatments that are not significantly different from each other.

2.2.2. Insecticidal Activity of **MT**, **OMT**, **MT/OMT** (8/2), 1,8-Cineole (**CN**) and **MT/OMT** (8/2)/**CN** against *P. xylostella* by the Leaf-Dipping Method

The procedure for evaluation of larvicidal activity of 1,8-cineole (**CN**), **MT/OMT** (8/2), and **MT/OMT** (8/2)/**CN** against *P. xylostella* (Figure 2) was as follows: The solutions of 1,8-cineole (**CN**, 0.4 and 0.5 mg/mL), **MT/OMT** (8/2, 0.5 mg/mL), and **MT/OMT** (8/2, 0.5 mg/mL)/**CN** (C_{CN}= 0.4 mg/mL) were prepared in acetone (acetone was used as CK). For each compound, 45 robust third instar larvae of *P. xylostella* were selected (15 insects per group). The cabbage leaf disc (surface area: 0.25 cm^2) was dipped into the corresponding solution for 3 s and taken out. The treated ones were added to three dishes during 48 h (15 insects per dish) (temperature: 25 ± 2 °C; RH: 70 ± 10%; photoperiod: light/dark = 16/8 h). Their CMR values were calculated in the same way as mentioned above [53].

The procedure for the determination of LC_{50} values of **MT**, **OMT**, **MT/OMT** (8/2), and **MT/OMT** (8/2)/**CN** at 48 h against *P. xylostella* was as follows (Table 2): Firstly, five different concentrations (2, 1, 0.5, 0.25, and 0.125 mg/mL) of **MT**, **OMT**, **MT/OMT** (8/2), and **MT/OMT** (8/2)/**CN** were prepared in acetone (acetone was used as CK). Five concentrations of β-cypermethrin (a positive control) were set as 0.5, 0.25, 0.125, 0.0625, and 0.03125 mg/mL in acetone, respectively. For each concentration, 45 robust third instar larvae of *P. xylostella* were selected (15 insects per group). The cabbage leaf disc (surface area: 0.25 cm^2) was dipped into the corresponding solution for 3 s and taken out. The treated ones were added to three dishes during 48 h (15 insects per dish) (temperature: 25 ± 2 °C; RH: 70 ± 10%; photoperiod: light/dark = 16/8 h). Their CMR values at 48 h were calculated in the same way as mentioned above. Finally, their 48 h median lethal concentration (LC_{50}) values were calculated upon the different concentrations and CMRs (Table 3).

Table 2. Corrected mortality rates at 48 h at different concentrations of **MT**, **OMT**, **MT/OMT** (8/2), and **MT/OMT** (8/2)/1,8-cineole (**CN**) against *P. xylostella*.

Compound	Corrected Mortality Rate (%) at Concentration (mg/mL) [a]				
	2 (0.5) [b]	1 (0.25) [b]	0.5 (0.125) [b]	0.25 (0.0625) [b]	0.125 (0.03125) [b]
MT	65.1 ± 3.8	53.4 ± 2.2	39.5 ± 2.2	23.2 ± 0	13.9 ± 2.2
OMT	60.4 ± 4.4	48.8 ± 2.2	32.5 ± 4.4	18.6 ± 2.2	11.6 ± 2.2
MT/OMT (8/2)[c]	72.4 ± 3.3	58.6 ± 0	41.3 ± 1.6	27.5 ± 2.8	15.5 ± 1.6
MT/OMT (8/2)/**CN**[d]	81.0 ± 1.6	65.5 ± 1.6	53.4 ± 2.8	37.9 ± 0	24.1 ± 1.6
β-cypermethrin	83.7 ± 2.2	55.8 ± 5.8	37.2 ± 3.8	25.5 ± 2.2	13.9 ± 2.2

[a] Values are mean ± SE of three replicates. [b] Number in parentheses is the tested concentration of β-cypermethrin. [c] Mass ratio; [d] **MT/OMT** (mass ratio: 8/2) containing **CN** (C_{CN}= 0.4 mg/mL) in acetone.

Table 3. LC_{50} values of **MT**, **OMT**, **MT/OMT** (8/2), and **MT/OMT** (8/2)/**CN** at 48 h against *P. xylostella*.

Compound	Linear Regression Equation[a]	LC_{50} (mg/mL)	Confidence Interval 95% (mg/mL)	r	χ^2	df	*P*	Co-Toxicity Coefficient
MT	Y = 0.057 + 1.240X	0.899	0.636~1.455	0.997	0.171	3	0.982	/
OMT	Y = −0.08 + 1.258X	1.158	0.806~2.045	0.993	0.155	3	0.984	/
MT/OMT (8/2) [b]	Y = −0.201 + 1.338X	0.708	0.539~0.970	0.998	0.025	3	0.999	133
MT/OMT (8/2)/ **CN** [c]	Y = 0.456 + 1.280X	0.441	0.324~0.585	0.999	0.161	3	0.984	213
β-cypermethrin	Y = 1.281 + 1.652X	0.168	0.128~0.226	0.980	1.845	3	0.605	/

[a] Regression analysis by IBM SPSS Statistics 20.0 ($p < 0.05$); [b] Mass ratio; [c] **MT/OMT** (mass ratio: 8/2) containing **CN** (C_{CN} = 0.4 mg/mL) in acetone.

The co-toxicity coefficient (CTC) values of the binary complexes were further evaluated according to Sun's formula [54]. The value of CTC is used to determine whether the efficiency is increased: when CTC > 120, it is synergistic; when CTC < 80, it is antagonistic; when 80 < CTC < 120, it is additive. A significant synergistic effect is observed when the value of CTC is 200.

2.2.3. Control Efficiency of **MT/OMT** (8/2), and **MT/OMT** (8/2)/**CN** against *P. xylostella* in the Greenhouse

The solutions of **MT/OMT** (8/2), **MT/OMT** (8/2)/**CN** and β-cypermethrin were prepared at 0.2 mg/mL in 0.1% aq. Tween-80, respectively (Table 4). Each cabbage seedling was infested with 20 third instar larvae of *P. xylostella* prior to spraying. One cabbage seedling was chosen for one group, and each treatment was three replicates. An airbrush was used to spray 10 mL of the corresponding solution for each treatment. The cabbage seedlings treated with 0.1% aq. Tween-80 alone were used as CK (temperature: 25 ± 2 °C; RH: 70 ± 10%; photoperiod: light/dark = 16/8 h). Their control effects on the 1st, 3rd, and 5th days were calculated in the same way as mentioned above [55].

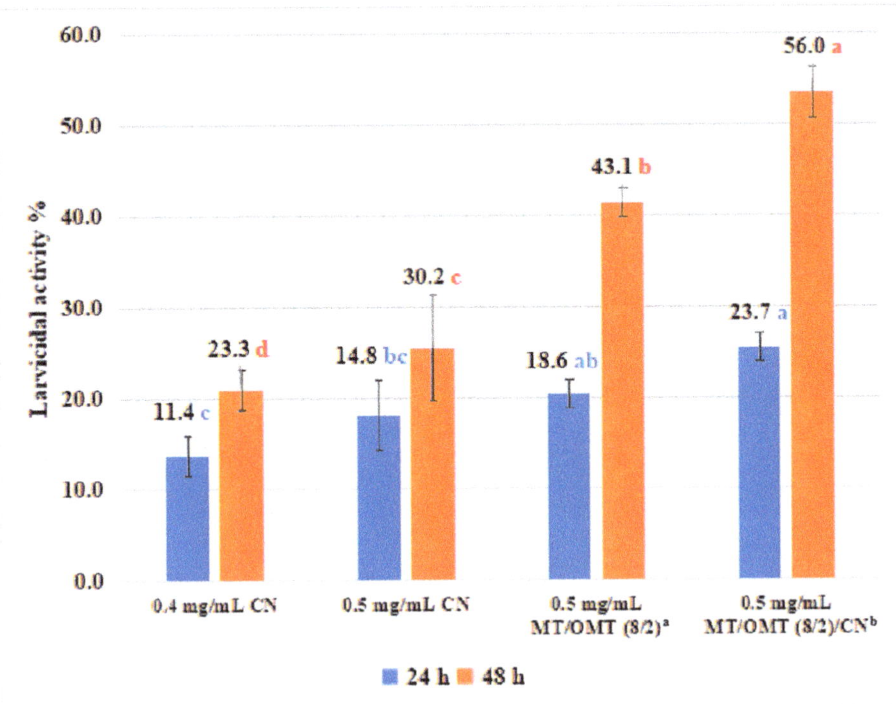

Figure 2. Larvicidal activity of 1,8-cineole (**CN**, 0.4 and 0.5 mg/mL), **MT/OMT** (8/2, 0.5 mg/mL) [a], and **MT/OMT** (8/2, 0.5 mg/mL)/**CN** [b] against *P. xylostella.* [a] Mass ratio; [b] **MT/OMT** (mass ratio: 8/2) were dissolved in the solution of **CN** (C_{CN}= 0.4 mg/mL) in acetone. Multiple range test using Duncan's test ($p < 0.05$). The same letters denote treatments that are not significantly different from each other (different compounds at the same time).

Table 4. Control efficiency of **MT/OMT** (8/2), and **MT/OMT** (8/2)/**CN** against *P. xylostella* in the greenhouse tests at a concentration of 0.2 mg/mL.

Compound	Control Efficiency (%) [c]		
	1st Day	3rd Day	5th Day
MT/OMT (8/2) [a]	10.1 ± 1.6 b [d]	25.8 ± 1.6 b	36.8 ± 0 c
MT/OMT (8/2)/**CN** [b]	15.5 ± 4.4 b	32.1 ± 3.3 b	49.0 ± 3.3 b
β-cypermethrin	40.6 ± 4.4 a	58.6 ± 2.8 a	71.9 ± 1.6 a

[a] Mass ratio; [b] **MT** and **OMT** (mass ratio: 8/2) containing **CN** (C_{CN} = 0.04 mg/mL) in 0.1% aq. Tween-80; [c] Values are mean ± SE of three replicates; [d] Multiple range test using Duncan's test ($p < 0.05$). The same letters denote treatments that are not significantly different from each other (different compounds at the same time).

2.2.4. Acaricidal Activity of **MT**, **OMT**, **CN**, and Their Mixtures against *T. urticae* by the Slide-Dipping Method

The solutions of **MT**, **OMT**, their binary complexes (with different mass ratio), and spirodiclofen (a positive control) (treated by 0.1 g/L of aq. Tween-80 as CK) were prepared at 0.5 mg/mL in Tween-80 in water (0.1 g/L), respectively (Table 5). For each compound, 90–120 healthy and size-consistency female adults of mites (30–40 ones per group) were selected out. Slides affixed with mites were dipped into the corresponding solution for 5 s and taken out (temperature: 26 ± 1 °C; RH: 60–80%; photoperiod: light/dark = 14/10 h). Their mortalities at 48 and 72 h were calculated as follows: corrected mortality rate (%) = $(T - C) \times 100/(100\% - C)$; C is the mortality rate of CK, and T is the mortality rate of the treated *T. urticae* [56].

Table 5. Acaricidal activity of **MT**, **OMT**, **CN**, and their mixtures against *T. urticae* treated at a concentration of 0.5 mg/mL.

Compound	Corrected Mortality Rate (Mean ± SE, %)	
	48 h	72 h
MT	7.0 ± 2.5	23.5 ± 1.7 cde [d]
MT/OMT (9/1) [a]	10.5 ± 1.6	20.4 ± 3.2 cdef
MT/OMT (8/2)	6.9 ± 0.7	24.3 ± 2.3 cde
MT/OMT (7/3)	4.5 ± 1.0	21.9 ± 5.8 cdef
MT/OMT (6/4)	10.3 ± 0.5	16.0 ± 3.4 ef
MT/OMT (5/5)	10.9 ± 1.6	24.6 ± 4.4 cde
MT/OMT (4/6)	8.7 ± 2.0	27.1 ± 3.4 bcde
MT/OMT (3/7)	9.4 ± 3.2	32.2 ± 0.9 bc
MT/OMT (2/8)	9.8 ± 2.2	28.9 ± 1.6 bcd
MT/OMT (1/9)	4.0 ± 1.2	25.2 ± 6.3 cde
OMT	6.0 ± 1.3	18.2 ± 2.9 def
1,8-cineole (CN) [b]	6.5 ± 0.2	11.2 ± 0.4 f
MT/OMT (3/7)/**CN** [c]	17.8 ± 0.3	38.0 ± 3.2 b
spirodiclofen	47.7 ± 4.7	88.6 ± 2.3 a

[a] Mass ratio; [b] at a concentration of 0.048 mg/mL; [c] **MT/OMT** (mass ratio: 3/7) containing CN (C_{CN}= 0.048 mg/mL) in Tween-80 in water (0.1 g/L); [d] Multiple range test using Duncan's test ($p < 0.05$). The same letters denote treatments that are not significantly different from each other (different compounds at the same time).

According to the above results, the acaricidal activity of **CN** and **MT/OMT** (3/7)/**CN** against *T. urticae* (Table 5) was tested as follows: The solution of **CN** was prepared at 0.048 mg/mL in Tween-80 in water (0.1 g/L), and the solution of **MT/OMT** (3/7)/**CN** was prepared at 0.5 mg/mL in Tween-80 in water (0.1 g/L) containing **CN** (C_{CN}= 0.048 mg/mL). The next procedure was completed in the same way as mentioned above.

2.2.5. LC_{50} of **MT**, **OMT**, **MT/OMT** (3/7), and **MT/OMT** (3/7)/**CN** against *T. urticae*

Firstly, five different concentrations (6, 3, 1.5, 0.75, and 0.375 mg/mL) of **MT** and **OMT** were prepared in Tween-80 in water (0.1 g/L) (treated by 0.1 g/L of aq. Tween-80 as CK) (Table 6). Six concentrations of **MT/OMT** (3/7) and **MT/OMT** (3/7)/**CN** were set as 4, 2, 1, 0.5, 0.25, and 0.125 mg/mL in Tween-80 in water (0.1 g/L), respectively. Five concentrations of spirodiclofen were set as 0.5, 0.25, 0.125, 0.0625, and 0.03125 mg/mL in Tween-80 in water (0.1 g/L), respectively. For each concentration, 90–120 healthy and size-consistency female adults of mites (30–40 ones per group) were selected out. Slides affixed with mites were dipped into the corresponding solution for 5 s and taken out (temperature: 26 ± 1 °C; RH: 60–80%; photoperiod: light/dark = 14/10 h). Their CMR values at 72 h were calculated in the same way as mentioned above. Finally, LC_{50} values were calculated by the linear regressions of 72 h CMRs (%) and concentrations (Table 7) [56]. The co-toxicity coefficient (CTC) values of the binary complexes were further evaluated according to Sun's formula [54].

2.2.6. Control Efficiency of **MT/OMT** (3/7), and **MT/OMT** (3/7)/**CN** against *T. urticae* in the Greenhouse

The solutions of **MT/OMT** (3/7), **MT/OMT** (3/7)/**CN**, and spirodiclofen were all prepared at 0.2 mg/mL in 0.1% aq. Tween-80, respectively. Asparagus bean plants were infested with the female adults of *T. urticae* prior to spraying. Three plants were chosen for one group, and each treatment was three replicates. An airbrush was used to spray 10 mL of the corresponding solution for three replicates. The plants treated with 0.1% aq. Tween-80 alone were used as CK (temperature: 26 ± 1 °C; RH: 60–80%; photoperiod:

light/dark = 14/10 h). Their control effects on the 1st, 3rd, and 5th days were calculated in the same way as mentioned above (Table 8) [56].

Table 6. Corrected mortality rates at 72 h at different concentrations of **MT**, **OMT**, **MT/OMT** (3/7), and **MT/OMT** (3/7)/**CN** against *T. urticae*.

Compound	Corrected Mortality Rate (%) at Concentration (mg/mL) [a]					
	6	3	1.5	0.75	0.375	/
MT	80.3 ± 0.9	56.6 ± 1.6	40.5 ± 2.4	24.1 ± 2.7	15.6 ± 3.9	/
OMT	75.1 ± 2.4	53.9 ± 2.2	38.5 ± 2.1	20.5 ± 3.5	10.3 ± 1.0	/
Compound	Corrected mortality rate (%) at concentration (mg/mL) [a]					
	4 (0.5) [b]	2 (0.25) [b]	1 (0.125) [b]	0.5 (0.0625) [b]	0.25 (0.03125) [b]	0.125
MT/OMT (3/7) [c]	72.6 ± 0.4	53.2 ± 1.3	40.4 ± 2.7	32.2 ± 0.9	21.7 ± 0.6	13.4 ± 1.5
MT/OMT (3/7)/**CN** [d]	80.1 ± 0.8	61.0 ± 0.5	49.6 ± 2.2	38.0 ± 3.2	29.6 ± 3.5	17.4 ± 1.6
spirodiclofen	90.1 ± 0.5	72.5 ± 0.2	54.5 ± 0.8	36.8 ± 2.6	26.0 ± 1.2	/

[a] Values are mean ± SE of three replicates. [b] Number in parentheses is the tested concentration of spirodiclofen. [c] Mass ratio; [d] **MT** and **OMT** (mass ratio: 3/7) containing **CN** (C_{CN} = 0.048 mg/mL) in Tween-80 in water (0.1 g/L).

Table 7. LC_{50} values of **MT**, **OMT**, **MT/OMT** (3/7), and **MT/OMT** (3/7)/**CN** at 72 h against *T. urticae*.

Compound	Linear Regression Equation [a]	LC_{50} (mg/mL)	Confidence Interval 95% (mg/mL)	r	χ^2	df	*P*	Co-Toxicity Coefficient
MT	Y = −0.471 + 1.549X	2.014	1.667~2.451	0.987	2.004	3	0.572	/
OMT	Y = −0.604 + 1.606X	2.377	1.970~2.924	0.994	0.398	3	0.941	/
MT/OMT (3/7) [b]	Y = −0.160 + 1.083X	1.405	1.108~1.858	0.986	1.681	4	0.794	160
MT/OMT (3/7)/**CN** [c]	Y = 0.054 + 1.103X	0.894	0.713~1.134	0.992	2.166	4	0.705	252
spirodiclofen	Y = 1.597 + 1.558X	0.094	0.077~0.113	0.996	2.198	3	0.532	/

[a] Regression analysis by IBM SPSS Statistics 20.0 ($p < 0.05$); [b] Mass ratio; [c] **MT** and **OMT** (mass ratio: 3/7) containing **CN** (C_{CN} = 0.048 mg/mL) in Tween-80 in water (0.1 g/L).

Table 8. Control efficiency of **MT/OMT** (3/7), and **MT/OMT** (3/7)/**CN** against *T. urticae* in the greenhouse tests at a concentration of 0.2 mg/mL.

Compound	Control Efficiency (%) [c]		
	1st Day	3rd Day	5th Day
MT/OMT (3/7) [a]	7.7 ± 0.8 b [d]	19.6 ± 1.9 c	31.3 ± 1.1 c
MT/OMT (3/7)/**CN** [b]	13.8 ± 1.8 b	31.9 ± 1.0 b	42.3 ± 1.2 b
spirodiclofen	25.0 ± 2.4 a	58.3 ± 1.0 a	73.9 ± 1.5 a

[a] Mass ratio; [b] **MT** and **OMT** (mass ratio: 3/7) containing **CN** (C_{CN} = 0.04 mg/mL) in 0.1% aq. Tween-80; [c] Values are mean ± SE of three replicates; [d] Multiple range test using Duncan's test ($p < 0.05$). The same letters denote treatments that are not significantly different from each other (different compounds at the same time).

2.3. *Enzyme Activity Assay against P. xylostella*

2.3.1. Sample Preparation Using Leaf-Dipping Method

According to the above-mentioned leaf-dipping method, 180 robust 3rd instar larvae of *P. xylostella* were treated with **CN**, **MT/OMT** (8/2) and **MT/OMT** (8/2)/**CN** at 0.4 mg/mL, respectively (treated by acetone as CK). Then, the 30 surviving larvae in the treated group were collected at 12, 24, 36, and 48 h, respectively. They were then snap-frozen in liquid nitrogen and stored at −80 °C for subsequent enzyme activity analysis [53].

2.3.2. Preparation of Homogenous Liquid

The homogenization treatment (including ten larvae: Weight (g)/Volume (mL) = 1/10) was performed in an ice bath. The homogenous liquid was obtained for carboxylesterase

(CarE) activity assay when the sample was centrifuged at 12,000× *g* for 30 min at 4 °C. The homogenous liquid was obtained for glutathione-*S*-transferase (GST) activity assay when the sample was centrifuged at 8000× *g* for 10 min at 4 °C.

2.3.3. CarE and GST Activity Assay According to Bradford's Method

The absorption values over time of CarE and GST were tested by using CarE (α-naphthyl acetate (α-NA) as a substrate) and GST (1-chloro-2,4-dinitrobenzene and reduced glutathione as substrates) assay kits (Suzhou Keming Biotechnology Co., Ltd., China), respectively. Total protein concentration was determined according to the Bradford method (using bovine serum albumin (BSA) as a standard). The protein content was tested by a BCA protein quantitative assay kit (Shaanxi Zhonghui Hecai Biomedical Technology Co., Ltd., China). Finally, the enzymes activity values were obtained according to the absorption value and the protein content. Each treatment was replicated three times [57].

2.4. Analysis of Morphology of Cuticles

2.4.1. Pretreatment of Mites

The complex (**MT/OMT** (3/7)/**CN**) was prepared at 0.894 mg/mL in 0.1 g/L of aq. Tween-80 (0.1 g/L of aq. Tween-80 as CK). Female adult mites with the same physiological status and good growth conditions were selected to cowpea leaves (5 cm in length and 3 cm in width; 30 ones/leaf). After 4 h, the leaves were immersed in the above solution for 5 s, the excess solution was absorbed by filter paper, and the leaves were placed in a Petri dish with a moist sponge (three replicates/treatment). Then, Petri dishes were placed in a light incubator (temperature: 26 ± 1 °C; RH: 60–80%; photoperiod: light/dark = 14/10 h). Finally, the dead mites were collected at 24, 48 and 72 h after treatment by the complex for scanning electron microscope (SEM) analysis [56,58].

2.4.2. Observation of Morphology and Structural Changes of Cuticles in Spider Mites

The collected samples were fixed with 2.5% glutaraldehyde under ice bath conditions, incubated at 4 °C for 4 h, rinsed three times with 0.1 mol/L of phosphate buffer saline (PBS), dehydrated with different concentrations of ethanol, and freeze-dried for 3 h. The mites were then placed on a sample table and sprayed with gold under vacuum conditions. The morphology was observed and photographed by S-3400N SEM [59].

2.5. Statistic Analysis

Mortality data were corrected with Abbott's formula and analyzed by a multiple range test using Duncan's test ($p < 0.05$). The median lethal concentration (LC_{50}) values were calculated on log-concentration versus probit (% mortality) regression analysis. The values of r, χ^2, df, and *P* were obtained on regression analysis by IBM SPSS Statistics 20.0 ($p < 0.05$).

3. Results and Discussion

3.1. Insecticidal Activity

First, the insecticidal activities of two compounds (**MT** and **OMT**) and their binary mixtures at different mass ratios against *P. xylostella* were tested. As shown in Table 1, to binary mixtures, when the mass ratio of **MT** and **OMT** was 8:2, the corresponding 48 h corrected mortality rate (CMR) was 27.9%, which was higher than those of **MT** (20.9%) and **OMT** (16.2%). The 48 h CMRs were all 20.9% when the mass ratio of **MT/OMT** was 7:3 or 5:5. The 48 h CMR was 16.2% when the mass ratio of **MT/OMT** was 1:9, which was similar to that of **OMT**. However, when the mass ratio of **MT** and **OMT** was 3:7, the corresponding 48 h CMR was decreased to 13.9%, which was lower than those of **MT** and **OMT**. Obviously, for **MT** and **OMT** against *P. xylostella*, the best mass ratio of **MT** and **OMT** was 8:2. As shown in Figure 2, against *P. xylostella*, 48 h CMRs of 1,8-cineole (**CN**, 0.4 and 0.5 mg/mL), **MT/OMT** (8/2, 0.5 mg/mL), and **MT/OMT** (8/2, 0.5 mg/mL)/**CN** were 23.3%, 30.2%, 43.1%, and 56.0%, respectively. It demonstrated that when **CN** (0.4

mg/mL) was combined with **MT/OMT**(8/2, 0.5 mg/mL), the larvicidal activity of the corresponding complex was significantly improved. Then, 48 h LC_{50} values of **MT**, **OMT**, **MT/OMT** (8/2), and **MT/OMT** (8/2)/**CN** against *P. xylostella* were further evaluated according to their CMRs at different concentrations (Table 2). As described in Table 3, 48 h LC_{50} values of **MT/OMT** (8/2), and **MT/OMT** (8/2)/**CN** were 0.708 and 0.441 mg/mL, respectively; while 48 h LC_{50} values of **MT** and **OMT** were 0.899 and 1.158 mg/mL, respectively. That is, the larvicidal activity of **MT/OMT** (8/2)/**CN** was 2.0–2.6 fold those of **MT** and **OMT**. Moreover, the co-toxicity coefficient (CTC) values of **MT/OMT** (8/2) and **MT/OMT** (8/2)/**CN** were 133 and 213, respectively. Clearly, their CTCs were all > 120, so the above two mixtures showed the synergistic effect. Subsequently, the control effects of **MT/OMT** (8/2) and **MT/OMT** (8/2)/**CN** against *P. xylostella* in the greenhouse at 0.2 mg/mL were tested. As shown in Table 4, the control effects of **MT/OMT** (8/2) and **MT/OMT** (8/2)/**CN** against *P. xylostella* after 5 days were 36.8% and 49.0%, respectively. Symptoms at the fifth day of the cabbage seedling leaves treated with **MT/OMT** (8/2) and **MT/OMT** (8/2)/**CN** against *P. xylostella* are described in Figure 3. In the **MT/OMT** (8/2)- and **MT/OMT** (8/2)/**CN**-treated groups, the surfaces of the leaves were smooth and had almost no small puncture; however, there were many small punctures eaten by *P. xylostella* on the leaves in the CK-treated group. Photographs of control effects of those complexes against *P xylostella* are shown in Figures S1–S3.

Figure 3. Fifth-day symptoms of cabbage seedling leaves treated with **MT/OMT** (8/2, 0.2 mg/mL) and **MT/OMT** (8/2, 0.2 mg/mL)/**CN** against *P. xylostella*.

3.2. Acaricidal Activity

The 48 and 72 h acaricidal results of **MT**, **OMT**, and their binary mixtures at 0.5 mg/mL against the female adults of *T. urticae* are shown in Table 5. Among nine binary complexes, the 72 h CMRs of six complexes were higher than those of **MT** (23.5%) and **OMT** (18.2%); especially when the mass ratio of **MT** to **OMT** was 3:7, the corresponding 72 h CMR was 34.8%. At 0.048 mg/mL, the 72 h CMR of **CN** against *T. urticae* was only 11.2%; interestingly, when **MT** and **OMT** (mass ratio: 3/7) were dissolved in the solution of **CN** (C_{CN} = 0.048 mg/mL) in 0.01% aq. Tween-80, and $C_{MT/OMT\,(3/7)}$ was 0.5 mg/mL, the corresponding CMR was increased to 38.0%. It may be related to the strong permeability of **CN** as an essential oil [38]. Then, 72 h LC_{50} values of **MT**, **OMT**, **MT/OMT** (3/7), and **MT/OMT** (3/7)/**CN** against *T. urticae* were further calculated according to their CMRs at different concentrations (Table 6). As shown in Table 7, the 72 h LC_{50} values of **MT/OMT** (3/7) and **MT/OMT** (3/7)/**CN** (C_{CN}= 0.048 mg/mL) were 1.405 and 0.894 mg/mL, respectively, while 72 h LC_{50} values of **MT** and **OMT** were 2.014 and 2.377 mg/mL, respectively. That is, the acaricidal activity of **MT/OMT** (3/7)/**CN** (C_{CN}= 0.048 mg/mL) was increased to 2.3–2.7 folds when compared with those of **MT** and **OMT**. Meanwhile, the CTCs of **MT/OMT** (3/7) and **MT/OMT** (3/7)/**CN** (C_{CN}= 0.048 mg/mL) were 160 and 252, respectively. So, the above two mixtures showed the synergistic effect especially for the combination of **MT/OMT** (3/7)/**CN**. On the other hand, the control efficiency of **MT/OMT** (3/7) and **MT/OMT** (3/7)/**CN** (C_{CN} = 0.04 mg/mL in 0.1% aq. Tween-80) at 0.2 mg/mL against *T. urticae* in the greenhouse was evaluated. As shown in Table 8, the control effects of **MT/OMT** (3/7) and **MT/OMT** (3/7)/**CN** after 5 days were 31.3% and 42.3%, respectively. Fifth-day symptoms of asparagus bean seedling leaves treated with **MT/OMT** (3/7) and **MT/OMT** (3/7)/**CN** were described in Figure 4. There were lots of white spots destroyed by *T. urticae* on the seedling leaves in the CK-treated group, whereas in the **MT/OMT** (3/7)- and **MT/OMT** (3/7)/**CN**-treated groups, almost no small white spots were on the seedling leaves. These findings were the same as our previous report [58]. Photographs of control effects of those complexes against *T. urticae* are shown in Figures S4–S6.

Figure 4. Fifth-day symptoms of asparagus bean seedling leaves treated with **CK** (**a**), **MT/OMT** (3/7, 0.2 mg/mL) (**b**), and **MT/OMT** (3/7, 0.2 mg/mL)/**CN** (**c**) against *T. urticae*.

3.3. Changes of Detoxification Enzymes Activities

Subsequently, the changes of detoxification enzymes (CarE and GST) activities in *P. xylostella*, treated with **CN**, **MT/OMT** (8/2) and **MT/OMT** (8/2)/**CN** at 0.4 mg/mL after 12, 24, 36, and 48 h were depicted in Figures 5 and 6, respectively. The enzymatic activities of CarE and GST are always in dynamic change. As described in Figure 5, the CarE activity values in the treated groups at 36 and 48 h were much lower than those of the control group. The CarE activity value in the **CN**-treated group at 12 h (82.1 U/g) was higher than that of the control group (69.4 U/g). It suggested that **CN** was toxic to *P. xylostella* and stimulated the detoxification ability of the CarE in *P. xylostella* at 12 h. At 24 h, the CarE activity value in the **MT/OMT** (8/2)/**CN**-treated group (67.7 U/g) was higher than that of the control group (52.4 U/g). It indicated that **MT/OMT** (8/2)/**CN** stimulated the detoxification ability of the CarE in *P. xylostella* at 24 h. After that, the detoxification ability of the CarE in *P. xylostella* in all treated groups was inhibited. The time for reaching the lowest points of the CarE activities in the treated groups was different. For example, in the **CN**-treated group, the lowest point of the CarE activity was reached at 36 h, whereas the lowest points of the CarE activity in the **MT/OMT** (8/2)- and **MT/OMT** (8/2)/**CN**-treated group were reached at 48 h.

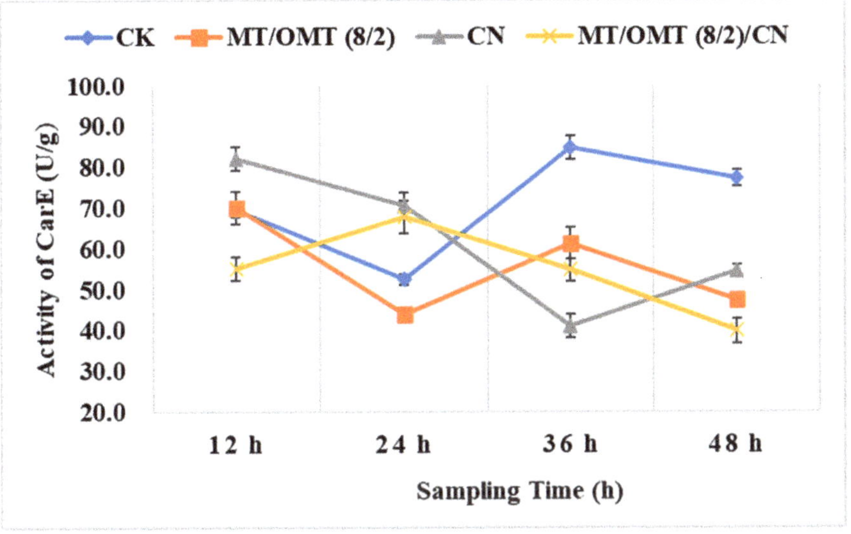

Figure 5. Effects of **CN**, **MT/OMT** (8/2) and **MT/OMT** (8/2)/**CN** at 0.4 mg/mL on CarE in *P. xylostella* at different sampling times..

As illustrated in Figure 6, the GST activity value in the **MT/OMT** (8/2)/**CN**-treated group at 12 h (3910 nmol/min/g) was higher than that of the control group (2698 nmol/min/g). Similarly, it suggested that the mixture **MT/OMT** (8/2)/**CN** stimulated and enhanced the detoxification ability of the GST in *P. xylostella* at 12 h. At 24 h, the GST activity value in the **CN**-treated group was slightly higher than that of the control group. Afterwards, the GST activity in the treated groups decreased significantly when compared with that of the control, and it reached the lowest point at 36 h in all treated groups. The GST activity values at 36 h in **CN**-, **MT/OMT** (8/2)- and **MT/OMT** (8/2)/**CN**-treated groups were 2208, 2591 and 1932 nmol/min/g, respectively, and they were decreased 1.4–1.9 folds of that of the control group (3603 nmol/min/g). Obviously, the GST activity of *P. xylostella* in the treated groups was largely inhibited at 36 h.

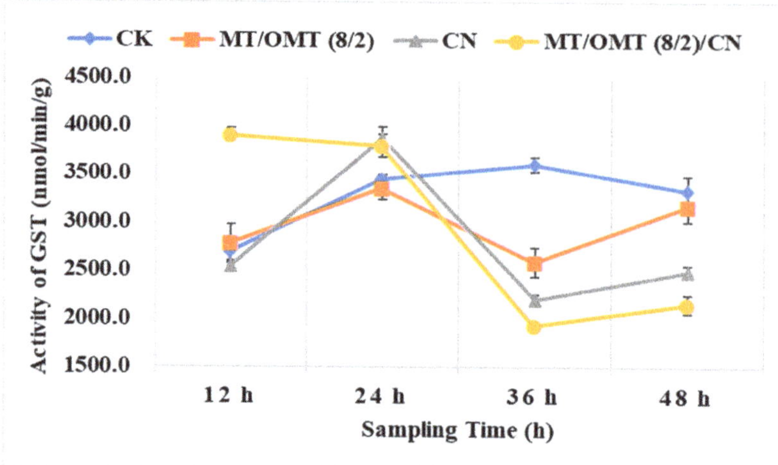

Figure 6. Effects of **CN**, **MT/OMT** (8/2) and **MT/OMT** (8/2)/**CN** at 0.4 mg/mL on GST in *P. xylostella* at different sampling times.

3.4. Toxicological Study of Structural Changes of Cuticles by SEM

The penetration of insecticide to pest epidermis is the premise of its toxic effect. The cuticle serves as the initial barrier of defense between the body and the outside world, effectively blocking pesticide penetration. The reduction in epidermal penetration can delay the time of pesticide reaching the target site, and pests have sufficient time and faster speed to metabolize the pesticide entering the body [59]. As shown in Figure 7, in the CK-treated group, the cuticles of *T. urticae* are flat and have an entire structure, with neat and continuous skin texture; whereas in the **MT/OMT** (3/7)/**CN**-treated group, the structure of the cuticles was damaged with visible wrinkles and an uneven arrangement of the inner ridges. The cuticles of wounded mites were thinner, softer, and more asymmetrically bent than those of normal mites, losing their barrier-like function against acaricide penetration. As a result, **CN** may enhance the penetration ability of this complex on *T. urticae* [38,60].

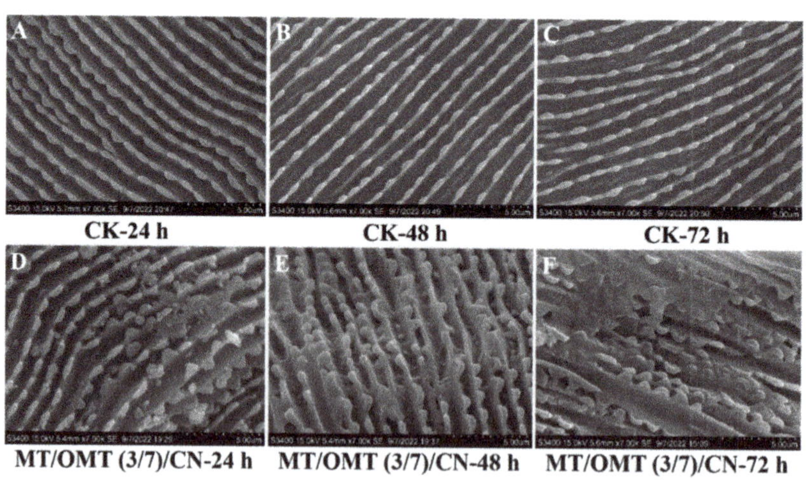

Figure 7. Scanning electron microscope images of cuticles of *T. urticae* treated with **MT/OMT** (3/7)/**CN** (**D–F**) and aq. Tween-80 (0.1 g/L) (CK) (**A–C**) after 24, 48 and 72 h, respectively. Bars: 5.0 μm.

4. Conclusions

In summary, the significant synergistic effects were observed when **CN** was combined with **MT** and **OMT** as pesticidal agents: against *P. xylostella*, the CTC of **MT/OMT** (8/2)/**CN** was 213; against *T. urticae*, the CTC of **MT/OMT** (3/7)/**CN** was 252. Furthermore, these mixtures displayed good control efficiency against *P. xylostella* and *T. urticae* in the greenhouse. Importantly, the enzymes activity changes over time of CarE and GST in *P. xylostella* treated with **CN**, **MT/OMT** (8/2) and **MT/OMT** (8/2)/**CN** were explored. Due to the accumulation of toxicant leading to a decrease in the detoxification ability, the time for reaching the lowest point of the CarE activity in the treated groups was different, whereas the time for reaching the lowest point of the GST activity in the treated groups was the same (at 36 h). Notably, by SEM analysis, the toxicology study suggested that the destruction of the cuticle layer crest of *T. urticae* by **MT/OMT** (3/7)/**CN** may be the main cause of their death. These results will pave the way for the future study of different combinations of **MT–OMT–CN** and the application of **CN** as a synergist with other bioactive natural products as pesticidal candidates in crop protection.

Supplementary Materials: The following supporting information can be downloaded at: https://www.mdpi.com/article/10.3390/toxins15040240/s1, Biological assay; Pictures of control effects of different complexes against *P. xylostella*; Pictures of control effects of different complexes against *T. urticae*. Figure S1. Pictures of control efficiency of **CK** after 1st day (a), 3rd day (b) and 5th day (c) against *P. xylostella* in the greenhouse; Figure S2. Pictures of control efficiency of **MT/OMT** (8/2, 0.2 mg/mL) after 1st day (a), 3rd day (b) and 5th day (c) against *P. xylostella* in the greenhouse; Figure S3. Pictures of control efficiency of **MT/OMT** (8/2, 0.2 mg/mL)/**CN** after 1st day (a), 3rd day (b) and 5th day (c) against *P. xylostella* in the greenhouse; Figure S4. Pictures of control efficiency of **CK** after 1st day (a), 3rd day (b) and 5th day (c) against *T. urticae* in the greenhouse; Figure S5. Pictures of control efficiency of **MT/OMT** (3/7, 0.2 mg/mL) after 1st day (a), 3rd day (b) and 5th day (c) against *T. urticae* in the greenhouse; Figure S6. Pictures of control efficiency of **MT/OMT** (3/7, 0.2 mg/mL)/**CN** after 1st day (a), 3rd day (b) and 5th day (c) against *T. urticae* in the greenhouse.

Author Contributions: J.X., S.F., Y.W., H.W. and S.Z. who have made substantial contributions to the present work are named in the manuscript. M.L. and H.X. designed the work, analyzed the data, and wrote and revised the paper. All authors have read and agreed to the published version of the manuscript.

Funding: This research was funded by the National Natural Science Foundation of China (Project No. 32272592).

Institutional Review Board Statement: Not applicable.

Informed Consent Statement: Not applicable.

Data Availability Statement: The dataset utilized in this study is available upon request.

Acknowledgments: H.X. want to thank the National Natural Science Foundation of China (Project No. 32272592).

Conflicts of Interest: The authors declare no conflict of interest.

References

1. Fu, S.; Liu, Z.; Chen, J.; Sun, G.; Jiang, Y.; Li, M.; Xiong, L.; Chen, S.; Zhou, Y.; Asad, M.; et al. Silencing arginine kinase/integrin β1 subunit by transgenic plant expressing dsRNA inhibits the development and survival of *Plutella xylostella*. *Pest Manag. Sci.* **2020**, *76*, 1761–1771. [CrossRef] [PubMed]
2. Vaschetto, L.M.; Beccacece, H.M. The emerging importance of noncoding RNAs in the insecticide tolerance, with special emphasis on *Plutella xylostella* (Lepidoptera: Plutellidae). *Wiley Interdiscip. Rev. RNA* **2019**, *10*, e1539. [CrossRef] [PubMed]
3. Banazeer, A.; Afzal, M.B.S.; Hassan, S.; Ijaz, M.; Shad, S.A.; Serrão, J.E. Status of insecticide resistance in *Plutella xylostella* (Linnaeus) (Lepidoptera: Plutellidae) from 1997 to 2019: Cross-resistance, genetics, biological costs, underlying mechanisms, and implications for management. *Phytoparasitica* **2021**, *50*, 465–485. [CrossRef]
4. Abdelgaleil, S.A.; Badawy, M.E.; Mahmoud, N.F.; Marei, A.E.-S.M. Acaricidal activity, biochemical effects and molecular docking of some monoterpenes against two-spotted spider mite (*Tetranychus urticae* Koch). *Pestic. Biochem. Physiol.* **2019**, *156*, 105–115. [CrossRef]

5. Ndiaye, S.G.; Welty, C. Augmentation and conservation biological control of *Tetranychus urticae* on hops in Ohio. *Biol. Control.* **2022**, *173*, 104980. [CrossRef]
6. Mavridis, K.; Papapostolou, K.M.; Ilias, A.; Michaelidou, K.; Stavrakaki, M.; Roditakis, E.; Tsagkarakou, A.; Bass, C.; Vontas, J. Next-generation molecular diagnostics (TaqMan qPCR and ddPCR) for monitoring insecticide resistance in *Bemisia tabaci*. *Pest Manag. Sci.* **2022**, *78*, 4994–5001. [CrossRef]
7. Shan, J.; Sun, X.; Li, R.; Zhu, B.; Liang, P.; Gao, X. Identification of ABCG transporter genes associated with chlorantraniliprole resistance in *Plutella xylostella* (L.). *Pest Manag. Sci.* **2021**, *77*, 3491–3499. [CrossRef]
8. Yamanaka, T.; Kitabayashi, S.; Jouraku, A.; Kanamori, H.; Kuwazaki, S.; Sudo, M. A feasibility trial of genomics-based diagnosis detecting insecticide resistance of the diamondback moth. *Pest Manag. Sci.* **2022**, *78*, 1573–1581. [CrossRef]
9. Poonsri, W.; Pluempanupat, W.; Chitchirachan, P.; Bullangpoti, V.; Koul, O. Insecticidal alkanes from *Bauhinia scandens* var. horsfieldii against Plutella xylostella L. (Lepidoptera: Plutellidae). *Ind. Crop. Prod.* **2015**, *65*, 170–174. [CrossRef]
10. Venzon, M.; Togni, P.H.B.; Perez, A.L.; Oliveira, J.M. Control of two-spotted spider mites with neem-based products on a leafy vegetable. *Crop Prot.* **2020**, *128*, 105006. [CrossRef]
11. Fan, Q.; Li, X.; Wei, C.; Wang, P.; Sun, H.; Zheng, S.; Li, Y.; Tian, Z.; Liu, J.; Zhang, Y. Extraction, structure characterization and biological activity determination of (S)-(-)-palasonin from *Butea monosperma* (Lam.) Kuntze seeds. *Ind. Crop. Prod.* **2022**, *187*, 115393. [CrossRef]
12. Li, T.; Lv, M.; Wen, H.; Wang, Y.; Thapa, S.; Zhang, S.; Xu, H. Synthesis of piperine-based ester derivatives with diverse aromatic rings and their agricultural bioactivities against *Tetranychus cinnabarinus* Boisduval, *Aphis citricola* Van der Goot, and *Eriosoma lanigerum* Hausmann. *Insects* **2023**, *14*, 40. [CrossRef]
13. Hao, M.; Xu, J.; Wen, H.; Du, J.; Zhang, S.; Lv, M.; Xu, H. Recent advances on biological activities and structural modifications of dehydroabietic acid. *Toxins* **2022**, *14*, 632. [CrossRef]
14. Yang, R.; Lv, M.; Xu, H. Synthesis of piperine analogs containing isoxazoline/pyrazoline scaffold and their pesticidal bioactivities. *J. Agric. Food Chem.* **2018**, *66*, 11254–11264. [CrossRef]
15. Qu, H.; Lv, M.; Yu, X.; Lian, X.; Xu, H. Discovery of some piperine-based phenylsulfonylhydrazone derivatives as potent botanically narcotic agents. *Sci. Rep.* **2015**, *5*, 13077. [CrossRef]
16. Koul, O.; Singh, R.; Kaur, B.; Kanda, D. Comparative study on the behavioral response and acute toxicity of some essential oil compounds and their binary mixtures to larvae of *Helicoverpa armigera*, *Spodoptera litura* and *Chilo partellus*. *Ind. Crop Prod.* **2013**, *49*, 428–436. [CrossRef]
17. Huang, J.-L.; Lv, M.; Xu, H. Semisynthesis of some matrine ether derivatives as insecticidal agents. *RSC Adv.* **2017**, *7*, 15997–16004. [CrossRef]
18. Li, S.; Sun, Z.; Zhang, B.; Lv, M.; Xu, H. Non-food bioactive products: Semisynthesis, biological activities, and mechanisms of action of oximinoether derivatives of matrine from *Sophora flavescens*. *Ind. Crop. Prod.* **2019**, *131*, 134–141. [CrossRef]
19. Xu, J.; Sun, Z.; Hao, M.; Lv, M.; Xu, H. Evaluation of biological activities, and exploration on mechanism of action of matrine-cholesterol derivatives. *Bioorg. Chem.* **2020**, *94*, 103439. [CrossRef]
20. Zhang, B.; Xu, H. Research progress of agricultural bioactivities and structural modifications of matrine and its analogues. *Chin. J. Pestic. Sci.* **2019**, *21*, 609–626.
21. Lv, M.; Ma, Q.; Zhang, S.; Xu, H. Agrochemical properties evaluation of some imines alkaloids of matrine/oxymatrine. *Bioorganic Med. Chem. Lett.* **2021**, *48*, 128246. [CrossRef] [PubMed]
22. Xu, H.; Xu, M.; Sun, Z.; Li, S. Preparation of matrinic/oxymatrinic amide derivatives as insecticidal/acaricidal agents and study on the mechanisms of action against *Tetranychus cinnabarinus*. *J. Agric. Food Chem.* **2019**, *67*, 12182–12190. [CrossRef] [PubMed]
23. Ding, Y.; Li, N.; Sun, J.; Zhang, L.; Guo, J.; Hao, X.; Sun, Y. Oxymatrine inhibits bocavirus MVC replication, reduces viral gene expression and decreases apoptosis induced by viral infection. *Virol. Sin.* **2019**, *34*, 78–87. [CrossRef] [PubMed]
24. Lan, X.; Zhao, J.; Zhang, Y.; Chen, Y.; Liu, Y.; Xu, F. Oxymatrine exerts organ- and tissue-protective effects by regulating in-flammation, oxidative stress, apoptosis, and fibrosis: From bench to bedside. *Pharmacol. Res.* **2020**, *151*, 104541. [CrossRef]
25. Chen, K.; Zhu, P.; Ye, J.; Liao, Y.; Du, Z.; Chen, F.; Juanjuan, H.; Zhang, S.; Zhai, W. Oxymatrine inhibits the migration and invasion of hepatocellular carcinoma cells by reducing the activity of MMP-2/-9 via regulating p38 signaling pathway. *J. Cancer* **2019**, *10*, 5397–5403. [CrossRef]
26. Arena, J.S.; Merlo, C.; Defagó, M.T.; Zygadlo, J.A. Insecticidal and antibacterial effects of some essential oils against the poultry pest *Alphitobius diaperinus* and its associated microorganisms. *J. Pest Sci.* **2019**, *93*, 403–414. [CrossRef]
27. Sun, Y.; Cai, X.; Cao, J.; Wu, Z.; Pan, D. Effects of 1,8-cineole on carbohydrate metabolism related cell structure changes of salmonella. *Front. Microbiol.* **2018**, *9*, 1078. [CrossRef]
28. Meng, C.; Zeng, W.; Lv, J.; Wang, Y.; Gao, M.; Chang, R.; Li, Q.; Wang, X. 1,8-cineole ameliorates ischaemic brain damage via TRPC6/CREB pathways in rats. *J. Pharm. Pharmacol.* **2021**, *73*, 979–985. [CrossRef]
29. de Albuquerque Lima, T.; de Queiroz Baptista, N.M.; de Oliveira, A.P.; da Silva, P.A.; de Gusmão, N.B.; dos Santos Correia, M.T.; Napoleão, T.H.; da Silva, M.V.; Paiva, P.M. Insecticidal activity of a chemotype VI essential oil from *Lippia alba* leaves collected at Caatinga and the major compound (1,8-cineole) against *Nasutitermes corniger* and *Sitophilus zeamais*. *Pestic. Biochem. Physiol.* **2021**, *177*, 104901. [CrossRef]
30. Kumrungsee, N.; Pluempanupat, W.; Koul, O.; Bullangpoti, V. Toxicity of essential oil compounds against diamondback moth, *Plutella xylostella*, and their impact on detoxification enzyme activities. *J. Pest Sci.* **2014**, *87*, 721–729. [CrossRef]

31. Liao, M.; Li, S.; Wu, H.; Gao, Q.; Shi, S.; Huang, Y.; Cao, H. Transcriptomic analysis of *Sitophilus zeamais* in response to limonene fumigation. *Pest. Manag. Sci.* **2022**, *78*, 4774–4782. [CrossRef]
32. Born, F.D.; da Camara, C.A.; de Melo, J.P.; de Moraes, M.M. Acaricidal property of the essential oil from *Lippia gracilis* against *Tetranychus urticae* and a natural enemy, *Neoseiulus californicus*, under greenhouse conditions. *Exp. Appl. Acarol* **2018**, *75*, 491–502. [CrossRef]
33. Kavallieratos, N.G.; Nika, E.P.; Skourti, A.; Boukouvala, M.C.; Ntalaka, C.T.; Maggi, F.; Spinozzi, E.; Petrelli, R.; Perinelli, D.R.; Benelli, G.; et al. *Carlina acaulis* essential oil nanoemulsion as a new grain protectant against different developmental stages of three stored-product beetles. *Pest Manag. Sci.* **2022**, *78*, 2434–2442. [CrossRef]
34. Ebadollahi, A. Chemical composition, acaricidal and insecticidal effects of essential oil from *Achillea filipendulina* against two arthropod pests; *Oryzaephilus surinamensis* and *Tetranychus urticae*. *Toxin Rev.* **2016**, *36*, 132–137. [CrossRef]
35. Ebadollahi, A.; Sendi, J.J.; Aliakbar, A. Efficacy of Nanoencapsulated *Thymus eriocalyx* and *Thymus kotschyanus* Essential Oils by a Mesoporous Material MCM-41 Against *Tetranychus urticae* (Acari: Tetranychidae). *J. Econ. Èntomol.* **2017**, *110*, 2413–2420. [CrossRef]
36. Ebadollahi, A.; Sendi, J.J. Acaricidal potentials of the terpene-rich essential oils of two Iranian *Eucalyptus* species against *Tetranychus urticae* Koch. *J. Oleo Sci.* **2017**, *66*, 307–314. [CrossRef]
37. Huang, X.; Huang, Y.; Yang, C.; Liu, T.; Liu, X.; Yuan, H. Isolation and insecticidal activity of essential oil from *Artemisia la-vandulaefolia* DC. against *Plutella xylostella*. *Toxins* **2021**, *13*, 842. [CrossRef]
38. Tak, J.H.; Isman, M.B. Enhanced cuticular penetration as the mechanism for synergy of insecticidal constituents of rosemary essential oil in *Trichoplusia ni*. *Sci. Rep.* **2015**, *5*, 12690. [CrossRef]
39. Aungtikun, J.; Soonwera, M.; Sittichok, S. Insecticidal synergy of essential oils from *Cymbopogon citratus* (Stapf.), *Myristica fragrans* (Houtt.), and *Illicium verum* Hook. f. and their major active constituents. *Ind. Crop. Prod.* **2021**, *164*, 113386. [CrossRef]
40. Sun, Z.; Lv, M.; Huang, W.; Li, T.; Xu, H. Development of botanical pesticides: Exploration on the phenotype of vestigial wings of insect pests induced by plant natural products or their derivatives by blocking tyrosine phosphorylation of insulin receptor 1. *J. Agric. Food Chem.* **2022**, *70*, 2117–2126. [CrossRef]
41. Cui, G.; Yuan, H.; He, W.; Deng, Y.; Sun, R.; Zhong, G. Synergistic effects of botanical curcumin-induced programmed cell death on the management of *Spodoptera litura* Fabricius with avermectin. *Ecotoxicol. Environ. Saf.* **2021**, *229*, 113097. [CrossRef] [PubMed]
42. Gao, Q.; Shi, Y.; Liao, M.; Xiao, J.; Li, X.; Zhou, L.; Liu, C.; Liu, P.; Cao, H. Laboratory and field evaluation of the aphidicidal activity of moso bamboo (*Phyllostachys pubescens*) leaf extract and identification of the active components. *Pest Manag. Sci.* **2019**, *75*, 3167–3174. [CrossRef] [PubMed]
43. Lu, X.; Liu, J.; Weng, H.; Ma, Z.; Zhang, X. Efficacy of binary combinations between methyl salicylate and carvacrol against thrips *Anaphothrips obscurus*: Laboratory and field trials. *Pest Manag. Sci.* **2019**, *76*, 589–596. [CrossRef] [PubMed]
44. Nong, Q.-Y.; Liu, Y.-A.; Qin, L.-T.; Liu, M.; Mo, L.-Y.; Liang, Y.-P.; Zeng, H.-H. Toxic mechanism of three azole fungicides and their mixture to green alga *Chlorella pyrenoidosa*. *Chemosphere* **2020**, *262*, 127793. [CrossRef] [PubMed]
45. Tak, J.-H.; Jovel, E.; Isman, M.B. Comparative and synergistic activity of *Rosmarinus officinalis* L. essential oil constituents against the larvae and an ovarian cell line of the cabbage looper, *Trichoplusia ni* (Lepidoptera: Noctuidae). *Pest Manag. Sci.* **2015**, *72*, 474–480. [CrossRef]
46. Tak, J.H.; Isman, M.B. Acaricidal and repellent activity of plant essential oil-derived terpenes and the effect of binary mixtures against *Tetranychus urticae* Koch (Acari: Tetranychidae). *Ind. Crop Prod.* **2017**, *108*, 786–792. [CrossRef]
47. Benelli, G.; Pavela, R.; Iannarelli, R.; Petrelli, R.; Cappellacci, L.; Cianfaglione, K.; Afshar, F.H.; Nicoletti, M.; Canale, A.; Maggi, F. Synergized mixtures of Apiaceae essential oils and related plant-borne compounds: Larvicidal effectiveness on the filariasis vector *Culex quinquefasciatus* Say. *Ind. Crop. Prod.* **2017**, *96*, 186–195. [CrossRef]
48. Yuan, L.; Yang, X.; Yu, X.; Wu, Y.; Jiang, D. Resistance to insecticides and synergistic and antagonistic effects of essential oils on dimefluthrin toxicity in a field population of *Culex quinquefasciatus* Say. *Ecotoxicol. Environ. Saf.* **2018**, *169*, 928–936. [CrossRef]
49. Zhou, C.; Yang, H.; Wang, Z.; Long, G.-Y.; Jin, D.-C. Protective and detoxifying enzyme activity and ABCG subfamily gene expression in *Sogatella furcifera* under insecticide stress. *Front. Physiol.* **2019**, *9*, 1890. [CrossRef]
50. Li, Y.; Sun, H.; Tian, Z.; Li, Y.; Ye, X.; Li, R.; Li, X.; Zheng, S.; Liu, J.; Zhang, Y. Identification of key residues of carboxylesterase PxEst-6 involved in pyrethroid metabolism in *Plutella xylostella* (L.). *J. Hazard. Mater.* **2021**, *407*, 124612. [CrossRef]
51. Zhu, W.; Duan, Y.; Chen, J.; Merzendorfer, H.; Zou, X.; Yang, Q. SERCA interacts with chitin synthase and participates in cu-ticular chitin biogenesis in *Drosophila*. *Insect. Biochem. Mol. Biol.* **2022**, *145*, 103783. [CrossRef]
52. Xu, H.; Zhang, K.; Lv, M.; Hao, M. Construction of cholesterol oxime ether derivatives containing isoxazoline/isoxazole fragments and their agricultural bioactive properties/control efficiency. *J. Agric. Food Chem.* **2021**, *69*, 8098–8109. [CrossRef]
53. Wu, C.; Yu, X.; Wang, B.; Liu, J.; Meng, F.; Zhao, Y.; Xiong, L.; Yang, N.; Li, Y.; Li, Z. Synthesis, insecticidal evaluation, and 3D-QASR of novel anthranilic diamide derivatives containing N-arylpyrrole as potential ryanodine receptor activators. *J. Agric. Food Chem.* **2020**, *68*, 9319–9328. [CrossRef]
54. Sun, Y.P.; Johnson, E.R. Analysis of joint action of insecticides against house flies. *J. Econ. Entomol.* **1960**, *53*, 887–892. [CrossRef]
55. Nian, X.; He, Y.; Lu, L.; Zhao, R. Evaluation of alternative *Plutella xylostella* control by two *Isaria fumosorosea* conidial formu-lations-oil-based formulation and wettable powder-combined with *Bacillus thuringiensis*. *Pest Manag. Sci.* **2015**, *71*, 1675–1684. [CrossRef]

56. Hao, M.; Sun, Z.; Xu, J.; Lv, M.; Xu, H. Semisynthesis and pesticidal activities of derivatives of the diterpenoid andrographolide and investigation on the stress response of *Aphis citricola* Van der Goot (Homoptera: Aphididae). *J. Agric. Food Chem.* **2020**, *68*, 4131–4143. [CrossRef]
57. Bradford, M.M. A rapid and sensitive method for the quantitation of microgram quantities of protein utilizing the principle of protein-dye binding. *Anal. Biochem.* **1976**, *72*, 248–254. [CrossRef]
58. Li, T.; Lv, M.; Wen, H.; Wang, J.; Wang, Z.; Xu, J.; Fang, S.; Xu, H. High value-added application of natural plant products in crop protection: Construction and pesticidal activities of piperine-type ester derivatives and their toxicology study. *J. Agric. Food Chem.* **2022**, *70*, 16126–16134. [CrossRef]
59. Zhou, H.; Liu, S.; Wan, F.; Jian, Y.; Guo, F.; Chen, J.; Ning, Y.; Ding, W. Graphene oxide–acaricide nanocomposites advance acaricidal activity of acaricides against *Tetranychus cinnabarinus* by directly inhibiting the transcription of a cuticle protein gene. *Environ. Sci. Nano* **2021**, *8*, 3122–3137. [CrossRef]
60. Togawa, T.; Dunn, A.; Emmons, A.C.; Willis, J.H. CPF and CPFL, two related gene families encoding cuticular proteins of *Anopheles gambiae* and other insects. *Insect Biochem. Mol. Biol.* **2007**, *37*, 675–688. [CrossRef]

toxins

MDPI

Article

Insecticidal Activities of *Sophora flavescens* Alt. towards Red Imported Fire Ants (*Solenopsis invicta* Buren)

Yongqing Tian and Zhixiang Zhang *

Key Laboratory of Natural Pesticide and Chemical Biology, Ministry of Education, South China Agricultural University, Guangzhou 510642, China
* Correspondence: zdsys@scau.edu.cn

Abstract: The red imported fire ant (*Solenopsis invicta*) is a worldwide invasive and dangerous insect that is controlled mainly by chemical insecticides. Plant-derived insecticidal compounds are generally better than synthetic insecticides for environmental compatibility and the biosafety of non-targets. The toxicity of the ethanol extract of *Sophora flavescens* roots against *S. invicta* was evaluated under laboratory conditions. The ethanol extract showed toxicity against minor and medium workers of *S. invicta* with 7-day LC_{50} values of 1426.25 and 2292.60 mg/L, respectively. By bioactivity-directed chromatographic separations using the minor worker as the test insect, two active compounds, matrine and sophocarpine, were isolated from the *S. flavescens* total alkaloids; their chemical structure was identified by ^{13}C NMR data. Matrine showed toxicities against minor and medium workers with 7-day LC_{50} values of 46.77 and 71.49 mg/L, respectively, and for sophocarpine, 50.08 and 85.87 mg/L, respectively. The two compounds could substantially reduce the foraging response, food consumption, and aggregation of *S. invicta* workers at a sublethal concentration of 15 mg/L. The present research suggests that *S. flavescens* roots have potential as a natural control agent for red imported fire ants.

Keywords: *Sophora flavescens*; *Solenopsis invicta*; matrine; sophocarpine

Key Contribution: This research has shown that the *Sophora flavescens* Alt. root was toxic to the red imported fire ant (*Solenopsis invicta* Buren), and the two insecticidal compounds, matrine and sophocarpine, were elucidated. The results suggest that *S. flavescens* roots could potentially be used as a botanical pesticide for controlling red imported fire ants.

1. Introduction

The red imported fire ant (RIFA), *Solenopsis invicta* Buren, is one of the most notoriously known insects in the world [1]. This ant is native to South America and is a global invasive insect [2], and at present, it is distributed in many countries, including the United States, Mexico, Australia, New Zealand, China, Malaysia, Singapore, and the West Indies. This insect can cause harm to humans, native ecosystems, and economic activities worldwide [3]. In infested areas, RIFAs are commonly found in gardens, lawns, parks, school yards, golf courses, and on roadsides. This ant is very harmful to human health; they will attack a person who tramples or touches their nests by stings. The venom can cause intense burning, itching, blistering, or pustules, which will last for several weeks [4,5]. The broken or scratched pustules can harbor secondary bacterial infections. In worse cases, severe allergic reactions can result in anaphylactic shock and even death [6]. The ants reproduce rapidly in invaded areas and negatively impact native ecosystems and wildlife [7]. These ants can also inflict structural damage; for example, ant colonies in electrical circuitry can cause shorted air conditioners, and they may also inhabit telephone junction boxes, traffic and light control boxes, and transformers [3].

Management of RIFAs mostly relies on synthetic insecticide treatment in nests and surrounding areas. Although synthetic insecticides can be used to rapidly control RIFAs, their drawbacks cannot be ignored, such as environmental contamination, killing non-target organisms. Therefore, to solve these problems, sustainable approaches have been explored and integrated into pest management projects. Plant-derived compounds are one of the most attractive alternatives to synthetic insecticides [8], because they have many desirable attributes, such as an eco-friendly nature, availability, safety, acceptability, and minimal side effect on beneficial organisms, thus plants could be used as alternatives to synthetic chemical pesticides [9]. In the past, many scholars studied the activities of plants towards RIFAs, and these results were summarized by Chen and Oi [8]. According to this review, the active compounds of plant origin included potential fire ant bait active ingredients, contact toxins, repellants, and fumigants. Among these active compounds, two conventional botanical insecticides, pyrethrins, and rotenone, together with seven plant essential oils, namely d-limonene, clove oil, cotton seed oil, lemongrass oil, peppermint oil, pine oil, rosemary oil, and turpentine, have been registered for use against fire ants in the U.S.

In our studies on insecticidal plants, we found that a crude ethanol extract from the root of *Sophora flavescens* Alt., a herbaceous subshrub grown in China, was toxic to RIFA workers. In this research, we tested the bioactivities of total alkaloids from the root ethanol extract under laboratory conditions, then isolated two active compounds, matrine and sophocarpine, using bioactivity-guided separations. We also found that the alkaloids in the *S. flavescens* root had sublethal effects towards RIFAs, including a reduced foraging response, food consumption, and aggregation. Our study demonstrated that *S. flavescens* has potential as a biopesticide for controlling RIFAs.

2. Results

2.1. Structure Determination of Insecticidal Compounds

By bioactivity-directed separation, two active compounds were isolated and identified to be matrine and sophocarpine (Figure 1) based on the ^{13}C NMR data (Figures 2 and 3; Table 1) which were in agreement with the reported data [10,11].

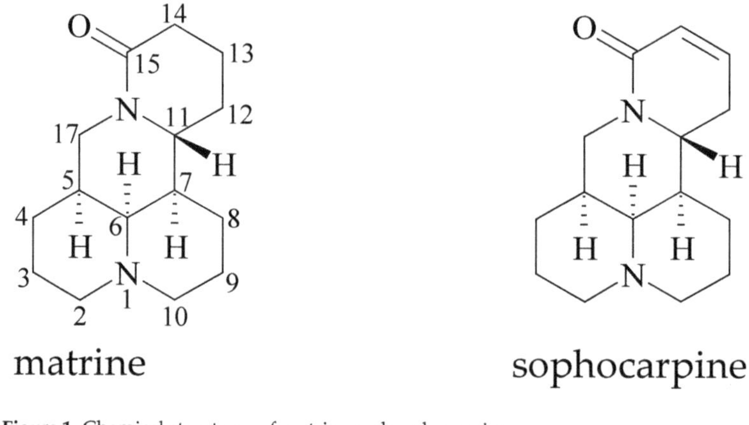

Figure 1. Chemical structures of matrine and sophocarpine.

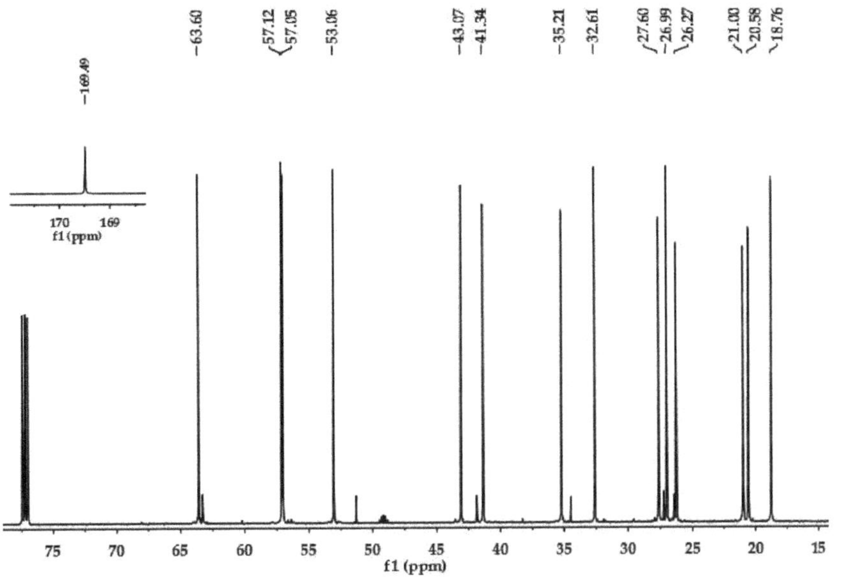

Figure 2. ^{13}C NMR spectrum of matrine (in $CDCl_3$).

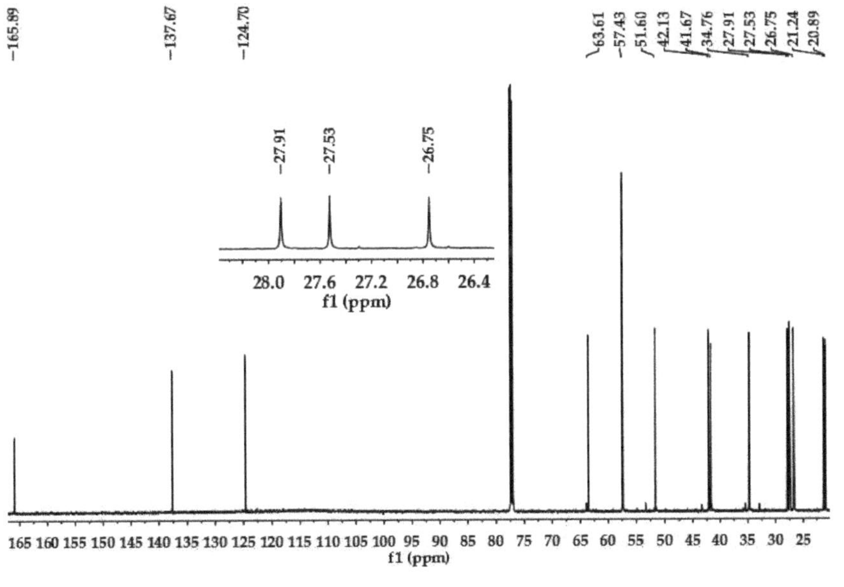

Figure 3. ^{13}C NMR spectrum of sophocarpine (in $CDCl_3$).

Table 1. ^{13}C NMR assignments of matrine and sophocarpine in $CDCl_3$.

Assignment	Matrine (δ)		Sophocarpine (δ)	
	Observed	Reference [10]	Observed	Reference [11]
2	57.12	57.6	57.43	57.3
3	21.00	21.4	21.24	21.1
4	27.60	27.9	27.91	27.8
5	35.21	35.6	34.76	34.7
6	63.60	64.0	63.61	63.5
7	43.07	43.4	41.67	41.6
8	26.27	26.6	26.75	26.6
9	20.58	21.0	20.89	20.8
10	57.05	57.4	57.43	57.3
11	53.06	53.4	51.60	51.5
12	26.99	27.4	27.53	27.4
13	18.76	19.2	137.67	137.3
14	32.61	33.0	124.70	124.5
15	169.49	169.7	165.89	165.4
17	41.34	41.7	42.13	41.9

2.2. Toxicities of Active Compounds

A bioassay showed that *S. flavescens* root ethanol extract possessed activities against minor and medium workers of RIFAs; the 7-day LC_{50} values were 1426.25 and 2292.60 mg/L (Tables 2 and 3), respectively. Because alkaloids are the typical compounds contained in *S. flavescens*, the total alkaloids were separated from the ethanol extract using an acid solution and alkali isolation procedure; the subsequent bioassay showed the total alkaloids had higher activities than those of the ethanol extract; the 7-day LC_{50} values towards minor and medium RIFA workers were 214.67 and 371.49 mg/L, respectively, while the rest of the substance possessed no activity, which suggested that the active compounds were in the total alkaloids. Therefore, the total alkaloids were further separated, guided by the bioassay using the minor workers, and finally two active compounds, matrine and sophocarpine, were obtained. The bioactivities of matrine and sophocarpine were much higher than those of the root ethanol extract and total alkaloids; the 7-day LC_{50} values of matrine towards the minor and medium workers were 46.77 and 71.49 mg/L respectively, and for sophocarpine, 50.08 and 85.87 mg/L (Tables 2 and 3), respectively.

Table 2. LC_{50} values (mg/L) of the *S. flavescens* ethanol extract, total alkaloids, matrine, and sophocarpine against RIFA minor workers.

Treatment	Regression Equations	7-Day LC_{50} (95% Confidence Interval)	Correlation Coefficient (R^2)
S. flavescens ethanol extract	Y = −0.3859 + 1.7075 x	1426.25 (1201.21~1693.45)	0.9838
Total alkaloids	Y = 1.7000 + 1.4152 x	214.67 (169.25~272.27)	0.9982
Matrine	Y = 1.9605 + 1.8201 x	46.77 (39.74~55.04)	0.9953
Sophocarpine	Y = 2.1387 + 1.6835 x	50.08 (42.12~59.54)	0.9969
Fipronil (control)	Y = 5.1185 + 2.2215 x	0.88 (0.76~1.03) [1]	0.9919

[1] 1-day LC_{50} value.

Table 3. LC_{50} values (mg/L) of the *S. flavescens* ethanol extract, total alkaloids, matrine, and sophocarpine against RIFA medium workers.

Treatment	Regression Equation	7-Day LC_{50} (95% Confidence Interval)	Correlation Coefficient (R^2)
S. flavescens ethanol extract	Y = 1.2697 + 1.1101 x	2292.60 (1740.50~3019.84)	0.9965
Total alkaloids	Y = 1.5613 + 1.3380 x	371.49 (297.38~464.07)	0.9900
Matrine	Y = 1.7937 + 1.7292 x	71.49 (60.28~84.77)	0.9952
Sophocarpine	Y = 1.6500+ 1.7323 x	85.87 (72.18~102.15)	0.9858
Fipronil (control)	Y = 4.7842 + 1.6863 x	1.34 (1.13~1.60) [1]	0.9920

[1] 1-day LC_{50} value.

2.3. Effects of Active Compounds on the Foraging Response of RIFAs

As shown in Figure 4, at sublethal concentrations, alkaloids in the *S. flavescens* root could affect the foraging response of RIFAs. After being treated with matrine, there were no significant differences in feeding rates among the three treatments during the first two days. On day 3, the feeding rate of 15 mg/L matrine was significantly lower than that of the control, but that of 7.5 mg/L matrine and the control did not differ significantly. Starting from day 4, the feeding rate of the three treatments became significantly different, and the differences continued to day 5. On day 5, the feeding rates of the control and 7.5 and 15 mg/L matrine were 96.67%, 82.35%, and 71.13%, respectively (Figure 4A). After being treated with sophocarpine, the pattern of the feeding rate was similar to that of matrine. On day 5, there were significant differences in the feeding rates among the control and 7.5 and 15 mg/L sophocarpine; the corresponding feeding rates were 96.67%, 84.73%, and 73.69%, respectively (Figure 4B). As for those of the total-alkaloids-treated groups, a slightly different pattern was observed. During the first three days, there were no significant differences in the feeding rates among the three treatments. On day 5, the feeding rates of 7.5 mg/L (88.34%) and 15 mg/L (84.17%) of total alkaloids were similar but significantly less than that of the control (96.67%) (Figure 4C).

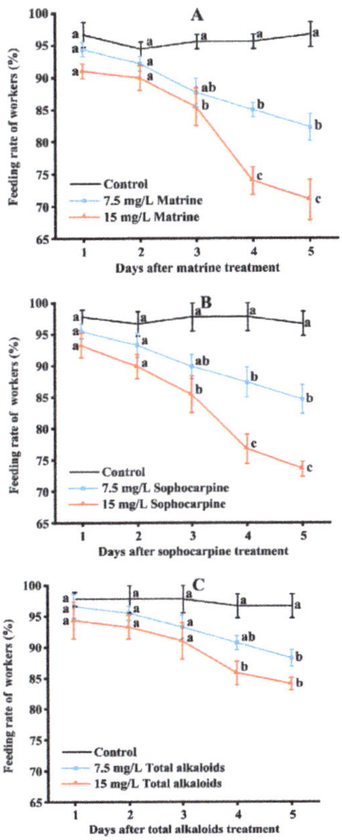

Figure 4. Foraging response of RIFA minor workers after being treated with matrine (**A**), sophocarpine (**B**), and total alkaloids (**C**). Data are presented as mean ± S.E. Different letters on each sampling day indicate significant differences per parameter among treatments at the $p < 0.05$ level based on Tukey's HSD test ($n = 3$).

2.4. Effects of Active Compounds on Food Consumption of RIFAs

The consumption of *T. molitor* (mealworm) by the ant workers treated with matrine, sophocarpine, and total alkaloids were examined. After being treated with matrine, there were no significant differences in the amount of *T. molitor* consumption among the three treatments on day 0 and day 3. On day 6, the consumption amount of workers treated with 15 mg/L matrine was significantly lower than that of the control, but there was no significant difference in the 7.5 mg/L treated group and the control. On day 9, significant differences could be seen among the three treated groups, and these differences continued to day 12. On day 12, the consumption rates of the control and the 7.5- and 15 mg/L matrine-treated groups were 0.14, 0.09, and 0.06 mg per ant, respectively (Figure 5A). After being treated with sophocarpine, the results were similar to those of matrine. On day 12, the consumption rates of the control and 7.5- and 15 mg/L sophocarpine treated groups were 0.16, 0.11, and 0.07 mg per ant, respectively (Figure 5B). As for the total alkaloids-treated groups, there were no significant differences among the three treatments during the first six days. On day 9, the *T. molitor* consumption of the 15 mg/L total-alkaloids-treated group was significantly less than that of the control, but the consumption of 7.5 mg/L total alkaloids and the control did not differ significantly. On day 12, there were significant differences among the control and the 7.5- and 15 mg/Ltotal alkaloids; the corresponding *T. molitor* consumption rates were 0.16, 0.13, and 0.09 mg per ant, respectively (Figure 5C).

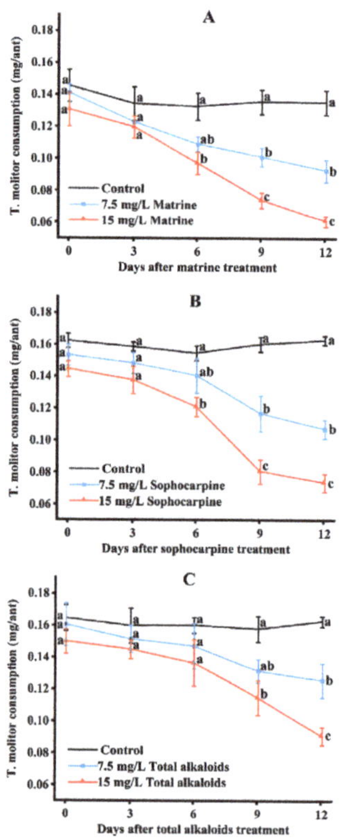

Figure 5. The amount of *T. molitor* consumed by RIFA minor workers after treatment with matrine (**A**), sophocarpine (**B**), and total alkaloids (**C**). Data are presented as mean ± S.E. Different letters on each sampling day indicate significant differences per parameter among treatments at the $p < 0.05$ level based on Tukey's HSD test ($n = 3$).

2.5. Effects of Active Compounds on RIFA Aggregation

Alkaloids in the *S. flavescens* root could affect aggregation of RIFAs. After being treated with matrine (Figure 6A), no differences were recorded on day 0. From day 3 to day 6, the aggregation rate of the 15 mg/L matrine-treated group was significantly lower than the control, but the 7.5 mg/L matrine-treated group was not significantly different from the control. From day 9 to 12, the 7.5 mg/L matrine-treated group also showed significant differences to the control. During the whole experiment period, there were no differences between the 7.5 and 15 mg/L matrine-treated groups. The aggregation rate of workers treated with sophocarpine showed a similar tendency (Figure 6B). As for the total-alkaloids-treated groups, the 15 mg/L-treated group was significantly lower than the control from day 3 to day 12, but no significant differences occurred between the 7.5 mg/L-treated group and the control during the 12 days of evaluation (Figure 6C).

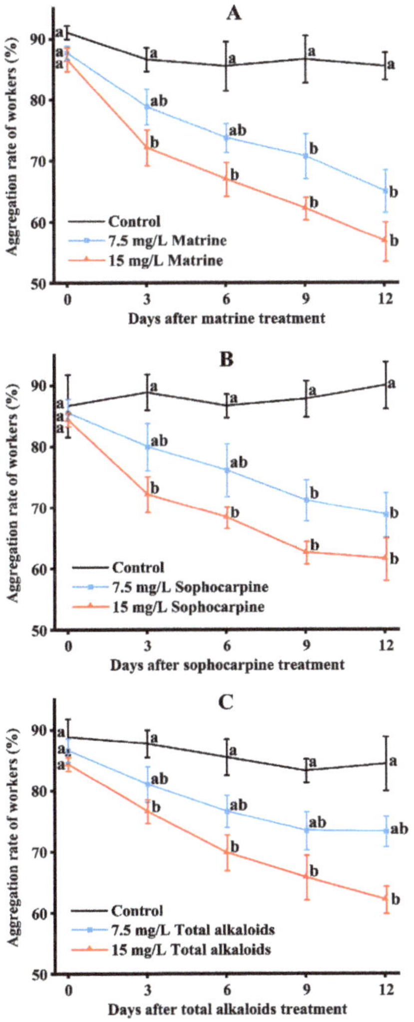

Figure 6. The aggregating rate of RIFA minor workers fed with matrine (**A**), sophocarpine (**B**), and total alkaloids (**C**). Data are presented as mean ± S.E. Different letters on each sampling day indicate significant differences per parameter among treatments at the $p < 0.05$ level based on Tukey's HSD test ($n = 3$).

3. Discussion

The reported methods for controlling RIFAs included the use of chemical insecticides, such as hydramethylnon, indoxacarb [12], and fipronil [13], or biological control agents, such as entomopathogenic fungus *Beauveria bassiana* [14]. Plant-derived chemicals have attracted researchers' attention because of their biodegradability, less harm to non-target organisms, and minimum effects on pest resistance [15]. Several plants have been studied that show activities towards RIFAs, such as *Syzygium aromaticum* [16], *Acorus calamus* [17], and *Cinnamomum osmophloeum* [18]. As far as is known, there has been no report on the use of *S. flavescens* for controlling this type of ant. This research showed that the ethanol extract of the *S. flavescens* root had activity towards RIFAs; the main active compounds were two alkaloids, matrine and sophocarpine. The two compounds could not only kill the RIFA workers, but also, at sublethal concentrations, they resulted in harmful chronic effects, including reduced foraging responses, consumption of *T. molitor*, and reduced aggregation. These chronic effects could affect their ability to attack other organisms and colony survival, which suggested another way of controlling this type of ant.

Approximately 300 compounds have been identified in *S. flavescens* [19,20], with most of them belonging to alkaloids. *S. flavescens* possesses diverse pharmacological properties including antianaphylaxis and antimicrobial, immunoregulatory, and cardioprotective activities [21]. These therapeutic effects of *S. flavescens* may be derived from complex interactions among its compounds. Inspired by the therapeutic effects, we conducted this research, and found the total alkaloids had bioactivity against RIFA workers, while the rest of the substance of the *S. flavescens* root ethanol extract showed no activity. Generally, the total alkaloids contained many compounds, but only those with high contents could be easily isolated, and it was difficult to isolate the minor constituents owing to their low contents. Thus, only two active compounds with high contents were isolated from the *S. flavescens* root in this work. According to the existing phytochemical research results, compounds in the same plant always have similar chemical structures and it has been suggested that they have similar bioactivities. Therefore, the minor alkaloids in the *S. flavescens* root may also have bioactivities against red fire ants, which needs further research.

The dried roots of *S. flavescens*, a traditional Chinese medicine [22], have been widely used in Korea, Japan, and China for the treatment of diarrhea, inflammation, abscesses, dysentery, and fever in East Asian countries [23]. Modern research has proven that compounds in this plant have clinical significance [24], such as treating diabetic retinopathy, tumour or cancer [25,26], and aerobic vaginitis [27]. These results show the *S. flavescens* compounds are safe for humans to some extent.

At present, the botanical pesticides are mainly maded from plant materials. *S. flavescens* can grow in more than ten provinces in China, and also distribute in Korea, Japan, India, etc., which make it possible to provide enough material for pesticide production. According to the above discussion, it can be preliminarily concluded that *S. flavescens* is a promising plant for controlling RIFAs.

4. Conclusions

In summary, this research demonstrated that the alkaloids in *S. flavescens* roots are lethal to RIFAs; this toxicity is mainly attributed to matrine and sophocarpine. At sublethal concentrations, the alkaloids could result in harmful chronic effects, including reduced foraging responses, food consumption, and aggregation, which could affect their ability to attack other organisms and colony survival. Thus, the alkaloids in *S. flavescens* roots have potential as natural insecticides for controlling RIFAs by either directly killing them at a high concentration or them reducing viability at a lower concentration. However, this work is only the beginning of developing *S. flavescens* alkaloids as an ideal control agent; the applicability and stability need to be further investigated.

5. Materials and Methods

5.1. Plant and Insects

Plant material: The roots of *S. flavescens* were purchased as a commercially available product from a standard drugstore, because it is a very common traditional Chinese medicine.

The origin and rearing of the RIFA workers: Workers of RIFAs were collected from the suburbs of Guangzhou city and were classified as minor (2.8–3.0 mm body length and 0.6–0.7 mm head width) and medium (3.5–3.7 mm body length and 0.8–0.9 mm head width) workers. The collected ants were fed with a mixture of 10% honey and ham sausage. A test tube (25 mm × 200 mm) partially filled with water and plugged with cotton was used as a water source. The ants were maintained in the laboratory at 25 ± 2 °C.

5.2. Chemicals and Reagents

Silica gel (60–100 and 100–200 mesh) and TLC precoated plates (GF_{254}) were bought from Qingdao Marine Chemical Ltd., Qingdao, China. Sephadex LH-20 was bought from GE Healthcare Bio-Sciences AB, Uppsala, Sweden. $CDCl_3$ was bought from Cambridge Isotope Laboratories, Inc., Andover, MA, USA. The organic solvents used in this work, ethanol, MeOH, EtOAc, CH_2Cl_2, and $CHCl_3$, were all of analytical grade.

5.3. Extraction and Compound Isolation

Extraction and bioactivity-directed isolation of matrine and sophocarpine: The powdered dry roots of *S. flavescens* (6 kg) was extracted by the method of immersion with ethanol three times at room temperature; the ethanol solutions were combined and concentrated in vacuo. This crude ethanol extract (900.9 g) was subsequently dissolved in aqueous hydrochloric acid (to pH = 2) and filtered. The filtered aqueous acid layer was made alkaline to pH = 9 with $NH_3 \cdot H_2O$, then extracted three times with EtOAc to give 1.3 g total alkaloids. The total alkaloids and the rest of the substance were bioassayed using minor RIFA workers, which showed that the active compounds were in the total alkaloids and the rest of the substance possessed no activity. The total alkaloids were chromatographed over a silica gel column (100–200 mesh) and eluted with CH_2Cl_2–MeOH mixtures of increasing polarities (60:1–60:3), to yield seven fractions (I–VII) according to the TLC analysis. The activities of the fractions I–VII against *S. invicta* minor workers were tested at concentration of 500 mg/L. Fractions III and V, obtained on elution with CH_2Cl_2–MeOH (60:1–60:2), demonstrated strong activities towards *S. invicta* workers and caused 86.67% and 90.00% mean mortalities after 7-day treatment, respectively. The mean mortality of fraction IV was 38.89% and the rest of the fractions did not display any activity. Fractions III and V were respectively separated by a Sephadex LH-20 column, eluted with $CHCl_3$–MeOH (1:1), to give two active insecticidal compounds, sophocarpine (70.8 mg) and matrine (90.2 mg).

The ^1H (600 MHz) and ^{13}C (150 MHz) NMR spectra of the isolated compounds were recorded in $CDCl_3$ on a Bruker AV-600 instrument using TMS as an internal reference. For column chromatography, silica gel and Sephadex LH-20 were used. TLC was performed on precoated plates with detection under a fluorescent (λ = 254 nm) light, in iodine vapor and using Dragendroff's reagent.

5.4. Toxicities of Active Compounds

The activity of the crude ethanol extract of the *S. flavescens* root, the total alkaloids, and the active compounds against RIFA workers were evaluated using the "water tube" method [28] with slight modifications. The water source was a test tube (10 mm × 30 mm) filled approximately two-thirds full with solution and tightly plugged with a saturated cotton. The solution in the test tube contained the abovementioned crude ethanol extract, total alkaloids, and the active compounds dissolved in a DMSO/water (1:99, *v*/*v*) mixture. The water test tube and worker ants were placed in a disposable plastic cup (top/bottom/height: 62 mm/40 mm/60 mm) whose vertical wall was precoated with a fluon emulsion and allowed to dry for 24 h to prevent the ants from escaping. Each treatment was replicated three times, and each replicate included 30 worker ants. The test tube in the control group was filled with a DMSO/water (1:99, *v*/*v*) mixture. The treated workers were maintained at 24 °C to 26 °C and 60% to 80% relative humidity and were fed with ham sausage. The average mortalities of the three replications at each concentration were recorded on the 7th day after treatment and were corrected by Abbott's formula [29]. The median of the lethal concentration value (LC_{50}), which was defined as the concentration causing 50% mortality, was determined using probit analysis.

For activity-directed separation, the mortalities of fractions from the subsequent separation procedure towards RIFA workers were assessed at the concentration of 500 or 1000 mg/L using the method described above.

5.5. Effects of Active Compounds on the Foraging Response of RIFAs

RIFA workers were treated in the same manner as those mentioned above at sublethal concentrations, 7.5 and 15 mg/L. Minor workers, about 3 mm in length [18,30], were used for recording the foraging responses. A total of 30 workers were tested for each treatment and the control group, and each treatment had three replicates. The foraging responses were observed at the 24th, 48th, 72nd, 96th, and 120th hour after treatment. Workers were considered to possess foraging abilities if they could feed on the fresh sausage within 10 min after the fresh sausage was added. The foraging response of the RIFAs was evaluated by the feeding rate; the equation was as follows: feeding rate (%) = (number of ants possessing a feeding ability/number of alive workers) × 100.

5.6. Effects of Active Compounds on the Food Consumption of RIFAs

RIFA minor workers were treated in the same manner as the above method at sublethal concentrations. On the 0, 3rd, 6th, 9th, and 12th day, *T. molitors* was weighed and then added to the plastic cup as ant food. After 24 h of feeding, the *T. molitors* was removed from the cup and weighed to determine the consumption of *T. molitor* (mg *T. molitor* per ant), then the average consumption was calculated. A total of 60 minor workers were tested for each treatment and the control group, and each treatment had three replicates. The consumption of *T. molitor* was calculated using the following equation: consumption of *T. molitor* (%) = (weight loss of *T. molitor*/number of alive ants) × 100.

5.7. Effects of Active Compounds on the Aggregating Rate of RIFAs

RIFA workers were treated in the same manner as the above method at sublethal concentrations. A total of 30 minor workers were tested for each treatment and the control group, and all treatments were replicated thrice. Aggregation was measured every 3 days up to 12 days of treatment; aggregation was defined as the gathering of more than five workers. The aggregation rate was calculated using the following equation: aggregation rate (%) = (number of aggregated workers/number of alive workers) × 100.

5.8. Data Analysis

The collected data from the aforementioned experiments were subjected to one-way analysis of variance (ANOVA) using SPSS Statistics, Version 17.0, 2009 (International Business Machines Corporation, Armonk, NY, USA). If significant differences occurred among the treatments, the means were separated by Tukey's honestly significant difference (HSD) test at the $p < 0.05$ level. The means were presented in graphs with standard error bars which were drawn using OriginPro (Version 2021, OriginLab Corporation, Northampton, MA, USA).

Author Contributions: Conceptualization, Y.T. and Z.Z.; methodology, Y.T. and Z.Z.; validation, Y.T. and Z.Z.; formal analysis, Y.T.; investigation, Y.T. and Z.Z.; resources, Y.T. and Z.Z.; data curation, Y.T. and Z.Z.; writing—original draft preparation, Y.T.; writing—review and editing, Y.T. and Z.Z.; visualization, Y.T. and Z.Z.; supervision, Z.Z.; project administration, Y.T. and Z.Z.; funding acquisition, Y.T. All authors have read and agreed to the published version of the manuscript.

Funding: This work was supported by Science and Technology Planning Project of Guangdong Provinvce (2019B030316021).

Institutional Review Board Statement: Not applicable.

Informed Consent Statement: Not applicable.

Data Availability Statement: Data are contained within the article.

Conflicts of Interest: The authors declare no conflict of interest.

References

1. Zhang, R.Z.; Li, Y.C.; Liu, N.; Porter, S.D. An overview of the red imported fire ant (Hymenoptera: Formicidae) in mainland China. *Fla. Entomol.* **2007**, *90*, 723–731. [CrossRef]
2. Wetterer, J.K. Exotic spread of *Solenopsis invicta* Buren (Hymenoptera: Formicidae) beyond North America. *Sociobiology* **2013**, *60*, 50–55. [CrossRef]
3. Vinson, S.B. Impact of the invasion of the imported fire ant. *Insect Sci.* **2013**, *20*, 439–455. [CrossRef] [PubMed]
4. Fu, J.T.; Tang, L.; Li, W.S.; Wang, K.; Cheng, D.M.; Zhang, Z.X. Fumigant toxicity and repellence activity of camphor essential oil from *Cinnamonum camphora* Siebold against *Solenopsis invicta* workers (Hymenoptera: Formicidae). *J. Insect Sci.* **2015**, *15*, 129–134. [CrossRef] [PubMed]
5. Addesso, K.M.; Oliver, J.B.; O'Neal, P.A.; Youssef, N. Efficacy of nootka oil as a biopesticide for management of imported fire ants (Hymenoptera: Formicidae). *J. Econ. Entomol.* **2017**, *110*, 1547–1555. [CrossRef]
6. Wang, L.; Lu, Y.Y.; Li, R.P.; Zeng, L.; Du, J.L.; Huang, X.; Xu, Y.J. Mental health effects caused by red imported fire ant attacks (*Solenopsis invicta*). *PLoS ONE* **2018**, *13*, e0199424. [CrossRef]
7. Epperson, D.M.; Allen, C.R. Red imported fire ant impacts on up-land arthropods in southern Mississippi. *Am. Midl. Nat.* **2010**, *163*, 54–63. [CrossRef]
8. Chen, J.; Oi, D.H. Naturally occurring compounds/materials as alternatives to synthetic chemical insecticides for use in fire ant management. *Insects* **2020**, *11*, 758. [CrossRef]
9. Isman, M.B. Botanical insecticides: For richer, for poorer. *Pest Manag. Sci.* **2008**, *64*, 8–11. [CrossRef]
10. Bai, G.Y.; Wang, D.Q.; Ye, C.H.; Liu, M.L. ^{1}H and ^{13}C chemical shift assignments and stereochemistry of matrine and oxymatrine. *Appl. Magn. Reson.* **2002**, *23*, 113–121. [CrossRef]
11. Xiao, P.; Kubo, H.; Komiya, H.; Higashiyama, K.; Yan, Y.N.; Li, J.S.; Ohmiya, S. Lupin alkaloids from seeds of *Sophora viciifolia*. *Phytochemistry* **1999**, *50*, 189–193. [CrossRef]
12. Allen, H.R.; Miller, D.M. Field evaluations of broadcast, and individual mound treatments for red imported fire ant, *Solenopsis invicta* Buren, (Hymenoptera: Formicidae) control in Virginia, USA. *Sociobiology* **2015**, *62*, 10–17. [CrossRef]
13. Ipser, R.M.; Gardner, W.A. Impact of broadcast applications of hydramethylnon and fipronil on red imported fire ant (Hymenoptera: Formicidae) and nontarget arthropods. *J. Entomol. Sci.* **2010**, *45*, 58–61. [CrossRef]
14. Kafle, L.; Wu, W.J.; Kao, S.S.; Shih, C.J. Efficacy of *Beauveria bassiana* against the red imported fire ant, *Solenopsisinvicta* (Hymenoptera: Formicidae), in Taiwan. *Pest Manag. Sci.* **2011**, *67*, 1434–1438. [CrossRef]
15. Issakul, K.; Araya, J.; Pawelzik, E.; Jatisatienr, C. Potential of *Mammea siamensis* as a botanical insecticide: Its efficiency on diamondback moth and side effects on non-target organisms. *J. Med. Plant Res.* **2011**, *5*, 2149–2156.
16. Kafle, L.; Shih, C.J. Toxicity and repellency of compounds from clove (*Syzygium aromaticum*) to red imported fire ants *Solenopsis invicta* (Hymenoptera: Formicidae). *J. Econ. Entomol.* **2013**, *106*, 131–135. [CrossRef]
17. Kafle, L.; Shih, C.J. Insecticidal activities of compounds from sweet flag (*Acorus calamus*) against red imported fire ants *Solenopsis invicta* (Hymenoptera: Formicidae). *Sociobiology* **2017**, *64*, 398–403. [CrossRef]
18. Cheng, S.S.; Liua, J.Y.; Lina, C.Y.; Hsui, Y.R.; Lu, M.C.; Wu, W.J.; Chang, S.T. Terminating red imported fire ants using *Cinnamomum osmophloeum* leaf essential oil. *Biores. Technol.* **2008**, *99*, 889–893. [CrossRef]
19. Li, F.; Yang, Y.G.; Gu, L.H.; Yang, L.; Wang, Z.T. Advances in chemical constituents and bioactivities of *Sophorae flavescentis* Radix. *Shanghai J. Tradit. Chin. Med.* **2021**, *55*, 84–100.
20. Zhang, Y.B.; Luo, D.; Yang, L.; Cheng, W.; He, L.J.; Kuang, G.K.; Li, M.M.; Li, Y.L.; Wang, G.C. Matrine-type alkaloids from the roots of *Sophora flavescens* and their antiviral activities against the hepatitis B virus. *J. Nat. Prod.* **2018**, *81*, 2259–2265. [CrossRef]
21. He, X.; Fang, J.; Huang, L.; Wang, J.; Huang, X. *Sophora flavescens* Alt.: Traditional usage, phytochemistry and pharmacology of an important traditional Chinese medicine. *J. Ethnopharmacol.* **2015**, *172*, 10–29. [CrossRef] [PubMed]
22. Weng, Z.B.; Guo, S.; Qian, D.W.; Zhu, Z.H.; Zhang, S.; Li, A.P.; Lei, Z.H.; Duan, J.N. *Sophora flavescens* seed as a promising high potential by-product: Phytochemical characterization and bioactivity evaluation. *Ind. Crops. Prod.* **2017**, *109*, 19–26. [CrossRef]
23. Yim, D.; Kim, M.J.; Shin, Y.; Lee, S.J.; Shin, J.G.; Kim, D.H. Inhibition of cytochrome p450 activities by *Sophora flavescens* extract and its prenylated flavonoids in human liver microsomes. *Evid.-Based Compl. Alt.* **2019**, *2019*, 2673769. [CrossRef] [PubMed]
24. Sun, P.; Zhao, W.J.; Wang, Q.; Chen, L.L.; Sun, K.K.; Zhan, Z.S.; Wang, J.F. Chemical diversity, biological activities and traditional uses of and important Chinese herb *Sophora*. *Phytomedicine* **2022**, *100*, 154054. [CrossRef] [PubMed]
25. Chen, M.H.; Gu, Y.Y.; Zhang, A.L.; Sze, D.M.Y.; Mo, S.L.; May, B.H. Biological effects and mechanisms of matrine and other constituents of *Sophora flavescens* in colorectal cancer. *Pharmacol. Res.* **2021**, *171*, 105778. [CrossRef]

26. Long, G.Q.; Wang, D.D.; Wang, J.; Jia, J.M.; Wang, A.H. Chemical constituents of *Sophora flavescens* Alt. and cytotoxic activities of two new compounds. *Nat. Prod. Res.* **2022**, *36*, 108–113. [CrossRef]
27. Wang, X.; Li, J.C.; Hu, Y.Z.; Chen, W.Y.; Jin, Y.G. Effect of *Sophora flavescens* alkaloid on aerobic vaginitis in gel form for local treatment. *J. Tradit. Chin. Med.* **2017**, *37*, 314–320.
28. Zhang, Z.X.; Zhou, Y.; Song, X.N.; Xu, H.H.; Cheng, D.M. Insecticidal activity of the whole grass extract of *Typha angustifolia* and its active component against *Solenopsis invicta*. *Sociobiology* **2013**, *60*, 362–366. [CrossRef]
29. Abbott, W.S. A method of computing the effectiveness of an insecticide. *J. Econ. Entomol.* **1925**, *18*, 265–267. [CrossRef]
30. Zhang, N.; Tang, L.; Hu, W.; Wang, K.; Zhou, Y.; Li, H.; Huang, C.L.; Chun, J.; Zhang, Z.X. Insecticidal, fumigant, and repellent activities of sweet wormwood oil and its individual components against red imported fire ant workers (Hymenoptera: Formicidae). *J. Insect Sci.* **2014**, *14*, 241. [CrossRef]

agronomy

Article

Identifying a Detoxifying Uridine Diphosphate Glucosyltransferase (UGT), MdUGT83K2, Which Can Glycosylate the Aryloxyphenoxypropionate Herbicide

Pan Li, Aijuan Zhao, Ru Li, Shibo Han, Na Li, Lusha Ji * and Kang Lei

School of Pharmaceutical Sciences, Liaocheng University, Liaocheng 252000, China
* Correspondence: jilu2060@163.com

Abstract: Glycosylation is a common modification reaction in plants. The products obtained upon glycosylation have different biological functions, making glycosylation an important mechanism affecting and regulating the balance of plant growth and metabolism. In this study, we first speculated that Group I in the apple glycosyltransferase family may have a predicted function like UGT83A1, according to gene chip data published online. Subsequently, by real-time PCR (polymerase chain reaction), we analyzed whether the expression of nine glycosyltransferase genes in Group I was induced by our previously reported ACCase (Acetyl-CoA carboxylase) inhibition-based herbicide QPP ((R)-ethyl·2-(4-((6-fluoro-3-methyl-4-oxo-3,4-dihydroquinazolin-2-yl)oxy) phenoxy) propanoate). It was found that expression of the MdUGT83K2 gene in Group I was significantly increased by QPP. In order to determine whether MdUGT83K2 can glycosylate QPP, we confirmed the enzymatic reaction of MdUGT83K2 in vitro and the presence of QPP glycosides in MdUGT83K2 transgenic apple seedlings by HPLC (High Performance Liquid Chromatography), and found that MdUGT83K2 can transfer glucose to QPP in vivo, which is glycosylated. In this work, we identified a novel apple glycosyltransferase, MdUGT83K2, which functions to glycosylate the ACCase-inhibiting herbicide QPP and may be involved in plant detoxification. Key Contribution: A novel apple glycosyltransferase, MdUGT83K2, was identified, which may be involved in plant detoxification by glycosylation modification of the ACCase-inhibiting herbicide.

Keywords: glycosylation; MdUGT83K2; detoxification; ACCase; herbicide

1. Introduction

Glycosyltransferase (GT, EC 2.4.x.y) is a class of activated glycosyl donors as substrates, which catalyzes the glycosylation of various substances, such as proteins, lipids, hormones, and phenylpropane compounds. The glycosylation reaction catalyzed by glycosyltransferase is a modification reaction necessary for the growth and development of plant cells and metabolic balance while also playing an important role in other life processes [1,2]. Glycosyltransferases are divided into 110 different families (CAZy, http://www.cazy.org (accessed on 15 January 2020)) according to amino acid sequence similarity, substrate specificity, and catalytic specificity, among which glycosyltransferase family 1 (GT family 1, GT1) has the largest number of gene members and plays the most important roles. The GT1 family is often referred to as UDP-glycosyltransferase (UGT) because it mainly catalyses the transfer of UDP-sugar to specific receptors (e.g., proteins, nucleic acids, antibiotics, alkaloids and plant hormones), which can regulate plant signaling pathways and cellular homeostasis, with broad roles [3].

Bronzel I, the first gene of the plant UGT family, was discovered by Dooner and Nelson when they were studying the genetic traits of maize (Zea mays L) transposons [4]. The gene encodes a protein with UGT activity, which can synthesize flavonoids and regulate melanin accumulation in maize kernels [4]. To date, genes encoding the plant UGT family have been found in wheat (Triticum aestivum L.) [5], soybean (Glycine max (L.) Merr.) [6], and rice (Oryza sativa L.) [7].

On the other hand, the application of herbicides has gradually become an indispensable part of agricultural production [8]. The potential toxic effects of herbicides threaten ecosystem [9]. It is thus very important to study the reproductive and genotoxicity of herbicides to select safe species for the protection of ecosystems. Among the known herbicides, aryloxyphenoxypropionate (APP) is a class of herbicides that inhibits fatty acid synthesis by inhibiting the activity of acetyl-CoA carboxylase and destroys membrane structures to achieve its herbicidal effect [10,11]. In our recent study, we discovered the compound QPP could be used as a potential lead structure for the further development of novel ACCase-inhibiting herbicides (Figure 1) [12]. In this study, we used biological methods to find glycosyltransferases that can glycosylate QPP so as to achieve the purpose of plant detoxification.

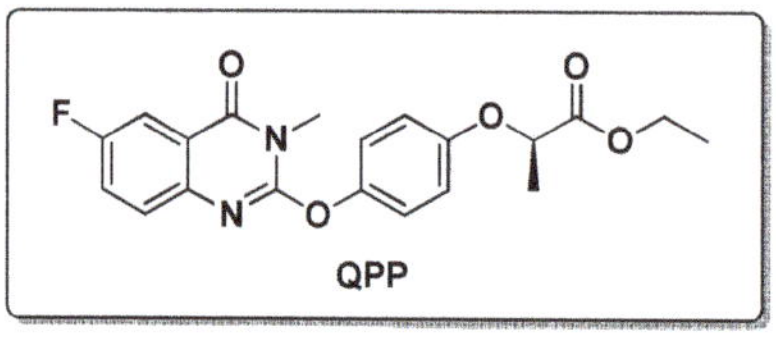

Figure 1. The structure of ACCase-inhibiting herbicide QPP.

2. Results

2.1. MdUGT83K2 Candidate Glycosyltransferase Identification

According to the gene chip database published online and a phylogenetic tree of apple (*Malus domestica*) glycosyltransferases [13], it was found that 7 glycosyltransferase genes *MdUGT83L3* (*MD09G1064900*), *MdUGT83L4* (*MD09G1064700*), *MdUGT83L5* (*MD17G1058200*), *MdUGT83L6* (*MD17G1058400*), *MdUGT83L7* (*MD17G1058100*), *MdUGT83L8* (*MD09G1065000*), and *MdUGT83K2* (*MD09G1065400*) in the apple glycosyltransferase Group I family may predicted function like UGT83A1 according to gene chip data published online [14]. The ACCase inhibition-based herbicide QPP was used to act on the wild-type "Gala" strain, RNA was extracted and reverse-transcribed into cDNA, confirmed by real-time quantitative PCR, and it was found that MdUGT83K2 is affected by QPP. Among the findings, when QPP treatment was applied for 12 h, the upregulation was up to 40-fold, which is significant difference ($p < 0.01$) (Figure 2). Therefore, we identified MdUGT83K2 as a candidate glycosyltransferase gene involved in QPP detoxification.

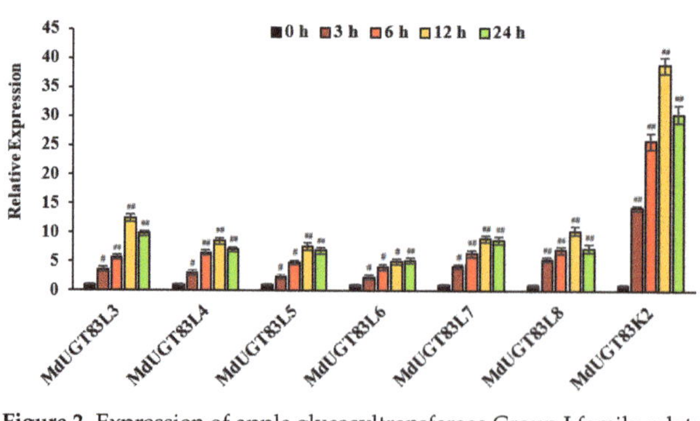

Figure 2. Expression of apple glycosyltransferase Group I family related genes induced by herbicide QPP. For the treatment of wild-type "Gala" seedlings (normal culture for 3 weeks), QPP synthesized was prepared in a solution with a final concentration of 10 μmol/L. After 12 h, the materials were collected and stored at −80 °C for subsequent RNA extraction. Note: $^{\#}$ $p < 0.05$, $^{\#\#}$ $p < 0.01$, $n = 3$.

2.2. Analysis of the Spatiotemporal Expression Pattern of MdUGT83K2

To further understand the expression pattern of MdUGT83K2, real-time PCR was further performed, the results of which are shown in Figure 3. The expression levels of MdUGT83K2 vary relatively widely across the different parts of a plant, with the lowest expression level in roots and the highest in young seeds.

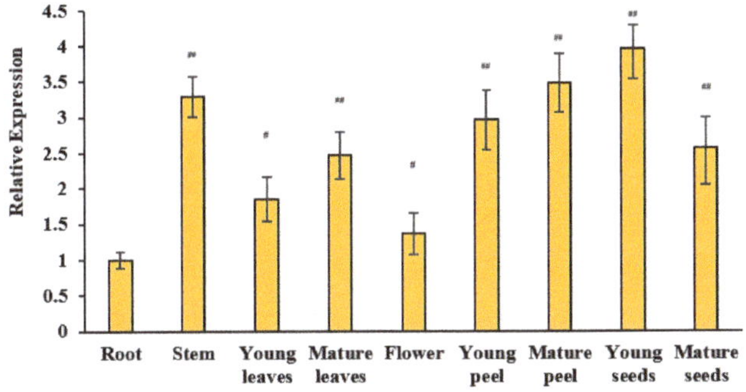

Figure 3. MdUGT83K2 spatiotemporal expression pattern. Material was collected from the root, stem, leaf, flower, and fruit of a 10-year-old "Gala" apple tree. Note: $^{\#}$ $p < 0.05$, $^{\#\#}$ $p < 0.01$, $n = 3$.

2.3. In Vitro Enzymatic Reaction to Determine MdUGT83K2 Glycosylation Modification of QPP

In the above study, we learned that MdUGT83K2 was induced to be upregulated by QPP. To further verify whether QPP was modified by MdUGT83K2 glycosylation, we constructed a MdUGT83K2 prokaryotic expression vector and expressed it into *E. coli* protein strain BL-21. After obtaining the active purified enzyme protein of MdUGT83K2, it was subjected to in vitro enzymatic reaction with QPP. HPLC detection showed that MdUGT83K2 could glycosylate QPP (Figure 4; a: HPLC image, b: mass spectrum). The specific enzyme activity of MdUGT74BP1 to QPP was 1.83 ± 0.32 (nkat/mg protein).

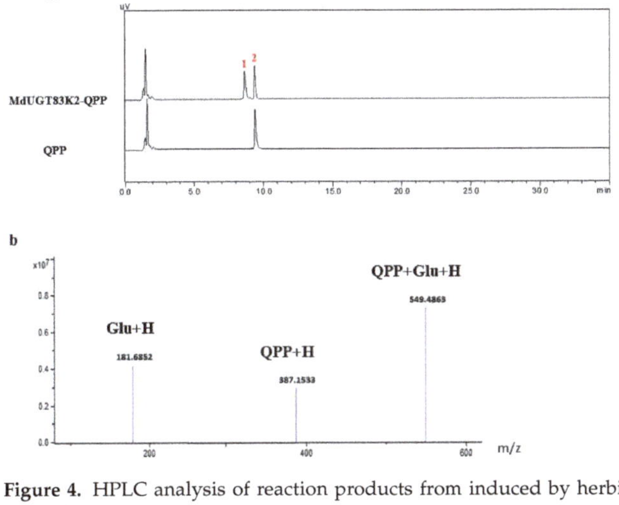

Figure 4. HPLC analysis of reaction products from induced by herbicide QPP by MdUGT83K2. The MdUGT83K2 prokaryotic expression vector was constructed and expressed into *E. coli* protein strain BL-21, and the MdUGT83K2 enzymatic protein was purified and in vitro enzymatic reaction demonstrated that MdUGT83K2 glycosylation modifies QPP. (**a**): HPLC image, (**b**): mass spectrum. 1, QPP glycosides; 2, QPP.

2.4. Acquisition of Transgenic MdUGT83K2 and Determination of QPP Glycoside Content

In the above results, we demonstrated that MdUGT83K2 can glycosylate and modify QPP in vitro, so in vivo to further determine whether MdUGT83K2 glycosylation modifies QPP in plants, we constructed a plant expression vector of MdUGT83K2 and obtained transgenic "Gala" lines OE6 and OE11 (Figure 5). After extracting QPP glycosides, HPLC analysis was performed. Compared to the level in the wild-type, the level of free QPP in OE6 and OE11 decreased, while the level of QPP glycoside increased, indicating that MdUGT83K2 can still glycosylate QPP in plants (Figure 6).

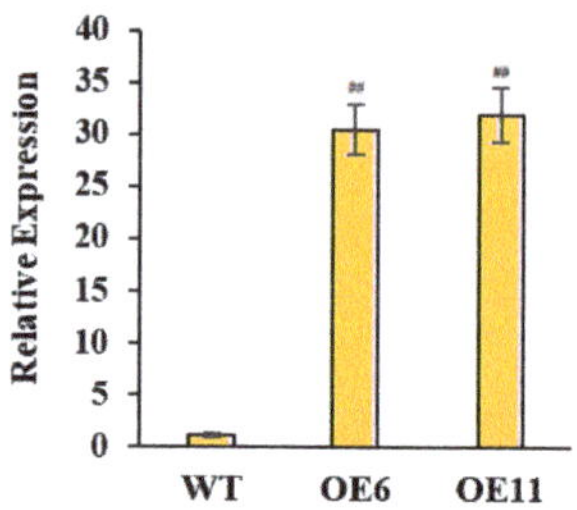

Figure 5. Expression intensity of MdUGT83K2 OE lines (normal culture for 3 weeks). Note: ## $p < 0.01$, $n = 3$.

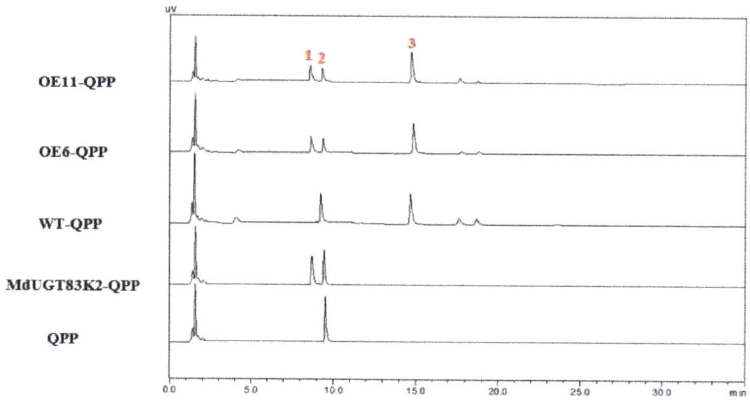

Figure 6. HPLC detection of QPP glycoside of various strains. For QPP glycoside extraction, tissue culture seedlings (normal culture for 3 weeks) of wild-type "Gala" and transgenic apple lines (OE6 and OE11) were treated with 10 μmol/L QPP for 12 h. 1, QPP glycosides; 2, QPP; 3, Internal reference "IBA".

3. Discussion

Herbicides play an important role in agricultural production, but they also bring new challenges to the breeding of crops for herbicide resistance and environmental restoration. Research on herbicide resistance of plants has increasingly become a hot filed. Most known plant resistance to herbicides is derived from the mutation or modification of herbicide target proteins [15–18], while less research has been performed on herbicide detoxification enzymes. Known herbicide detoxification enzymes include GST in maize and ketoaldehyde reductase in barnyard grass (Echinochloa crus galli (L.) Beauv.) [19]. Microbial herbicide detoxification enzymes include PAT, glyphosate oxidoreductase, and the products of genes such as bxn, glpA, glpB, atzA, atzB, atzC, and trzN. Mouse-derived cytochrome oxidase P450 has also been used for the detoxification of plant herbicides [20,21]. This is the focus of current research on herbicide detoxifying. This study found that the apple glycosyltransferase MdUGT83K2 can glycosylate QPP and participate in the process of plant detoxification. It can be used alone or in combination with other herbicide-resistance genes in crop herbicide-resistance breeding, which can expand the range of QPP applications. This work also provides new ideas and methods for research on herbicide selectivity, and it is expected to lead to the cultivation of new crop varieties that are resistant to herbicides. Therefore, the results of this study have potential value for applications in research on the herbicide resistance of crops.

Numerous studies have shown that plant glycosyltransferases can glycosylate small-molecule toxins to detoxify them [22,23]. Toxin molecules include not only biologically derived molecules, but also some non-biologically derived small-molecule substances, such as 2,4,5-trichlorophenol and 3,4-dichlorobenzamide. The identification of the plant glycosyltransferase MdUGT83K2 and its herbicide activity not only enriches the functional study of plant glycosyltransferases, but also provides a theoretical basis for an understanding of the functional evolution of glycosyltransferases. Glycosyltransferases first appeared in unicellular algae, and with the evolution of plants, the members of the glycosyltransferase family continued to expand in mosses, ferns, gymnosperms, and angiosperms, with the number increasing exponentially [24]. This reflects the growing importance of glycosyltransferases in adaptation to terrestrial life and their increasingly diverse functions in response to unfavorable internal and external environments. With the development of human social activities and industrial production, the external environment faced by plant growth is now constantly changing, and unfavorable factors continue to appear. It may be that plants must use existing glycosyltransferase family members to deal with emerging harmful substances and achieve detoxification through the functional evolution of enzymes so that plants can survive and develop better. Studies have shown that glycosyltransferases play an important role in the process of removing exogenous toxins, so the glycosylation and detoxification of herbicides may be a further manifestation of this function. Therefore, it is believed that the modification activity of glycosyltransferases on herbicides may be the result of their functional evolution. However, it cannot be ruled out that the glycosyltransferase MdUGT83K2 focused on in this study has other diverse substrates and exerts diverse functions enabling plants to cope with unfavorable environments. This is a topic that requires further exploration and research.

4. Conclusions

In this study, we first speculated that Group I in the apple glycosyltransferase family may predicted function like UGT83A1 to have possible detoxification effects according to gene chip data published online [14]. By qPCR technology, it was found that MdUGT83K2 in Group I was upregulated by the novel herbicide QPP. In vitro experiments, it was found that MdUGT83K2 can glycosylate the compound QPP, which indicated it may reduce the damage caused by herbicides to plants through the glycosylation of QPP. Therefore, in studies on transgenic plants, we further proved that MdUGT83K2 overexpression in "Gala" can indeed reduce free QPP and increase QPP glycosides.

This study found that the apple glycosyltransferase MdUGT83K2 can glycosylate QPP and participate in the process of plant detoxification. It can be used alone or in combination with other herbicide-resistance genes in crop herbicide-resistance breeding, which can expand the range of QPP applications. This work also provides new ideas and methods for research on herbicide selectivity, and it is expected to lead to the cultivation of new crop varieties that are resistant to herbicides. Therefore, the results of this study have potential value for applications in research on the herbicide resistance of crops.

5. Materials and Methods

5.1. Plant Material

Material for clarifying the spatiotemporal expression pattern of MdUGT83K2 was collected from the root, stem, leaf, flower, and fruit of a 10-year-old "Gala" apple (Malus × domestica Borkh.) tree planted at the School of Life Sciences, Shandong Agricultural University. After sampling, it was quick-frozen in liquid nitrogen and stored at −80 °C. Transgenic apple lines (OE6 and OE11) were obtained by transformation of the leaves of "Gala", using tissue culture plantlets of the clone "GL-3" apple. The growth conditions of tissue culture seedlings were 22 °C, 60 μmol m^{-2} s^{-1}, and a 14/10 h photoperiod. After subculture for 30 d and culture with rooting medium for 45 d, the seedlings were cultivated for 1 week, after which the tissue-cultured seedlings were transferred to plastic pots (8.5 × 8.5 × 7.5 cm) and then placed in a light incubator for cultivation.

5.2. Treatment of "Gala" Seedlings with QPP

For the treatment of wild-type "Gala" seedlings (normal culture for 3 weeks), QPP synthesized was prepared in a solution with a final concentration of 10 μmol/L. After 12 h, the materials were collected and stored at −80 °C for subsequent RNA extraction.

For the treatment of transgenic "Gala" seedlings, QPP synthesized in the early stage was prepared in a solution with a final concentration of 10 μmol/L. Then, the wild-type "Gala" and transgenic OE6 and OE11 lines were placed in the solution for 12 h of treatment. After that, the materials were collected and stored at −80 °C for subsequent extraction of QPP glycosides.

5.3. Real-Time PCR

For RNA extraction, 2 g of frozen material was taken and quickly ground into a fine powder in liquid nitrogen. The ground material was quickly transferred to a centrifuge containing 15 mL of CTAB extraction buffer (Weber Technology Co., Ltd., Guangzhou, China) in the tube, immediately vortexed vigorously for 30 s, and then warmed in a 65 °C water bath for 3–5 min. Subsequently, an equal volume of chloroform/isoamyl alcohol (24:1) was added, the sample was shaken evenly, centrifuged at 10,000 rpm and room temperature for 20 min, and then the supernatant was transferred to another centrifuge tube. The chloroform/isoamyl alcohol extraction was repeated one more time, and then the aqueous phase was transferred to another centrifuge tube. According to the solution volume, 1/3 volume of 8 mol/L LiCl was added to make the final concentration of LiCl 2 mol/L, followed by precipitation overnight at 4 °C, centrifugation at 4 °C (12,000 rpm) for 20 min, and then pouring off the supernatant. The precipitate was rinsed with 500 μL of 75% ethanol to remove LiCl, the precipitate was dissolved with 500 μL of SSTE buffer (Weber Technology Co., Ltd., Guangzhou, China), the solution was transferred to a 1.5 mL centrifuge tube, and then the extract was supplemented one more time with an equal volume of chloroform/isoamyl alcohol. Two volumes of 100% ethanol were added to the supernatant, followed by precipitation at −70 °C for 30 min or −20 °C for 2 h, and then centrifugation at full speed (12,000 rpm/min) and 4 °C for 20 min to precipitate RNA. The pellet was first rinsed with 400 μL of 70% ethanol, after which 400 μL of 100% ethanol was added to rinse the pellet. After drying the pellet, it was dissolved in 50 μL of DEPC-treated water.

For cDNA reverse transcription, PrimeScript RNA Reverse Transcriptase Reagent Kit with gDNA Eraser Kit (Code No. RR047A; Takara Biotechnology Co., Ltd., Dalian, China) was used in accordance with the kit instructions. The primer design was completed by Beijing Aoke Company (see Supplementary Materials Table S1).

For the real-time PCR, ChemoHS qPCR Mix (MonAmp™) and a real-time PCR instrument (ABI 7500) were used. The 10 μL reaction system included 0.4 μL each of upstream and downstream primers, 5 μL of MonAmp™ ChemoHS qPCR Mix, 0.1 μL of Low ROX Dye, 1–2 μL of cDNA template, and nuclease-free water to bring the volume to 10 μL. The PCR amplification program involved 95 °C for 5 min, followed by 40 cycles of 95 °C for 5 s of denaturation and 60 °C for 34 s of annealing and extension. Three replicates were set for each sample, melting curves were analyzed, and the relative expression levels of each gene were calculated.

5.4. Vector Construction

PCR amplification was performed using Ex Taq enzyme (RR030A; TaKaRa Biotechnology Co., Ltd., Dalian, China). Product recovery was performed using an agarose gel DNA recovery kit (ZY511-100T; Runye Biotechnology Co., Ltd., Shanghai, China). The gel-recovered product was connected to the pMD18-T vector and transformed into *E. coli* DH5α, and the recombinants were screened by the blue-white spot method. The recombinant plasmids (prokaryotic expression vector PGEX, plant expression vector PBI121) were double digested to identify positive clones and sequenced by UW Genetics (Shenzhen, China).

The PCR amplification system included 2 μL of Ex Taq (5 U/μL), 2.5 μL of 10 × Ex Taq Buffer (Mg^{2+} plus), 2 μL of dNTP mixture (2.5 mmol/L each), 0.5 μL each of upstream and downstream primers (20 μmol/L), 2 μL of template DNA (50 $ng \cdot L^{-1}$), and ddH_2O to make up the volume to 25 μL. PCR amplification conditions were pre-denaturation at 94 °C for 4 min; followed by 35 cycles of denaturation at 94 °C for 30 s, annealing at 55 °C for 40 s and extension at 72 °C for 1.5 min; and then final extension at 72 °C for 10 min. The reaction system included 0.5 μL of MdUGT83K2 restriction endonuclease I (BamHI), 0.5 μL of restriction endonuclease II (HindIII), 2 μL of 10 × buffer, 5 μL of plasmid, and sterile water to make the volume up to 20 μL. Incubation was performed at 37 °C for 1 h, followed by verification by electrophoresis.

5.5. In Vitro Enzymatic Reaction

The correctly validated recombinant plasmid MdUGT83K2-PGEX was transformed into E. coli BL-21 receptor state, selected 3–5 picked single bacteria were inoculated into LB liquid medium (Amp+) and incubated overnight at 37 °C in a 225 rpm/min shaker. The strains were inoculated in 30 mL LB medium containing antibiotics at a ratio of 1: 50 and incubated at 37 °C 225 rpm/min until the OD reached about 0.6. The OD value was then recorded, 1 mL was taken as the sample at 0 h, to which 0.5 $mol \cdot L^{-1}$ IPTG was added to make the final concentration 0.5 mmol. Samples were taken every 1 h until 5 h, so that the amount of bacteria in each sample was the same. Then, 2 mL of pre-chilled 1 spheroplast buffer (0.5 mM EDTA, 750 mM sucrose, 200 mM Tris-HCl; pH 8.0) was added to the collected cells, pipetting was performed to suspend them, followed by the addition of 14 mL of pre-chilled 1/2 concentration spheroplast buffer solution (containing 2 mg of bacterial lyase). Next, the samples were placed on ice for 30 min with shaking several times during this period. Subsequently, lysed cells were collected by centrifugation at 4500 rpm for 10 min at 4 °C, after which 5 mL of 1 PBS buffer solution was added to the pellet (2.7 mM KCl, 140 mM NaCl, 10 mM Na_2HPO_4, 1.8 mM KH_2PO_4) along with 10 μL of 100 mM PMSF(phenylmethyl sulfonylfluoride), followed by mixing by gentle pipetting and then placed in an ice bath for 30 min with shaking several times during this period. Samples were then centrifuged at 7000 rpm and 4 °C for 20 min; the supernatant was transferred to a new centrifuge tube, 100 μL of 50% glutathione Sepharose 4B (BH-S0637; Bohu, Shanghai, China) beads were added, shaking upside down was performed, followed by slowly shaking on a horizontal shaker for 1 h. Subsequently, centrifugation was performed at 500× *g* for 5 min, the supernatant was discarded, and the pipette tip was used to cut the pellets. The beads were aspirated and transferred to a 1.5 mL centrifuge tube, followed by washing three to four times with PBS solution. Centrifugation was again performed at 500× *g* for 5 min; the supernatant was discarded, and 50 μL of protein eluent was added, followed by shaking for 15 min. Another round of centrifugation was then performed at 500× *g* for 5 min, with the supernatant subsequently obtained after centrifugation being the target protein solution. The above steps were repeated to collect the protein several times. Ten microliters of protein were obtained, which was homogenized using 25 μL of 2 × protein loading buffer, followed by electrophoresis through an SDS-PAGE (sodium dodecyl sulfate polyacrylamide gel electropheresis) gel. The gel block was gently removed from the gel rack after electrophoresis and put in Coomassie staining solution, shaken for 30 min and then supplemented with decolorization solution or water for decolorizing. The obtained results were then observed in a gel imager after decolorization (Figure S1). Another part of the protein was concentrated and then subjected to an in vitro enzymatic reaction.

The enzymatic reaction system contained 10 μL of 0.5 M Tris-HCl (pH 8.0), 5 μL of 50 mM $MgSO_4$, 5 μL of 200 mM KCl, 2.5 μL of 0.1 M UDP-glucose, 1 μL of 10% β-mercaptoethanol, 1 μL of 100 mM QPP, 2 μL of MdUGT83K2 protein, and ddH_2O added to make up the volume to 100 μL.

5.6. Obtaining and Testing Genetically Modified Apples

The test material used for apple genetic transformation was the tissue culture seedlings of clone "GL-3", with strong regenerative ability and high transformation efficiency, obtained from the "Gala" apple. The leaves of "GL-3" were transformed with Agrobacterium carrying the MdUGT83K2 sequence, as described by the reported method [25]. MS medium was supplemented with 50 $mg \cdot L^{-1}$ kanamycin and 250 $mg \cdot L^{-1}$ cefotaxime sodium for the selection of regenerated shoots. During the screening process, the slowly growing and gradually yellowing regenerated buds were discarded, and the normally growing regenerated buds continued to be subcultured on the screening medium until the regenerated buds were cultivated into robust individuals with stem and leaf tissue.

Identification of transgenic lines at the DNA level was performed as follows. DNA from "GL-3" and transgenic apple plants was extracted using the Wolact Plant Genomic DNA Purification Kit (Wolact, Vicband Life Sciences Company Limited). The partial sequence of the CaMV 35S promoter on the vector was used as the upstream primer, and the downstream primer was designed with the sequence of the *MdUGT83K2* gene for PCR amplification. The primers are shown in Table S1.

mRNA level identification of transgenic lines was performed as follows. In accordance with the above RNA extraction method, RNA was extracted from MdUGT83K2 transgenic plants and non-transgenic plants and then reverse-transcribed into cDNA. The expression status of MdUGT83K2 in positive plants was detected by qRT-PCR, and the primers are shown in Table S1. The identified MdUGT83K2-positive transgenic plants and untransformed wild-type plants ("GL-3") were propagated and rooted for subsequent experimental treatments.

5.7. Extraction and HPLC Analysis of QPP Glycosides in Transgenic Apples

For QPP glycoside extraction, tissue culture seedlings of wild-type "Gala" and transgenic apple lines (OE6 and OE11) were treated with 10 μmol/L QPP for 12 h. Here, 10 g fresh samples were taken and ground into powder with liquid nitrogen; 10 mL of 80% methanol solution was added for extraction, and 2 μL of internal reference (Piclorom) was added at the same time. The samples were shaken well and placed at room temperature for 6 h. During this period, samples were continuously inverted and mixed. After vacuum evaporation in a rotary evaporator, the residual powder on the inner wall of the centrifuge tube was dissolved with 1 mL of methanol solution, centrifuged at 15,000 rpm for 20 min and the glycosides for subsequent HPLC detection.

HPLC analysis conditions involved the analytical instrument Shimadzu LC-20AT (Shimadzu, Japan). The main instruments included a workstation LC solution (Verl.21), diode array detector SPD-M20A, autosampler SIL-20A, system controller CBM-20A, and degasser DGM-20A3. The column was a reverse column. The detection wavelength of each species' eluted peak was between 190 and 430 nm.

Regarding the elution procedure, the mobile phase was acetonitrile, water (both containing 0.1% trifluoroacetic acid) and 10–100% acetonitrile, the flow rate was 1 mL/min, the elution time was 25 min, and the detection wavelength was 320 nm.

LC-MS (Liquid Chromatograph Mass Spectrometer) analysis conditions involved a Surveyor MSQ (USA) single quadrupole liquid chromatography mass spectrometer as the instrument, and the ESI (electrospray ionization) interface was used in this experiment. Here, 0.1% phosphoric acid or triethylamine acetic acid was replaced with 0.1% formic acid in the mobile phase to reduce ion suppression. The chromatographic conditions were the same as in the HPLC elution procedure, the detection was carried out in positive ion mode, and the surface-induced decomposition intensity was 30 eV.

5.8. Data Analysis

Statistical analysis was performed using SPSS Statistics for Windows version 17.0 (SPSS Inc., Chicago, IL, USA). Statistical significance was assessed by one-way ANOVA and Tukey's multiple-range test. Different letters indicate significant differences ($p < 0.05$) and extremely significant differences ($p < 0.01$).

Supplementary Materials: The following supporting information can be downloaded at: https://www.mdpi.com/article/10.3390/agronomy13020306/s1. Figure S1: MdUGT83K2 purified protein SDS-PAGE gel. 1, GST, 2, MdUGT83K2-GST. Table S1: Primer sequence.

Author Contributions: Conceptualization, L.J. and K.L.; methodology, P.L.; validation, A.Z. and R.L.; formal analysis, P.L. and S.H.; investigation, P.L. and N.L.; data curation, L.J.; writing—original draft preparation, L.J.; writing—review and editing, P.L.; project administration, L.J.; funding acquisition, P.L. and L.J. All authors have read and agreed to the published version of the manuscript.

Funding: This project was financially supported by National Natural Science Foundation of China (No. 32001982, 31701827), Shandong Provincial Natural Science Foundation of China (No. ZR202102180037), National Key Research and Development Program of China (No. SQ2020YFF0422322) and the China Postdoctoral Science Foundation (No. 2020M671984).

Data Availability Statement: Not applicable.

Acknowledgments: This project was supported by grants from the National Natural Science Foundation of China (Grant No. 32001982), Shandong Provincial Natural Science Foundation of China (Grant No. ZR202102180037), National Key Research and Development Program of China (SQ2020YFF0422322). Thank you very much for the plant materials provided by the College of Life Sciences of Shandong Agricultural University.

Conflicts of Interest: The authors declare no conflict of interest.

References

1. Wu, B.; Liu, X.; Xu, K.; Zhang, B. Genome-wide characterization, evolution and expression profiling of UDP-glycosyltransferase family in pomelo (Citrus grandis) fruit. *BMC Plant Biol.* **2020**, *20*, 459. [CrossRef] [PubMed]
2. Li, Y.; Liu, F.; Li, P.; Wang, T.; Zheng, C.; Hou, B.K. An *Arabidopsis* cytokinin-modifying glycosyltransferase UGT76C2 improves drought and salt tolerance in rice. *Front Plant Sci.* **2020**, *11*, 560696. [CrossRef] [PubMed]
3. Zhao, M.; Zhang, N.; Gao, T.; Jin, J.; Jing, T.; Wang, J.; Wu, Y.; Wan, X.; Schwab, W.; Song, C. Sesquiterpene glucosylation mediated by glucosyltransferase UGT91Q2 is involved in the modulation of cold stress tolerance in tea plants. *New Phytol.* **2020**, *226*, 362–372. [CrossRef] [PubMed]
4. Dooner, H.K.; Nelson, O.E. Controlling element-induced alterations in UDPglucose: Flavonoid glucosyltransferase, the enzyme specified by the bronze locus in maize. *Proc. Natl. Acad. Sci. USA* **1977**, *74*, 5623–5627. [CrossRef] [PubMed]
5. He, Y.; Ahmad, D.; Zhang, X.; Zhang, Y.; Wu, L.; Jiang, P.; Ma, H. Genome-wide analysis of family-1 UDP glycosyltransferases (UGT) and identification of UGT genes for FHB resistance in wheat (*Triticum aestivum* L.). *BMC Plant Biol.* **2018**, *18*, 67. [CrossRef] [PubMed]
6. Yano, R.; Takagi, K.; Tochigi, S.; Fujisawa, Y.; Nomura, Y.; Tsuchinaga, H.; Takahashi, Y.; Takada, Y.; Kaga, A.; Anai, T.; et al. Isolation and characterization of the soybean Sg-3 gene that is involved in genetic variation in sugar chain composition at the C-3 position in soyasaponins. *Plant Cell Physiol.* **2018**, *59*, 792–805. [CrossRef] [PubMed]
7. Dong, N.Q.; Sun, Y.; Guo, T.; Guo, T.; Shi, C.L.; Zhang, Y.M.; Kan, Y.; Xiang, Y.H.; Zhang, H.; Yang, Y.B.; et al. UDP-glucosyltransferase regulates grain size and abiotic stress tolerance associated with metabolic flux redirection in rice. *Nat. Commun.* **2020**, *11*, 2629. [CrossRef] [PubMed]
8. Tong, L. Acetyl-coenzyme A carboxylase: Crucial metabolic enzyme and attractive target for drug discovery. *Cell Mol. Life Sci.* **2005**, *62*, 1784–1803. [CrossRef]
9. Délye, C. Weed resistance to acetyl coenzyme A carboxylase inhibitors: An update. *Weed Sci.* **2005**, *53*, 728–746. [CrossRef]
10. Liu, W.; Harrison, D.K.; Chalupska, D.; Gornicki, P.; O'donnell, C.C.; Adkins, S.W.; Haselkorn, R.; Williams, R.R. Single-site mutations in the carboxyltransferase domain of plastid acetyl-CoA carboxylase confer resistance to grass-specific herbicides. *Proc. Natl. Acad. Sci. USA* **2007**, *104*, 3627–3632. [CrossRef]
11. Zhang, X.Q.; Powles, S.B. Six amino acid substitutions in the carboxyltransferase domain of the plastidic acetyl-CoA carboxylase gene are linked with resistance to herbicides in a Lolium rigidum population. *New Phytol.* **2006**, *172*, 636–645. [CrossRef]

12. Wang, C.C.; Chen, K.; Li, N.; Fu, S.Y.; Li, P.; Ji, L.S.; Liu, G.Y.; Wang, X.K.; Lei, K. Design, synthesis, mode of action and herbicidal evaluation of quinazolin-4(3*H*)-one derivatives based on aryloxyphenoxypropionate motif. *Agronomy* **2022**, *12*, 1840. [CrossRef]
13. Li, Y.J.; Li., P.; Zhang, L.; Shu, J.; Court, M.H.; Sun, Z.; Jiang, L.; Zheng, C.; Shu, H.; Ji., L.S.; et al. Genome-wide analysis of the apple family 1 glycosyltransferases identified a flavonoid-modifying UGT, MdUGT83L3, which is targeted by MdMYB88 and contributes to stress adaptation. *Plant Sci.* **2022**, *321*, 111314. [CrossRef]
14. Bai, S.; Zhao, Y.F.; Zhou, Y.M.; Wang, M.L.; Li, Y.H.; Luo, X.Y.; Li, L.X. Identification and expression of main genes involved in non-target site resistance mechanisms to fenoxaprop-p-ethyl in Beckmannia syzigachne. *Pest Manag. Sci.* **2020**, *76*, 2619–2626. [CrossRef]
15. Uzogara, S.G. The impact of genetic modification of human foods in the 21st century: A review. *Biotechnol Adv.* **2000**, *18*, 179–206. [CrossRef]
16. Dong, H.; Huang, Y.; Wang, K. The Development of herbicide resistance crop plants using CRISPR/Cas9-mediated gene dditing. *Genes* **2021**, *12*, 912. [CrossRef]
17. Hou, C.X.; Dirk, L.M.; Pattanaik, S.; Das, N.C.; Maiti, I.B.; Houtz, R.L.; Williams, M.A. Plant peptide deformylase: A novel selectable marker and herbicide target based on essential cotranslational chloroplast protein processing. *Plant Biotechnol. J.* **2007**, *5*, 275–281. [CrossRef]
18. Burns, E.E.; Keith, B.K.; Refai, M.Y.; Bothner, B.; Dyer, W.E. Constitutive redox and phosphoproteome changes in multiple herbicide resistant *Avena fatua L.* are similar to those of systemic acquired resistance and systemic acquired acclimation. *J. Plant Physiol.* **2018**, *220*, 105–114. [CrossRef]
19. Li, P.; Yu, Q.; Han, H.P.; Mao, L.; Nyporko, A.; Fan, L.; Bai, L.; Powles, S. Aldo-keto Reductase metabolizes glyphosate and confers glyphosate resistance in Echinochloa colona. *Plant Physiol.* **2019**, *181*, 1519–1534.
20. Schuler, M.A. Plant cytochrome P450 monooxygenases. *Crit. Rev. Plant Sci.* **1996**, *15*, 235–284. [CrossRef]
21. Inui, H.; Shiota, N.; Ishige, T.; Yasunobu, O.; Hideo, O. Herbicide metabolism and resistance of transgenic potato plantexpressing rat cytochrome P4501A1. *Breed. Sci.* **1998**, *48*, 135–143.
22. Pflugmacher, D.; Sandermann, H., Jr. The lipid/protein interface as a target site for general anesthetics: A multiple-site kinetic analysis of synaptosomal Ca^{2+}-ATPase. *Biochim. Biophys. Acta* **1998**, *1415*, 174–180. [CrossRef] [PubMed]
23. Jones, P.; Vogt, T. Glycosyltransferases in secondary plant metabolism: Tranquilizers andstimulant controllers. *Planta* **2001**, *213*, 164–174. [CrossRef] [PubMed]
24. Caputi, L.; Malnoy, M.; Goremykin, V.; Nikiforova, S.; Martens, S. A genome-wide phylogenetic reconstruction of family 1 UDP-glycosyltransferase revealed the expansion of the family during the adaptation of plants to life on land. *Plant J.* **2012**, *69*, 1030–1042. [CrossRef]
25. Dai, H.; Li, W.; Han, G.; Yang, Y.; Yue, M.; He, L.; Zhang, Z.H. Development of a seedling clone with high regeneration capacity and susceptibility to Agrobacterium in apple. *Sci. Hortic.* **2013**, *164*, 202–208. [CrossRef]

agronomy

MDPI

Article

Discovery of 3-(1-Amino-2-phenoxyethylidene)-6-methyl-2*H*-pyran-2,4(3*H*)-dione Derivatives as Novel Herbicidal Leads

Chao-Chao Wang [1], Ke Chen [2], Na Li [1], Xue-Kun Wang [1], Shi-Ben Wang [1], Pan Li [1], Xue-Wen Hua [3], Kang Lei [1,*] and Lu-Sha Ji [1]

[1] School of Pharmaceutical Sciences, Liaocheng University, Liaocheng 252000, China
[2] Department of Bioscience and Biotechnology, The University of Suwon, Whasung 18323, Republic of Korea
[3] College of Agricultural Science and Engineering, Liaocheng University, Liaocheng 252000, China
* Correspondence: leikang@lcu.edu.cn

Abstract: Natural products are one of the resources for discovering novel pesticide leads. Here, by molecular hybridization between the natural enamino diketone skeleton and the reported herbicide lead compound **I**, a series of 3-(1-aminoethylidene)-6-methyl-2*H*-pyran-2,4(3*H*)-dione derivatives (**APD**) were rationally designed, synthesized and tested for herbicidal activity in a greenhouse. The bioassay results showed that most of the target compounds possessed good herbicidal activity under pre-emergence conditions, of which the analog **APD-II-15** displayed good pre-emergent herbicidal activity against *Abutilon theophrasti Medicus*, *Amaranthus retroflexus L.*, *Echinochloa crus-galli*, *Eragrostis curvula (Schrad.) Nees*, *Avena fatua L.*, *Cyperus difformis L.*, *Chenopodium album L.*, *Ixeris denticulata*, *Plantago asiatica L.*, *Capsella bursa-pastoris (Linn.) Medic* and *Flaveria bidentis (L.) Kuntze* with >60% inhibition even at a dosage of 187.5 g ha^{-1}, and displayed good crop safety for wheat, soybean, millet and sorghum at a dosage of 375 g ha^{-1}. The preliminary study of the molecular mode of action by RNA sequencing suggested that a growth inhibition of weeds by **APD-II-15** might result from the disruptions of carbon metabolism and formation of a cytoskeleton. The present work indicated that **APD-II-15** might be used as a novel herbicidal lead compound for further optimization.

Keywords: enamino diketone; pyran-2,4(3*H*)-dione; synthesis; herbicide; RNA sequencing

1. Introduction

Weeds are the biggest competitor of crops, leading to enormous yield losses in agriculture. It is estimated that if these were left untreated, weeds would reduce global crop yields by 34% [1]. Herbicides, as the efficient weed control tool, play a very important role in protecting crops, increasing yields, and reducing labor costs in agriculture. However, an inevitable problem associated with the overuse of the same herbicide or several herbicides with the same mechanism of action is the evolvement of herbicide resistance [2–5]. To overcome this problem, developing efficient herbicides with novel structures or novel modes of action is necessary [6–8]. Biologically active natural products (NPs) are often served as the lead structures for novel herbicide discovery. To date, various NP-derived herbicides have been extensively developed and brought to the marketplace. For example, glufosinate, served as a commercial glutamine synthetase (GS) inhibitor, is derived from bialaphos obtained from fermentation cultures of the actinomycete *Streptomyces hygroscopis* [9]; mesotrione, served as a commercial 4-hydroxyphenylpyruvate dioxygenase (HPPD) inhibitor, is derived from the allelochemical leptospermone [10]. Furthermore, the introduction of a natural active skeleton is also an efficient method for novel agrichemical discovery [11,12]. For example, Yang et al. introduced the natural active quinazoline-2,4-dione skeleton into triketone HPPD-inhibiting herbicides and successfully developed quinotrione [13,14].

The enamino diketone skeleton is a basic structural feature that can be found in some natural products (Figure 1A) [15,16]. Over the last few decades, compounds with an enamino diketone skeleton (Figure 1B) have played an important role in pharmaceutical and agrochemical fields due to their antimicrobial, [17,18] antibacterial, [19–21] anticancer, [22–26] herbicidal, [27–29] antifungal [30–32] and anti-toxoplasma gondii activity [33]. Such fascinating structural and biological properties of the enamino diketone skeleton enable its application in the development of novel pesticides.

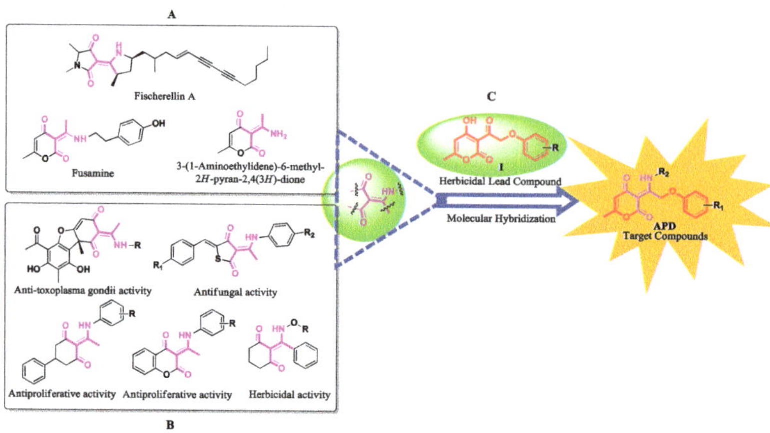

Figure 1. Design of target compound **APD** based on the molecular hybridization strategy. (**A**) natural product containing enamino diketone skeleton; (**B**) bioactive molecular containing enamino diketone skeleton; (**C**) the previous reported herbicidal lead compound I.

Recently, during our search for novel compounds that have potential use as herbicides, we have developed a promising herbicide lead: structure **I** (Figure 1C) [34]. It is easy to find that compound **I** has similar diketone moiety compared to the enamino diketone derivative. We envisaged that changing the carbonyl group on the acyl group of compound **I** to an imine and then isomerizing to enamine form can help construct compounds containing the natural enamino diketone skeleton, which should possess good herbicidal activity (Figure 1). Therefore, thirty-three 3-(1-aminoethylidene)-6-methyl-2*H*-pyran-2,4(3*H*)- dione derivatives (**APD**) were designed and synthesized, and their herbicidal activities were evaluated. Herein, we report the synthesis and structure–activity relationship (SAR) of **APD** together with the molecular mode of action study results.

2. Materials and Methods

2.1. General

In most cases, the reagents and solvents purchased from Energy Chemical or Tokyo Chemical Industry were analytical grade and used without further purification. Column chromatography purification was carried out using silica gel column chromatography (silica gel 200–300 mesh) (Qingdao Makall Group Co., Ltd., Qingdao, China). ^{1}H and ^{13}C NMR spectrum were obtained at 400 MHz and 100 MHz, respectively, using an AV-400 spectrometer (Bruker, Billerica, MA, USA) in $CDCl_3$ or DMSO-d_6 solution with tetramethylsilane (TMS) as the internal standard. The chemical shifts are reported as δ values relative to TMS. High-resolution mass spectra were conducted using an Ionspec 7.0 T spectrometer (Varian, Palo Alto, CA, USA) by the electrospray ionization fourier transform ion cyclotron resonance (ESI-FTICR) technique. The crystal structure was determined on a Saturn 724 CCD area-detector diffractometer (Rigaku, Tokyo, Japan).

2.2. Chemical Synthesis Procedures

The synthetic pathway used to prepare the target compounds **APD-I** and **APD-II** is outlined in Scheme 1. The yields were not optimized.

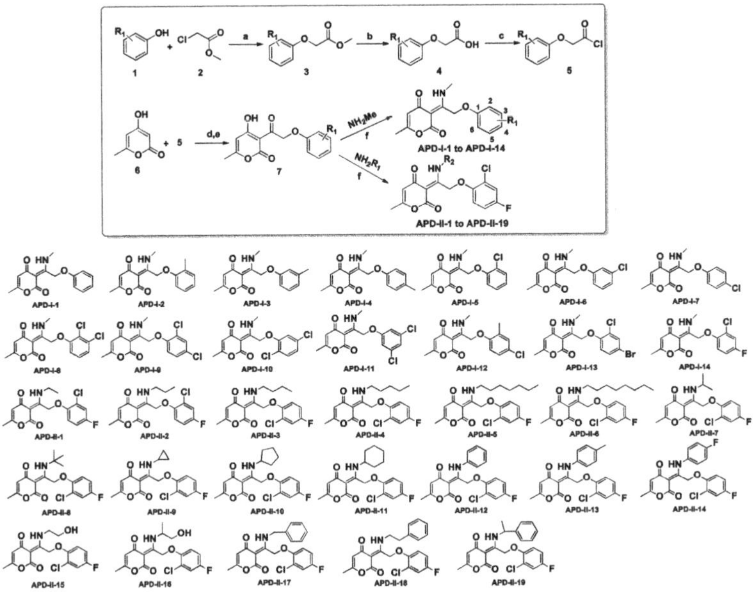

Scheme 1. Synthetic route for preparing target compounds **APD-I** and **APD-II**. Reagent and conditions: (a) K_2CO_3 DMF, 50 °C, 12 h; (b) NaOH, $MeOH/H_2O$, 70 °C, 2 h; (c) oxalyl chloride, DMF, DCM, 25 °C, 12 h; (d) Et_3N, DCM, 25 °C, 12 h; (e) Et_3N, KCN, 18-crown-6, DCM, 25 °C, 72 h; (f) Et_3N, EtOH, 80 °C, 12 h.

2.2.1. General Procedure for Preparing 4-hydroxyl-3-acetyl-pyran-2-one Derivatives 7

The 4-hydroxyl-3-acetyl-pyran-2-one derivatives **7** were prepared according to our previously reported method [34]. The appropriate phenol **1** (0.1 mol) and potassium carbonate (0.2 mol) were added successively to *N, N*-dimethylformamide (250 mL) in a 500 mL flask with stirring, and the mixture was heated to 50 °C for 1 h. Methyl chloroacetate **2** (0.1 mol) was added to the mixture, and the suspension was stirred at 50 °C for 12 h. Then, the mixture was cooled to room temperature, poured into water, and stirred for another 0.5 h. To the solution was added 300 mL of ethyl acetate; the ethyl acetate layer was separated and concentrated by rotary evaporation to give crude intermediate **3**.

Subsequently, 200 mL of methanol were added to dissolve the crude intermediate **3**, and lithium hydroxide (0.5 mol) and water (10 mL) were added to the solution with stirring. The solution was heated to 50 °C for 12 h. After completion of the reaction, 150 mL of methanol was removed by rotary evaporation, and the residue was poured into 200 mL of ice water. The solution was acidified by a hydrochloric acid solution (6 M) to pH =1. The resulting solid was collected by filtration, washed with water and dried in a vacuum to afford phenoxyacetic acid **4**.

Phenoxyacetic acid **4** (10 mmol) was dissolved in dichloromethane (100 mL) in a 250 mL flask; oxalyl chloride (20 mmol) and *N, N*-dimethylformamide (one drop) were added. The mixture was stirred at room temperature for 12 h. After completion of the reaction, dichloromethane was removed by rotary evaporation to provide phenoxyacetyl chloride **5**, which was used in the next step without purification.

Both 4-hydroxy-6-methyl-2*H*-pyran-2-one **6** (5 mmol) and 50 mL of dichloromethane were added to a 100 mL flask, and the solution was cooled to 0 °C. Phenoxyacetyl chloride **5** was added to the solution and stirred for another 30 min; then, triethylamine (7.5 mmol) and

4-dimethylaminopyridine (0.5 mmol) were added successively to the mixture at 0 °C, and the solution was stirred for 12 h. After completion of the reaction, an aqueous hydrochloric acid solution (1 M, 20 mL) was added. The dichloromethane layer was separated and washed with H_2O, saturated sodium chloride solution, dried by anhydrous sodium sulfate, concentrated by rotary evaporation, and the residue was dissolved in dichloromethane (20 mL) and added to triethylamine (5 mmol), 18-crown-6 (0.2 mmol) and potassium cyanide (1 mmol). The mixture was stirred at room temperature for 72 h. The mixture was poured into water and extracted with dichloromethane. The combined organic phase was dried with anhydrous sodium sulfate, filtered and removed by rotary evaporation. The residue was scratched from ethyl acetate and ethanol (1/1 by volume) to give intermediate **7** as a solid.

2.2.2. General Preparation of 3-(1-aminoethylidene)-6-methyl-2H-pyran-2,4(3H)-dione Derivatives (Compounds APD-I-1 to APD-I-14 and APD-II-1 to APD-II-19)

A measure of 4-Hydroxyl-3-acetyl-pyran-2-one derivative **7** (2.0 mmol) was dissolved in 20 mL of ethanol, followed by the addition of Et_3N (3.0 mmol) and the corresponding amine (2.2 mmol), and the mixture was refluxed at 80 °C for 12 h. After cooling, the precipitation was selected to give target compounds **APD-I-1** to **APD-I-14** and **APD-II-1** to **APD-II-19**.

Compound 6-Methyl-3-(1-(methylamino)-2-phenoxyethylidene)-2*H*-pyran-2,4(3*H*)-dione (**APD-I-1**): light yellow solid; yield 77.2%; ^{1}H NMR (400 MHz, $CDCl_3$) δ: 14.30 (s, 1H), 7.32–7.28 (m, 2H), 7.08–6.91 (m, 3H), 5.74 (s, 1H), 5.68 (s, 2H), 3.34 (d, J = 4.0 Hz, 3H), 2.15 (s, 3H); ^{13}C NMR (100 MHz, $CDCl_3$) δ: 185.0, 173.0, 163.4, 163.1, 157.4, 129.8, 122.0, 114.5, 107.4, 96.2, 63.2, 32.8, 19.9; HRMS: calcd for $C_{15}H_{16}NO_4^+$ [M + H]$^+$ 274.1074, found 274.1080; calcd for $C_{15}H_{15}NNaO_4^+$ [M + Na]$^+$ 296.0893, found 296.0901.

Compound 6-Methyl-3-(1-(methylamino)-2-(*o*-tolyloxy)ethylidene)-2*H*-pyran-2,4(3*H*)-dione (**APD-I-2**): light yellow solid; yield 73.0%; ^{1}H NMR (400 MHz, $CDCl_3$) δ: 14.29 (bs, 1H), 7.18 (t, J = 7.7 Hz, 1H), 6.88–6.73 (m, 3H), 5.74 (d, J = 0.8 Hz, 1H), 5.64 (s, 2H), 3.34 (d, J = 4.0 Hz, 3H), 2.33 (s, 3H), 2.15 (s, 3H); ^{13}C NMR (100 MHz, $CDCl_3$) δ: 185.0, 173.1, 163.4, 163.1, 157.4, 139.9, 129.5, 122.8, 115.4, 111.33, 107.4, 96.2, 63.3, 32.7, 21.5, 19.9; HRMS: calcd for $C_{16}H_{18}NO_4^+$ [M + H]$^+$ 288.1230, found 288.1238; calcd for $C_{16}H_{17}NNaO_4^+$ [M + Na]$^+$ 310.1050, found 310.1060.

Compound 6-Methyl-3-(1-(methylamino)-2-(*m*-tolyloxy)ethylidene)-2*H*-pyran-2,4(3*H*)-dione (**APD-I-3**): light yellow solid; yield 71.3%; ^{1}H NMR (500 MHz, $CDCl_3$) δ: 14.28 (s, 1H), 7.18 (t, J = 7.8 Hz, 1H), 6.85–6.74 (m, 3H), 5.73 (d, J = 0.7 Hz, 1H), 5.64 (s, 2H), 3.34 (d, J = 5.2 Hz, 3H), 2.33 (s, 3H), 2.14 (s, 3H); ^{13}C NMR (125 MHz, $CDCl_3$) δ: 185.0, 173.1, 163.4, 163.1, 157.5, 139.9, 129.5, 122.9, 115.4, 111.4, 107.4, 96.2, 63.3, 32.7, 21.5, 19.8; HRMS: calcd for $C_{16}H_{18}NO_4^+$ [M + H]$^+$ 288.1230, found 288.1241; calcd for $C_{16}H_{17}NNaO_4^+$ [M + Na]$^+$ 310.1050, found 310.1062.

Compound 6-Methyl-3-(1-(methylamino)-2-(p-tolyloxy)ethylidene)-2*H*-pyran-2,4(3*H*)-dione (**APD-I-4**): light yellow solid; yield 79.7%; ^{1}H NMR (400 MHz, $CDCl_3$) δ: 14.28 (bs, 1H), 7.10 (d, J = 8.4 Hz, 2H), 6.93–6.82 (m, 2H), 5.73 (s, 1H), 5.64 (s, 2H), 3.34 (d, J = 5.3 Hz, 3H), 2.29 (s, 3H), 2.14 (s, 3H); ^{13}C NMR (100 MHz, $CDCl_3$) δ: 185.0, 173.1, 163.4, 163.1, 157.4, 139.9, 129.5, 122.8, 115.4, 111.33, 107.4, 96.2, 63.3, 32.7, 21.5, 19.9; HRMS: calcd for $C_{16}H_{18}NO_4^+$ [M + H]$^+$ 288.1230, found 288.1237; calcd for $C_{16}H_{17}NNaO_4^+$ [M + Na]$^+$ 310.1050, found 310.1058.

Compound 3-(2-(2-Chlorophenoxy)-1-(methylamino)ethylidene)-6-methyl-2*H*-pyran-2,4(3*H*)-dione (**APD-I-5**): light yellow solid; yield 71.4%; ^{1}H NMR (400 MHz, $CDCl_3$) δ: 14.38 (bs, 1H), 7.40–7.37 (m, 1H), 7.25–7.21 (m, 1H), 7.14 (dd, J = 8.3, 1.2 Hz, 1H), 6.97 (td, J = 7.7, 1.3 Hz, 1H), 5.77 (s, 2H), 5.75 (s, 1H), 3.45 (d, J = 4.0 Hz, 3H), 2.15 (s, 3H); ^{13}C NMR (100 MHz, $CDCl_3$) δ: 185.0, 172.2, 163.6, 163.1, 152.9, 130.6, 128.2, 122.8, 114.1, 107.5, 96.2, 64.1, 33.1, 19.9; HRMS: calcd for $C_{15}H_{15}ClNO_4^+$ [M + H]$^+$ 308.0684, found 308.0695; calcd for $C_{15}H_{14}ClNNaO_4^+$ [M + Na]$^+$ 330.0504, found 330.0514.

Compound 3-(2-(3-Chlorophenoxy)-1-(methylamino)ethylidene)-6-methyl-2*H*-pyran-2,4(3*H*)-dione (**APD-I-6**): light yellow solid; yield 73.0%; ^{1}H NMR (400 MHz, $CDCl_3$) δ: 14.31 (bs, 1H), 7.23 (t, J = 8.1 Hz, 1H), 7.01–7.00 (m, 2H), 6.91–6.86 (m, 1H), 5.75 (s, 1H), 5.65 (s, 2H), 3.33 (d, J = 4.0 Hz, 3H), 2.15 (s, 3H); ^{13}C NMR (100 MHz, $CDCl_3$) δ: 185.0, 172.1, 163.4, 163.3, 158.1, 135.2, 130.6, 122.3, 115.5, 112.7, 107.4, 96.3, 63.3, 32.6, 19.9; HRMS: calcd for $C_{15}H_{15}ClNO_4^+$ $[M + H]^+$ 308.0684, found 308.0695; calcd for $C_{15}H_{14}ClNNaO_4^+$ $[M + Na]^+$ 330.0504, found 330.0514.

Compound 3-(2-(4-Chlorophenoxy)-1-(methylamino)ethylidene)-6-methyl-2*H*-pyran-2,4(3*H*)-dione (**APD-I-7**): light yellow solid; yield 78.1%; ^{1}H NMR (400 MHz, $CDCl_3$) δ: 14.28 (bs, 1H), 7.10 (d, J = 8.4 Hz, 2H), 6.87 (d, J = 8.6 Hz, 2H), 5.73 (s, 1H), 5.64 (s, 2H), 3.34 (d, J = 4.0 Hz, 3H), 2.15 (s, 3H); ^{13}C NMR (100 MHz, $CDCl_3$) δ: 185.0, 173.2, 163.4, 163.1, 155.6, 131.1, 127.2, 126.6, 121.7, 111.3, 107.4, 96.2, 63.4, 32.6, 19.9; HRMS: calcd for $C_{15}H_{15}ClNO_4^+$ $[M + H]^+$ 308.0684, found 308.0690; calcd for $C_{15}H_{14}ClNNaO_4^+$ $[M + Na]^+$ 330.0504, found 330.0510.

Compound 3-(2-(2,3-Dichlorophenoxy)-1-(methylamino)ethylidene)-6-methyl-2*H*-pyran-2,4(3*H*)-dione (**APD-I-8**): white solid; yield 70.0%; ^{1}H NMR (400 MHz, $CDCl_3$) δ: 14.39 (bs, 1H), 7.20–7.11 (m, 2H), 7.07 (dd, J = 7.8, 1.8 Hz, 1H), 5.79 (s, 2H), 5.76 (s, 1H), 3.44 (d, J = 4.0 Hz, 3H), 2.16 (s, 3H); ^{13}C NMR (100 MHz, $CDCl_3$) δ: 185.0, 171.7, 163.6, 163.2, 154.3, 134.2, 127.8, 123.7, 122.0, 112.0, 107.5, 96.2, 64.2, 33.0, 19.9; HRMS: calcd for $C_{15}H_{13}Cl_2NNaO_4^+$ $[M + Na]^+$ 364.0114, found 364.0125.

Compound 3-(2-(2,4-Chlorophenoxy)-1-(methylamino)ethylidene)-6-methyl-2*H*-pyran-2,4(3*H*)-dione (**APD-I-9**): white; yield 79.3%; ^{1}H NMR (400 MHz, $CDCl_3$) δ: 14.38 (bs, 1H), 7.38 (d, J = 2.5 Hz, 1H), 7.21 (dd, J = 8.8, 2.5 Hz, 1H), 7.11 (d, J = 8.8 Hz, 1H), 5.75 (bs, 3H), 3.44 (d, J = 4.0 Hz, 3H), 2.15 (s, 3H); ^{13}C NMR (100 MHz, $CDCl_3$) δ: 185.0, 171.6, 163.6, 163.2, 151.8, 130.3, 128.1, 127.4, 123.8, 115.2, 107.5, 96.2, 64.2, 32.9, 19.9; HRMS: calcd for $C_{15}H_{14}Cl_2NO_4^+$ $[M + H]^+$ 342.0294, found 342.0304; calcd for $C_{15}H_{13}Cl_2NNaO_4^+$ $[M + Na]^+$ 364.0114, found 364.0126.

Compound 3-(2-(2,5-Chlorophenoxy)-1-(methylamino)ethylidene)-6-methyl-2*H*-pyran-2,4(3*H*)-dione (**APD-I-10**): white solid; yield 79.0%; ^{1}H NMR (400 MHz, $CDCl_3$) δ: 14.38 (bs, 1H), 7.30 (d, J = 8.5 Hz, 1H), 7.17 (d, J = 2.2 Hz, 1H), 6.97 (dd, J = 8.5, 2.2 Hz, 1H), 5.76 (s, 1H), 5.72 (s, 2H), 3.42 (d, J = 4.0 Hz, 3H), 2.16 (s, 3H); ^{13}C NMR (100 MHz, $CDCl_3$) δ: 185.0, 171.2, 163.5, 163.3, 153.4, 133.5, 131.0, 122.9, 121.4, 114.8, 107.5, 96.3, 64.1, 32.8, 19.9; HRMS: calcd for $C_{15}H_{14}Cl_2NO_4^+$ $[M + H]^+$ 342.0294, found 342.0306; calcd for $C_{15}H_{13}Cl_2NNaO_4^+$ $[M + Na]^+$ 364.0114, found 364.0124.

Compound 3-(2-(3,5-Chlorophenoxy)-1-(methylamino)ethylidene)-6-methyl-2*H*-pyran-2,4(3*H*)-dione (**APD-I-11**): white solid; yield 72.1%; ^{1}H NMR (400 MHz, $CDCl_3$) δ: 14.32 (bs, 1H), 7.03 (s, 1H), 6.91 (d, J = 1.6 Hz, 2H), 5.75 (s, 1H), 5.62 (s, 2H), 3.32 (d, J = 4.0 Hz, 3H), 2.15 (s, 3H); ^{13}C NMR (100 MHz, $CDCl_3$) δ: 185.1, 171.2, 163.4, 163.3, 158.4, 135.7, 122.5, 113.8, 107.4, 96.3, 63.2, 32.4, 19.9; HRMS: calcd for $C_{15}H_{14}Cl_2NO_4^+$ $[M + H]^+$ 342.0294, found 342.0306; calcd for $C_{15}H_{13}Cl_2NNaO_4^+$ $[M + Na]^+$ 364.0114, found 364.0125.

Compound 3-(2-(4-Chloro-2-methylphenoxy)-1-(methylamino)ethylidene)-6-methyl-2*H*-pyran-2,4(3*H*)-dione (**APD-I-12**): light yellow solid; yield 76.6%; ^{1}H NMR (400 MHz, $CDCl_3$) δ: 14.34 (bs, 1H), 7.17–7.09 (m, 2H), 6.87 (d, J = 9.4 Hz, 1H), 5.75 (s, 1H), 5.65 (s, 2H), 3.35 (d, J = 4.0 Hz, 3H), 2.20 (s, 3H), 2.15 (s, 3H); ^{13}C NMR (100 MHz, $CDCl_3$) δ: 185.1, 172.6, 163.4, 163.2, 154.3, 130.8, 128.5, 126.8, 126.5, 112.6, 107.4, 96.2, 63.5, 32.5, 19.9, 16.3; HRMS: calcd for $C_{16}H_{17}ClNO_4^+$ $[M + H]^+$ 322.0841, found 322.0849; calcd for $C_{16}H_{16}ClNNaO_4^+$ $[M + Na]^+$ 344.0660, found 344.0669.

Compound 3-(2-(4-Bromo-2-chlorophenoxy)-1-(methylamino)ethylidene)-6-methyl-2*H*-pyran-2,4(3*H*)-dione (**APD-I-13**): white solid; yield 81.3%; ^{1}H NMR (400 MHz, $CDCl_3$) δ: 14.38 (bs, 1H), 7.52 (d, J = 2.4 Hz, 1H), 7.35 (dd, J = 8.8, 2.4 Hz, 1H), 7.05 (d, J = 8.8 Hz, 1H), 5.75 (bs, 3H), 3.43 (d, J = 4.0 Hz, 3H), 2.15 (s, 3H); ^{13}C NMR (100 MHz, $CDCl_3$) δ: 185.0, 171.6, 163.6, 163.2, 152.3, 133.0, 131.0, 124.1, 115.6, 114.3, 107.5, 96.2, 64.1, 32.9, 19.9; HRMS: calcd for $C_{15}H_{14}BrClNO_4^+$ $[M + H]^+$ 385.9789, found 385.9801; calcd for $C_{15}H_{13}BrClNNaO_4^+$ $[M + Na]^+$ 407.9609, found 407.9620.

Compound 3-(2-(2-Chloro-4-fluorophenoxy)-1-(methylamino)ethylidene)-6-methyl-2*H*-pyran-2,4(3*H*)-dione (**APD-I-14**): white solid; yield 77.7%; ^{1}H NMR (400 MHz, CDCl$_3$) δ: 14.37 (bs, 1H), 7.18–7.12 (m, 2H), 7.01–6.92 (m, 1H), 5.75 (s, 1H), 5.72 (s, 2H), 3.45 (d, *J* = 8.0 Hz, 3H), 2.15 (s, 3H); ^{13}C NMR (100 MHz, CDCl$_3$) δ: 185.0, 171.7, 157.3 (d, *J* = 244.1 Hz), 149.6 (d, *J* = 3.0 Hz), 123.8 (d, *J* = 10.5 Hz), 117.8 (d, *J* = 26.1 Hz), 115.6 (d, *J* = 8.8 Hz), 114.6 (d, *J* = 22.7 Hz), 107.5, 96.2, 64.6, 32.9, 19.9; HRMS: calcd for $C_{15}H_{14}ClFNO_4^+$ [M + H]$^+$ 326.0590, found 326.0600; calcd for $C_{15}H_{13}ClFNNaO_4^+$ [M + Na]$^+$ 348.0409, found 348.0419.

Compound 3-(2-(2-Chloro-4-fluorophenoxy)-1-(ethylamino)ethylidene)-6-methyl-2*H*-pyran-2,4(3*H*)-dione (**APD-II-1**): white solid; yield 67.5%; ^{1}H NMR (400 MHz, CDCl$_3$) δ: 14.39 (s, 1H), 7.19–7.10 (m, 2H), 7.01–6.90 (m, 1H), 5.74 (s, 1H), 5.68 (s, 2H), 3.91–3.81 (m, 2H), 2.15 (s, 3H), 1.42 (t, *J* = 7.3 Hz, 3H); ^{13}C NMR (100 MHz, CDCl$_3$) δ: 185.0, 170.3, 163.6, 163.1, 157.3 (d, *J* = 244.1 Hz), 149.7, 123.8 (d, *J* = 10.5 Hz), 117.7 (d, *J* = 26.2 Hz), 115.7 (d, *J* = 8.8 Hz), 114.6 (d, *J* = 22.6 Hz), 107.5, 95.8, 64.8, 41.2, 19.9, 15.1; HRMS: calcd for $C_{16}H_{16}ClFNO_4^+$ [M + H]$^+$ 340.0746, found 340.0759; calcd for $C_{16}H_{15}ClFNNaO_4^+$ [M + Na]$^+$ 362.0566, found 362.0576.

Compound 3-(2-(2-Chloro-4-fluorophenoxy)-1-(propylamino)ethylidene)-6-methyl-2*H*-pyran-2,4(3*H*)-dione (**APD-II-2**): light yellow solid; yield 72.0%; ^{1}H NMR (400 MHz, CDCl$_3$) δ: 14.45 (s, 1H), 7.18–7.11 (m, 2H), 7.01–6.91 (m, 1H), 5.74 (s, 1H), 5.68 (s, 2H), 3.79 (q, *J* = 12.0 Hz, 2H), 2.16 (s, 3H), 1.84–1.75 (m, 2H), 1.07 (t, *J* = 7.4 Hz, 3H); ^{13}C NMR (100 MHz, CDCl$_3$) δ: 185.0, 170.5, 163.6, 163.1, 157.3 (d, *J* = 244.0 Hz), 149.7 (d, *J* = 3.0 Hz), 123.8 (d, *J* = 10.5 Hz), 117.7 (d, *J* = 26.1 Hz), 115.7 (d, *J* = 8.7 Hz), 114.6 (d, *J* = 22.6 Hz), 107.5, 95.9, 64.8, 48.0, 23.0, 19.9, 11.4; HRMS: calcd for $C_{17}H_{18}ClFNO_4^+$ [M + H]$^+$ 354.0903, found 354.0915; calcd for $C_{17}H_{17}ClFNNaO_4^+$ [M + Na]$^+$ 376.0722, found 376.0732.

Compound 3-(1-(Butylamino)-2-(2-chloro-4-fluorophenoxy)ethylidene)-6-methyl-2*H*-pyran-2,4(3*H*)-dione (**APD-II-3**): white solid; yield 67.0%; ^{1}H NMR (400 MHz, CDCl$_3$) δ: 14.43 (s, 1H), 7.17–7.13 (m, 2H), 7.00–6.92 (m, 1H), 5.74 (s, 1H), 5.68 (s, 2H), 3.84–3.79 (m, 2H), 2.15 (s, 3H), 1.78–1.71 (m, 2H), 1.54–1.44 (m, 2H), 0.97 (t, *J* = 7.3 Hz, 3H); ^{13}C NMR (100 MHz, CDCl$_3$) δ: 185.0, 170.4, 163.6, 163.1, 157.3 (d, *J* = 244.2 Hz), 149.7 (d, *J* = 3.0 Hz), 123.8 (d, *J* = 10.6 Hz), 117.7 (d, *J* = 26.2 Hz), 115.8 (d, *J* = 8.8 Hz), 114.6 (d, *J* = 22.6 Hz), 107.5, 95.9, 64.8, 46.1, 31.5, 20.1, 19.9, 13.6; HRMS: calcd for $C_{18}H_{20}ClFNO_4^+$ [M + H]$^+$ 368.1059, found 368.1069; calcd for $C_{18}H_{19}ClFNNaO_4^+$ [M + Na]$^+$ 390.0879, found 390.0888.

Compound 3-(2-(2-Chloro-4-fluorophenoxy)-1-(pentylamino)ethylidene)-6-methyl-2*H*-pyran-2,4(3*H*)-dione (**APD-II-4**): white solid; yield 72.3%; ^{1}H NMR (400 MHz, CDCl$_3$) δ: 14.43 (s, 1H), 7.18–7.11 (m, 2H), 6.99–6.92 (m, 1H), 5.75 (s, 1H), 5.68 (s, 2H), 3.83–3.78 (m, 2H), 2.15 (s, 3H), 1.85–1.67 (m, 2H), 1.51–1.31 (m, 4H), 0.92 (t, *J* = 7.0 Hz, 3H); ^{13}C NMR (100 MHz, CDCl$_3$) δ: 185.0, 170.4, 163.6, 163.1, 157.3 (d, *J* = 244.0 Hz), 149.7 (d, *J* = 3.0 Hz), 123.8 (d, *J* = 10.5 Hz), 117.7 (d, *J* = 26.1 Hz), 115.8 (d, *J* = 8.8 Hz), 114.6 (d, *J* = 22.6 Hz), 107.5, 95.9, 64.7, 46.4, 29.3, 28.9, 22.3, 19.9, 13.9; HRMS: calcd for $C_{19}H_{22}ClFNO_4^+$ [M + H]$^+$ 382.1216, found 382.1227; calcd for $C_{19}H_{21}ClFNNaO_4^+$ [M + Na]$^+$ 404.1035, found 404.1044.

Compound 3-(2-(2-Chloro-4-fluorophenoxy)-1-(hexylamino)ethylidene)-6-methyl-2*H*-pyran-2,4(3*H*)-dione (**APD-II-5**): white solid; yield 70.0%; ^{1}H NMR (400 MHz, CDCl$_3$) δ: 14.42 (s, 1H), 7.20–7.09 (m, 2H), 7.03–6.88 (m, 1H), 5.74 (s, 1H), 5.68 (s, 2H), 3.83–3.78 (m, 2H), 2.15 (s, 3H), 1.80–1.70 (m, 2H), 1.51–1.40 (m, 2H), 1.36–1.26 (m, 4H), 0.89 (t, *J* = 6.7 Hz, 3H); ^{13}C NMR (100 MHz, CDCl$_3$) δ: 185.0, 170.4, 163.6, 163.1, 157.3 (d, *J* = 244.0 Hz), 149.7 (d, *J* = 2.9 Hz), 123.8 (d, *J* = 10.4 Hz), 117.7 (d, *J* = 26.1 Hz), 115.8 (d, *J* = 8.8 Hz), 114.6 (d, *J* = 22.7 Hz), 107.5, 95.9, 64.8, 46.4, 31.3, 29.5, 26.5, 22.5, 19.9, 14.0; HRMS: calcd for $C_{20}H_{24}ClFNO_4^+$ [M + H]$^+$ 396.1372, found 396.1385; calcd for $C_{20}H_{23}ClFNNaO_4^+$ [M + Na]$^+$ 418.1192, found 418.1201.

Compound 3-(2-(2-Chloro-4-fluorophenoxy)-1-(heptylamino)ethylidene)-6-methyl-2*H*-pyran-2,4(3*H*)-dione (**APD-II-6**): white solid; yield 71.1%; ^{1}H NMR (400 MHz, CDCl$_3$) δ: 14.42 (s, 1H), 7.17–7.13 (m, 2H), 7.00–6.90 (m, 1H), 5.74 (s, 1H), 5.68 (s, 2H), 3.83–3.78 (m, 2H), 2.15 (s, 3H), 1.81–1.70 (m, 2H), 1.48–1.39 (m, 2H), 1.36–1.19 (m, 6H), 0.88 (t, *J* = 6.8 Hz, 3H); ^{13}C NMR (100 MHz, CDCl$_3$) δ: 185.0, 170.4, 163.6, 163.1, 157.3 (d, *J* = 243.9 Hz), 149.7 (d, *J* = 3.0 Hz), 123.8 (d, *J* = 10.7 Hz), 117.7 (d, *J* = 26.1 Hz), 115.8 (d, *J* = 8.8 Hz),

114.6 (d, J = 22.7 Hz), 107.5, 95.9, 64.6, 46.4, 31.8, 29.6, 29.1, 29.1, 26.8, 22.6, 19.9, 14.1; HRMS: calcd for $C_{21}H_{26}ClFNO_4^+$ [M + H]$^+$ 410.1529, found 410.1542; calcd for $C_{21}H_{25}ClFNNaO_4^+$ [M + Na]$^+$ 432.1348, found 432.1357.

Compound 3-(2-(2-Chloro-4-fluorophenoxy)-1-(isopropylamino)ethylidene)-6-methyl-2*H*-pyran-2,4(3*H*)-dione (**APD-II-7**): white solid; yield 62.3%; ^{1}H NMR (400 MHz, $CDCl_3$) δ: 14.50 (s, 1H), 7.20–7.13 (m, 2H), 6.99–6.94 (m, 1H), 5.74 (s, 1H), 5.65 (s, 2H), 4.58–4.55 (m, 2H), 2.15 (s, 3H), 1.41 (s, 3H), 1.40 (s, 3H); ^{13}C NMR (100 MHz, $CDCl_3$) δ: 185.1, 168.5, 163.7, 163.0, 157.4 (d, J = 244.3 Hz), 149.9 (d, J = 2.8 Hz), 123.9 (d, J = 10.7 Hz), 117.7 (d, J = 26.3 Hz), 116.0 (d, J = 8.7 Hz), 114.6 (d, J = 22.6 Hz), 107.5, 95.5, 64.6, 48.3, 23.8, 19.9; HRMS: calcd for $C_{17}H_{18}ClFNO_4^+$ [M + H]$^+$ 354.0903, found 354.0916; calcd for $C_{17}H_{17}ClFNNaO_4^+$ [M + Na]$^+$ 376.0722, found 376.0732.

Compound 3-(1-(t-Butylamino)-2-(2-chloro-4-fluorophenoxy)ethylidene)-6-methyl-2*H*-pyran-2,4(3*H*)-dione (**APD-II-8**): white solid; yield 54.1%; ^{1}H NMR (400 MHz, DMSO-d_6) δ: 7.36 (dd, J = 8.4, 3.1 Hz, 1H), 7.05 (td, J = 8.8, 3.1 Hz, 1H), 6.72 (dd, J = 9.2, 5.0 Hz, 1H), 5.45 (s, 1H), 5.15 (s, 2H), 1.98 (s, 3H), 1.17 (s, 9H); ^{13}C NMR (100 MHz, DMSO-d_6) δ: 190.3, 181.5, 164.5, 160.0, 155.6 (d, J = 238.1 Hz), 151.5 (d, J = 2.5 Hz), 121.5 (d, J = 10.8 Hz), 117.2 (d, J = 26.2 Hz), 114.7 (d, J = 11.0 Hz), 114.5 (d, J = 2.5 Hz), 109.0, 100.6, 74.5, 51.1, 27.4, 19.6; HRMS: calcd for $C_{18}H_{20}ClFNO_4^+$ [M + H]$^+$ 368.1059, found 368.1069; calcd for $C_{18}H_{19}ClFNNaO_4^+$ [M + H]$^+$ 390.0879, found 390.0888.

Compound 3-(2-(2-Chloro-4-fluorophenoxy)-1-(cyclopropylamino)ethylidene)-6-methyl-2*H*-pyran-2,4(3*H*)-dione (**APD-II-9**): white solid; yield 63.2%; ^{1}H NMR (400 MHz, $CDCl_3$) δ: 14.41 (s, 1H), 7.19–7.09 (m, 2H), 6.96 (td, J = 8.8, 3.0 Hz, 1H), 5.76 (s, 2H), 5.72 (s, 1H), 2.15 (s, 3H), 1.05 (q, J = 6.9 Hz, 2H), 0.88 (q, J = 7.4 Hz, 2H); ^{13}C NMR (100 MHz, $CDCl_3$) δ: 185.0, 172.0, 157.3 (d, J = 243.9 Hz), 149.9 (d, J = 2.9 Hz), 123.9 (d, J = 10.5 Hz), 117.7 (d, J = 26.1 Hz), 115.7 (d, J = 8.8 Hz), 114.5 (d, J = 22.7 Hz), 107.4, 96.1, 65.0, 27.8, 19.9, 8.4; HRMS: calcd for $C_{17}H_{16}ClFNO_4^+$ [M + H]$^+$ 352.0746, found 352.0757; calcd for $C_{17}H_{15}ClFNNaO_4^+$ [M + Na]$^+$ 374.0566, found 374.0575.

Compound 3-(2-(2-Chloro-4-fluorophenoxy)-1-(cyclopentylamino)ethylidene)-6-methyl-2*H*-pyran-2,4(3*H*)-dione (**APD-II-10**): white solid; yield 57.7%; ^{1}H NMR (400 MHz, $CDCl_3$) δ: 14.63 (s, 1H), 7.20–7.08 (m, 2H), 6.96 (td, J = 8.9, 3.0 Hz, 1H), 5.73 (s, 1H), 5.67 (s, 2H), 4.72–4.61 (m, 1H), 2.15–2.12 (m, 5H), 1.87–1.83 (m, 2H), 1.78–1.64 (m, 4H); ^{13}C NMR (100 MHz, $CDCl_3$) δ: 185.0, 168.8, 163.6, 163.0, 157.3 (d, J = 244.0 Hz), 149.9 (d, J = 2.9 Hz), 123.9 (d, J = 10.4 Hz), 117.7 (d, J = 26.2 Hz), 116.0 (d, J = 8.7 Hz), 114.6 (d, J = 22.7 Hz), 107.5, 95.6, 64.8, 57.4, 34.3, 23.8, 19.8; HRMS: calcd for $C_{19}H_{20}ClFNO_4^+$ [M + H]$^+$ 380.1059, found 380.1064; calcd for $C_{19}H_{19}ClFNNaO_4^+$ [M + Na]$^+$ 402.0879, found 402.0882.

Compound 3-(2-(2-Chloro-4-fluorophenoxy)-1-(cyclohexylamino)ethylidene)-6-methyl-2*H*-pyran-2,4(3*H*)-dione (**APD-II-11**): white solid; yield 50.3%; ^{1}H NMR (400 MHz, $CDCl_3$) δ: 14.55 (bs, 1H), 7.21 (dd, J = 9.1, 4.8 Hz, 1H), 7.14 (dd, J = 7.9, 3.0 Hz, 1H), 6.96 (td, J = 9.0, 3.0 Hz, 1H), 5.74 (s, 1H), 5.66 (s, 2H), 4.28–4.19 (m, 1H), 2.15 (s, 3H), 2.02–1.99 (m, 2H), 1.83–1.80 (m, 2H) (dd, J = 9.1, 3.9 Hz, 2H), 1.65–1.25 (m, 6H); ^{13}C NMR (100 MHz, $CDCl_3$) δ: 185.1, 168.5, 163.7, 163.0, 157.4 (d, J = 244.1 Hz), 149.9 (d, J = 2.9 Hz), 123.9 (d, J = 10.5 Hz), 117.7 (d, J = 26.1 Hz), 116.3 (d, J = 8.8 Hz), 114.6 (d, J = 22.7 Hz), 107.5, 95.6, 64.6, 54.8, 33.6, 25.0, 24.1, 19.8; HRMS: calcd for $C_{20}H_{22}ClFNO_4^+$ [M + H]$^+$ 394.1216, found 394.1230; calcd for $C_{20}H_{21}ClFNNaO_4^+$ [M + Na]$^+$ 416.1035, found 416.1045.

Compound 3-(2-(2-Chloro-4-fluorophenoxy)-1-(phenylamino)ethylidene)-6-methyl-2*H*-pyran-2,4(3*H*)-dione (**APD-II-12**): white solid; yield 40.0%; ^{1}H NMR (400 MHz, $CDCl_3$) δ: 16.09 (s, 1H), 7.43–7.32 (m, 5H), 7.08 (dd, J = 8.0, 2.9 Hz, 1H), 6.99 (dd, J = 9.1, 4.9 Hz, 1H), 6.90 (td, J = 8.0, 4.9 Hz, 1H), 5.83 (s, 1H), 5.34 (s, 2H), 2.19 (s, 3H); ^{13}C NMR (100 MHz, $CDCl_3$) δ: 185.7, 168.4, 164.2, 162.6, 157.2 (d, J = 243.6 Hz), 150.1 (d, J = 2.9 Hz), 136.2, 129.6, 128.4, 125.1, 124.0 (d, J = 10.5 Hz), 117.5 (d, J = 26.1 Hz), 115.7 (d, J = 8.8 Hz), 114.2 (d, J = 22.7 Hz), 107.2, 97.1, 64.8, 20.1; HRMS: calcd for $C_{20}H_{16}ClFNO_4^+$ [M + H]$^+$ 388.0746, found 388.0755; calcd for $C_{20}H_{15}ClFNNaO_4^+$ [M + Na]$^+$ 410.0566, found 410.0573.

Compound 3-(2-(2-Chloro-4-fluorophenoxy)-1-(*p*-tolylamino)ethylidene)-6-methyl-2*H*-pyran-2,4(3*H*)-dione (**APD-II-13**): white solid; yield 42.8%; ^{1}H NMR (400 MHz, $CDCl_3$)

δ: 15.99 (s, 1H), 7.25 (d, J = 9.1 Hz, 2H), 7.20 (d, J = 8.4 Hz, 2H), 7.09 (dd, J = 8.0, 3.0 Hz, 1H), 7.00 (dd, J = 9.1, 4.9 Hz, 1H), 6.93–6.89 (m, 1H), 5.83 (s, 1H), 5.31 (s, 2H), 2.37 (s, 3H), 2.18 (s, 3H); ^{13}C NMR (100 MHz, $CDCl_3$) δ: 185.6, 168.2, 162.7, 150.2, 138.6, 133.4, 130.1, 124.9, 124.0 (d, J = 10.7 Hz), 117.5 (d, J = 26.0 Hz), 115.7 (d, J = 8.9 Hz), 114.2 (d, J = 22.7 Hz), 107.2, 101.4, 99.9, 97.1, 64.8, 21.1, 20.1; HRMS: calcd for $C_{21}H_{18}ClFNO_4^+$ [M + H]$^+$ 402.0903, found 402.0912; calcd for $C_{21}H_{17}ClFNNaO_4^+$ [M + Na]$^+$ 424.0722, found 424.0729.

Compound 3-(2-(2-Chloro-4-fluorophenoxy)-1-((4-fluorophenyl)amino)ethylidene)-6-methyl-2*H*-pyran-2,4(3*H*)-dione (**APD-II-14**): white solid; yield 35.5%; ^{1}H NMR (400 MHz, $CDCl_3$) δ: 16.06 (s, 1H), 7,37–7.33 (m, 2H), 7.13–7.07 (m, 3H), 7.00 (dd, J = 9.1, 4.8 Hz, 1H), 6.95–6.86 (m, 1H), 5.84 (s, 1H), 5.34 (s, 2H), 2.20 (s, 3H); ^{13}C NMR (100 MHz, $CDCl_3$) δ: 185.7, 168.8, 164.4, 162.6, 162.1 (d, J = 244.2 Hz), 160.9, 157.3 (d, J = 243.9 Hz), 149.9 (d, J = 2.9 Hz), 132.3, 126.9 (d, J = 8.7 Hz), 123.9 (d, J = 10.5 Hz), 117.6 (d, J = 26.1 Hz), 116.6 (d, J = 23.0 Hz), 115.7 (d, J = 8.8 Hz), 114.3 (d, J = 22.7 Hz), 97.2, 64.8, 20.1; HRMS: calcd for $C_{20}H_{15}ClF_2NO_4^+$ [M + H]$^+$ 406.0652, found 406.0659; calcd for $C_{20}H_{14}ClF_2NNaO_4^+$ [M + Na]$^+$ 428.0472, found 428.0474.

Compound 3-(2-(2-Chloro-4-fluorophenoxy)-1-((2-hydroxyethyl)amino)ethylidene)-6-methyl-2*H*-pyran-2,4(3*H*)-dione (**APD-II-15**): white solid; yield 70.1%; ^{1}H NMR (400 MHz, $CDCl_3$) δ: 14.49 (s, 1H), 7.18–7.10 (m, 2H), 7.01–6.88 (m, 1H), 5.72 (s, 3H), 3.99–3.93 (m, 4H), 2.14 (s, 3H); ^{13}C NMR (100 MHz, $CDCl_3$) δ: 184.9, 171.1, 163.5, 163.4, 157.3 (d, J = 244.2 Hz), 149.6 (d, J = 2.9 Hz), 123.7 (d, J = 10.4 Hz), 117.8 (d, J = 26.2 Hz), 115.6 (d, J = 8.8 Hz), 114.6 (d, J = 22.7 Hz), 107.6, 96.2, 64.9, 60.8, 48.4, 19.8; HRMS: calcd for $C_{16}H_{16}ClFNO_5^+$ [M + H]$^+$ 356.0696, found 356.0710; calcd for $C_{16}H_{15}ClFNNaO_5^+$ [M + Na]$^+$ 378.0515, found 378.0526.

Compound 3-(2-(2-Chloro-4-fluorophenoxy)-1-((1-hydroxypropan-2-yl)amino)ethylidene)-6-methyl-2*H*-pyran-2,4(3*H*)-dione (**APD-II-16**): white solid; yield 56.3%; ^{1}H NMR (400 MHz, $CDCl_3$) δ: 14.33 (s, 1H), 7.17–7.12 (m, 2H), 6.99–6.94 (m, 1H), 5.91 (d, J = 11.2 Hz, 1H), 5.67 (s, 1H), 5.48 (d, J = 12.3 Hz, 1H), 4.57–4.52 (m, 1H), 3.86 (dd, J = 11.5, 3.5 Hz, 1H), 3.66 (dd, J = 11.6, 7.2 Hz, 1H), 2.13 (s, 3H), 1.37 (d, J = 6.6 Hz, 3H); ^{13}C NMR (100 MHz, CDCl3) δ: 185.0, 170.2, 163.7, 163.5, 157.3 (d, J = 244.2 Hz), 149.8 (d, J = 3.0 Hz), 123.7 (d, J = 10.4 Hz), 117.7 (d, J = 26.2 Hz), 115.8 (d, J = 8.8 Hz), 114.6 (d, J = 22.6 Hz), 107.3, 95.8, 66.5, 64.7, 54.3, 19.8, 17.6; HRMS: calcd for $C_{17}H_{18}ClFNO_5^+$ [M + H]$^+$ 370.0852, found 370.0872; calcd for $C_{17}H_{17}ClFNNaO_5^+$ [M + Na]$^+$ 392.0671, found 392.0687.

Compound 3-(1-(Benzylamino)-2-(2-chloro-4-fluorophenoxy)ethylidene)-6-methyl-2*H*-pyran-2,4(3*H*)-dione (**APD-II-17**): white solid; yield 66.1%; ^{1}H NMR (400 MHz, $CDCl_3$) δ: 14.75 (s, 1H), 7.43–7.32 (m, 5H), 7.22–7.11 (m, 2H), 7.03–6.91 (m, 1H), 5.73 (s, 1H), 5.71 (s, 2H), 5.04 (d, J = 5.6 Hz, 2H), 2.15 (s, 3H); ^{13}C NMR (100 MHz, $CDCl_3$) δ: 185.1, 170.6, 163.5, 163.3, 157.4 (d, J = 244.4 Hz), 149.6 (d, J = 3.0 Hz), 135.2, 129.2, 128.4, 127.7, 124.0 (d, J = 10.5 Hz), 117.8 (d, J = 26.2 Hz), 115.9 (d, J = 8.9 Hz), 114.7 (d, J = 22.7 Hz), 107.5, 96.4, 65.0, 50.0, 19.9; HRMS: calcd for $C_{21}H_{18}ClFNO_4^+$ [M + H]$^+$ 402.0903, found 402.0916; calcd for $C_{21}H_{17}ClFNNaO_4^+$ [M + Na]$^+$ 424.0722, found 424.0733.

Compound 3-(2-(2-Chloro-4-fluorophenoxy)-1-(phenethylamino)ethylidene)-6-methyl-2*H*-pyran-2,4(3*H*)-dione (**APD-II-18**): white solid; yield 69.6%; ^{1}H NMR (400 MHz, $CDCl_3$) δ: 14.49 (s, 1H), 7.34–7.31 (m, 2H), 7.28–7.22 (m, 3H), 7.13 (dd, J = 7.9, 3.0 Hz, 1H), 7.08 (dd, J = 9.1, 4.8 Hz, 1H), 6.96–6.92 (m, 1H), 5.74 (s, 1H), 5.53 (s, 2H), 4.11–4.06 (m, 2H), 3.06 (t, J = 7.1 Hz, 1H), 2.14 (s, 3H); ^{13}C NMR (100 MHz, $CDCl_3$) δ: 185.0, 170.7, 163.5, 163.1, 157.4 (d, J = 244.2 Hz), 149.7 (d, J = 2.9 Hz), 137.3, 128.9 (d, J = 8.6 Hz), 127.2, 123.8 (d, J = 10.5 Hz), 117.7 (d, J = 26.2 Hz), 115.8 (d, J = 8.9 Hz), 114.6 (d, J = 22.7 Hz), 107.5, 96.0, 64.7, 47.7, 36.1, 19.9; HRMS: calcd for $C_{22}H_{20}ClFNO_4^+$ [M + H]$^+$ 416.1059, found 416.1071; calcd for $C_{22}H_{19}ClFNNaO_4^+$ [M + Na]$^+$ 438.0879, found 438.0889.

Compound 3-(2-(2-Chloro-4-fluorophenoxy)-1-((1-phenylethyl)amino)ethylidene)-6-methyl-2*H*-pyran-2,4(3*H*)-dione (**APD-II-19**): white solid; yield 48.6%; ^{1}H NMR (400 MHz, $CDCl_3$) δ: 14.93 (s, 1H), 7.39–7.28 (m, 5H), 7.16–7.07 (m, 2H), 6.98–6.89 (m, 1H), 6.03 (d, J = 12.7 Hz, 1H), 5.77 (s, 1H), 5.68–5.52 (m, 1H), 4.99 (d, J = 12.8 Hz, 1H), 2.15 (s, 3H), 1.71 (d, J = 6.8 Hz, 3H); ^{13}C NMR (100 MHz, $CDCl_3$) δ: 185.3, 169.5, 163.5, 163.3, 157.4 (d, J = 244.3 Hz), 149.8 (d, J = 2.8 Hz), 141.5, 129.2, 128.1, 125.9, 124.0 (d, J = 10.5 Hz),

117.7 (d, J = 26.1 Hz), 116.3 (d, J = 8.8 Hz), 114.6 (d, J = 22.7 Hz), 107.5, 96.2, 64.9, 55.6, 24.4, 19.9; HRMS: calcd for $C_{22}H_{20}ClFNO_4^+$ $[M + H]^+$ 416.1059, found 416.1071; calcd for $C_{22}H_{19}ClFNNaO_4+$ $[M + Na]^+$ 438.0879, found 438.0889.

2.3. X-ray Diffraction Analysis of Target Compound APD-I-9

The single crystal of **APD-I-9** was slowly cultivated from a mixture of dichloromethane and methanol (1/1 by volume). Crystallographic data of compound **APD-I-9** had been deposited in the Cambridge Crystallographic Data Centre as supplementary publications with the deposition number 2210257. The detail data can be acquired free of charge from http://www.ccdc.cam.ac.uk/. (accessed on 29 September 2022)

2.4. Evaluation of Herbicidal Activity

Herbicidal activity was evaluated based on the reported methods [34–37]. The commercial herbicide atrazine was selected as the positive control. The herbicidal activity of target compounds against two dicotyledonous species: *Abutilon theophrasti Medicus* (AM) and *Amaranthus retroflexus L.* (AR), and two monocotyledonous species: *Echinochloa crus-galli* (EC) and *Digitaria sanguinalis (L.) Scop.* (DS) was tested in a greenhouse. *Poa annua L.* (PA), *Eleusine indica* (EI), *Eragrostis curvula (Schrad.) Nees* (EN), *Avena fatua L.* (AF), *Setaria viridis (L.) Beauv.* (SB), *Cyperus difformis L.* (CD), *Leptochloa chinensis (L.) Nees* (LC), *Chenopodium album L.* (CA), *Ixeris denticulata* (ID), *Plantago asiatica L.* (PA), *Capsella bursa-pastoris (Linn.) Medic* (CB) and *Flaveria bidentis (L.) Kuntze* (FB) were also evaluated in herbicidal spectrum activity experiments. All of the target compounds were first dissolved in DMF and then diluted with Tween-80 to 100 g/L. The solutions were diluted with water to the appropriate concentrations and sprayed immediately after seed planting (pre-emergence treatment) or after the expansion of the first true leaf (post-emergence treatment). A mixture of the same amount of water, *N, N*-dimethylformamide and Tween 80 was sprayed as the control. The fresh weight of the above ground tissues was measured 14 days (pre-emergence treatment) or 21 days (post-emergence treatment) after treatment. The inhibition percent was used to describe the control efficiency of the compounds. The data represented the percentage displaying herbicidal damage as compared to the control, where complete control of the target is 100 and no control is 0. All of the bioassays were tested for three parallel experiments. SPSS 22.0 (SPSS, Chicago, IL, USA) was used as the statistical software program. Details of the experimental procedure are given in Supporting Information.

2.5. Crop Selectivity

The crops, i.e., wheat, soybean, millet, maize, cotton, peanut and sorghum, were planted in flowerpots (12 cm diameter) and grown at room temperature in the test soil. Crop safety experiments were conducted at a dosage of 375 g ha^{-1} when the crops had reached the four-leaf stage. After 14 days of treatment by compound **APD-II-15**, the crop selectivity was evaluated with three duplicates per experiment. The data represented the percentage displaying damage as compared to the control, where complete control of the target is 100 and no control is 0.

2.6. Phenotypic Study of Arabidopsis Thaliana

The *Arabidopsis thaliana* employed in this study is in the Columbia (Col-0) background. Surface-sterilized seeds were sown onto Murashige and Skoog (MS) plates at 4 °C for 3 days under darkness to favor vernalization and then grown at 22 °C for 2 days under LD (16 h light/8 h darkness). Subsequently, the plants were transferred to 1/2 MS plates supplemented with compound **APD-II-15** (0.01 μM) and atrazine (0.01 μM), respectively. The solution of compound **APD-II-15** and atrazine was prepared in 0.1% DMF and diluted in the 1/2 MS to obtain the tested concentrations, respectively. The solvent (0.1% DMF) was also added to control plates. The plants were grown on vertically oriented plates at 22 °C (air humidity of 40–60%) for 7 days under LD (16 h light/8 h darkness) for phenotypic investigation.

2.7. RNA Sequencing and Data Analysis

A.thaliana was cultured in an incubator with a temperature of 22 °C and a 16/8 h light/ dark photoperiod for 14 days. The compound APD-II-15 was dissolved in DMF and diluted in the deionized water to obtain the tested concentrations. *A. thaliana* seedlings with good growth were selected to be placed in a petri dish (filter-paper covered) and treated with **APD-II-15** (10 mM) as the experimental group. The distilled water (0.1% DMF) was added to the control plates. All treatments were conducted with three replicates. After treatment for 12 h, 0.1 g of fresh *A. thaliana* seedlings were weighed into a 15 mL centrifuge tube, quickly frozen and stored in a −80 °C refrigerator. Total RNA extraction and RNA-Seq analysis was performed by Beijing Novogene Biotechnology Co., Ltd. For all samples, total RNAs were extracted using a TRIzol reagent (Invitrogen, USA). RNA integrity was assessed using the RNA Nano 6000 Assay Kit of the Bioanalyzer 2100 system (Agilent Technologies, CA, USA). Total RNA was used as input material for the RNA sample preparations. The clustering of the index-coded samples was performed on a cBot Cluster Generation System using TruSeq PE Cluster Kitv3-cBot-HS (Illumia) according to the manufacturer's instructions. Differential expression analysis of two conditions/groups (dose and control) was performed using the DESeq2 R package (1.20.0), and genes with padj < 0.05 | log2FoldChange | >0.0 found by DESeq2 were assigned as differentially expressed. Gene Ontology (GO) and Kyoto Encyclopedia of Genes and Genomes (KEGG) enrichment analyses were implemented by the clusterProfiler R package (3.4.4) [38].

3. Results and Discussion

3.1. Chemistry

As depicted in Scheme 1, the target compounds **APD-I** and **APD-II** were easily prepared via a five-step synthetic route. Briefly, in the presence of potassium carbonate as a base, phenol **1** was reacted with methyl chloroacetate **2** to give the corresponding methyl aryloxyacetate **3**. The hydrolysis of methyl aryloxyacetate **3** with sodium hydroxide in the methanol/water system afforded the corresponding aryloxyacetyl acid **4**, which was reacted with oxalyl chloride in the presence of a catalytic amount of DMF to give the corresponding aryloxyacetyl chloride derivatives **5**. Subsequently, reacting compound **6** with aryloxyacetyl chloride **5** in the presence of trimethylamine afforded the key intermediate **7**. Finally, target compounds **APD-I** and **APD-II** were synthesized in 35.5–81.3% yield by the condensation of **7** with the corresponding amine in ethanol. The structures of all the target compounds were identified using ^{1}H and ^{13}C NMR spectroscopy, and HRMS. Furthermore, the structure of compound **APD-I-9** was confirmed using X-ray diffraction analysis (CCDC 2210257; Figure 2).

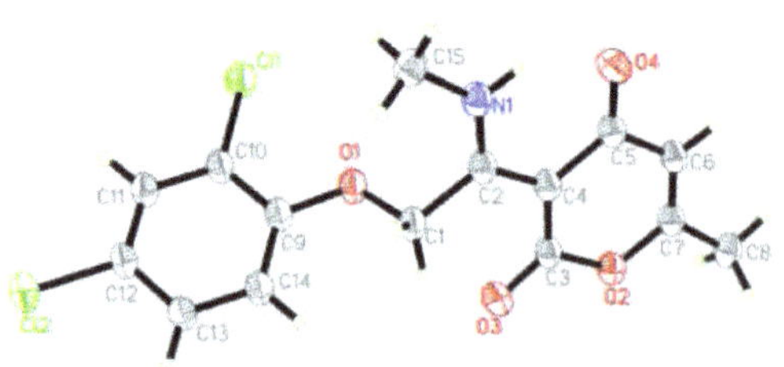

Figure 2. X-ray crystal structure of compound **APD-I-9**.

3.2. Herbicidal Activity of Target Compounds in Greenhouse Tests and SAR Study

As an initial work, the target compounds **APD-I-1** to **APD-I-14** were first synthesized, and their herbicidal activities were evaluated at a dosage of 1500 g ha^{-1} located in a greenhouse (data are given in Supporting Information, Table S1). Atrazine was selected as the positive control. As shown in Figure 3, some of the target compounds, such as **APD-I-9**, **APD-I-10**, **APD-I-12**, and **APD-I-14**, exhibited >200% sum inhibition against the weeds tested under pre-emergence conditions. Among them, compound **APD-I-14** exhibited excellent pre-emergent herbicidal activity with a sum inhibition rate 370.3%, which was comparable to the activity observed for atrazine. However, when the herbicidal activity of these target compounds was evaluated under post-emergence conditions, the result was not ideal. Only compound **APD-I-14** exhibited >200% sum inhibition against the weeds tested. It was found that the R_1 group on the benzene ring has a significant influence on herbicidal activity. Analyzing the pre-emergent herbicidal activity of these target compounds, it is easy to find that the compounds with a 2,4-di-substitution pattern on the benzene ring, such as **APD-I-9** and **APD-I-12** to **APD-I-14**, exhibited stronger herbicidal activity than that of compounds with mono-pattern or other di-substituted patterns. Among the 2,4-di-substituted analogues, compounds with 2-Cl-4-F substitution on the benzene ring, i.e., **APD-I-14**, possess the highest sum inhibition against the weeds tested, indicating that the 2-Cl-4-F substituent is an optimal orientation and that compound **APD-I-14** can be served as a potential lead compound for further optimization.

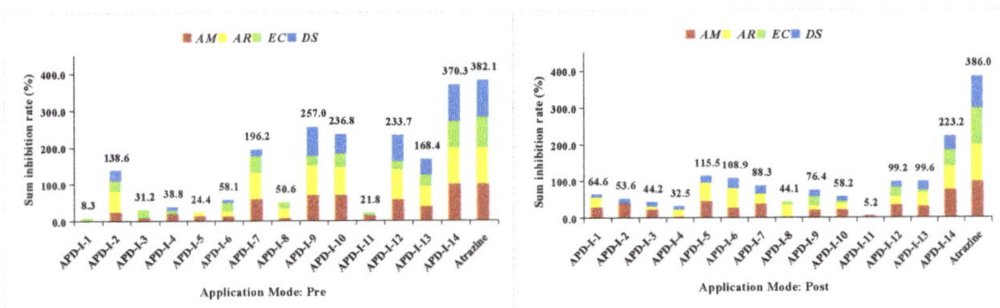

Figure 3. Effects (% inhibition) of target compounds **APD-I-1–APD-I-14** on loss of plant weight at a dosage of 1500 g ha^{-1} after treatment at 14 days (Pre) or 21 days (Post); Pre—pre-emergence condition; Post—post-emergence condition; *Abutilon theophrasti Medicus* (AM), *Amaranthus retroflexus L.* (AR), *Echinochloa crus-galli* (EC) and *Digitaria sanguinalis (L.) Scop.* (DS).

In order to explore the effect of the substituent at the nitrogen atom of **APD-I-14** on herbicidal activity, the target compounds **APD-II-1** to **APD-II-19** were synthesized and tested for herbicidal activity (data is given in Supporting Information, Table S2). Atrazine and compound **APD-I-14** were selected as positive controls. As shown in Figure 4, the target compounds **APD-II-1** to **APD-II-19** exhibited >300% sum inhibition against the weeds tested at a dosage of 1500 g ha^{-1} under pre-emergence conditions. Among them, compounds **APD-II-15** and **APD-II-16** exhibited excellent pre-emergent herbicidal activity, with sum inhibition rates of 398.1% and 394.8%, respectively, which were even better than that of atrazine. However, the herbicidal activity of compounds **APD-II-1** to **APD-II-19** was decreased under the post-emergence condition, which indicated that the target compounds possess enhanced herbicidal activity under pre-emergence conditions compared to under post-emergence conditions. This promising result encouraged us to further evaluate the herbicidal activity of these target compounds at a lower dosage under pre-emergence conditions.

Subsequently, the herbicidal activity of target compounds **APD-I-14** and **APD-II-1** to **APD-II-19** were tested under pre-emergence conditions by using a dose reduction with serial two-fold dilutions (data are given in Supporting Information, Table S3). With the dosage decreased from 750 g ha^{-1} to 187.5 g ha^{-1} (Figure 5), the herbicidal activity of these target compounds and atrazine became progressively lower. To our delight, when the dosage was decreased to 187.5 g ha^{-1}, compound **APD-II-15** still exhibited 270.9% sum inhibition against the tested weeds, which was superior to that of atrazine (sum inhibition rate = 232.8%) (Figure 6). Upon decreasing the dosage, the SAR gradually becomes obvious. It was found that the R_2 group at the nitrogen atom also has a significant influence on the herbicidal activity of the target compounds. When analyzing the herbicidal activity of these target compounds at a dosage of 187.5 g ha^{-1} (Figure 5C), it was found that among the compounds with straight-chain alkyl groups (i.e., **APD-I-14**, **APD-II-2** to **APD-II-6**), **APD-II-2** (R_2 = n-propyl) exhibited the highest herbicidal activity, with a sum inhibition rate 196.0%. With the extension of the carbon chain, compounds with a longer carbon chain progressively lost herbicidal activity. When a branched-chain alkyl groups was introduced on the nitrogen atom, these compounds exhibited lower herbicidal activity than that of the corresponding straight-chain compounds. For example, **APD-II-7** (R_2 = *i*-propyl, sum inhibition rate = 147%) < **APD-II-2** (R_2 = n-propyl, sum inhibition = 196.0%), and **APD-II-8** (R_2 = *t*-butyl, sum inhibition rate = 142.2%) < **APD-II-3** (R_2 = n-Butyl, sum inhibition = 175.5%). Cycloalkyl-substituted compounds (i.e., **APD-II-9** to **APD-II-11**) also exhibited lower herbicidal activity than that of the corresponding straight-chain compounds, and the herbicidal activity was decreased with the expansion of the ring. With the alkyl group on the nitrogen atom replaced by a benzene ring, compound **APD-II-12** displayed the worst control against the weeds tested. When a methyl or fluorine atom was introduced at the 4-rd of the benzene ring of compound **APD-II-12**, the herbicidal activity of **APD-II-13** and **APD-II-14** was slight improved, which indicated that the electronic effect of the substituent on the benzene ring had a certain influence on the herbicidal activity of the target compounds. The introduction of the alkyl chain between the benzene ring and the nitrogen atom also slightly improved the herbicidal activity, and the herbicidal activity of compounds **APD-II-12** and **APD-II-17** to **APD-II-19** could be placed in the following order: **APD-II-18** (R_2 = phenylethyl) > **APD-II-17** (R_2 = benzyl) > **APD-II-19** (R_2 = 1-phenylethyl) > **APD-II-12** (R_2 = phenyl). These findings demonstrate that the introduction of sterically bulky substituents at the at the nitrogen atom of **APD-I-14** is not conducive to improve herbicidal activity.

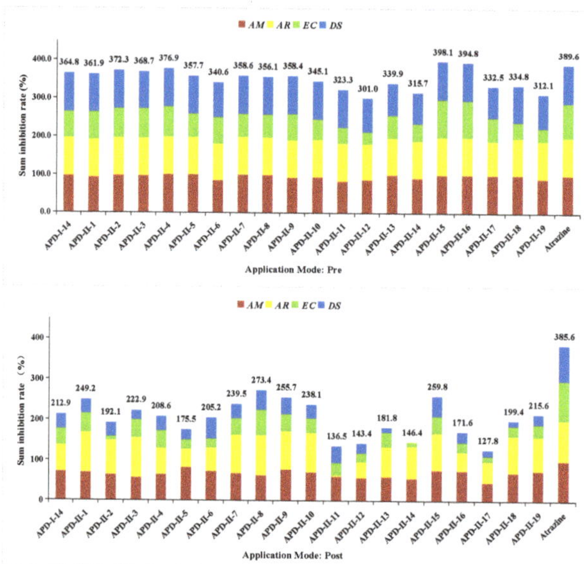

Figure 4. Effects (% inhibition) of compounds **APD-I-14**, **APD-II-1–APD-II-19** on loss of plant weight at a dosage of 1500 g ha^{-1} after treatment at 14 days (Pre) or 21 days (Post); Pre—pre-emergence condition; Post—post-emergence condition; *Abutilon theophrasti Medicus* (AM), *Amaranthus retroflexus L.* (AR), *Echinochloa crus-galli* (EC) and *Digitaria sanguinalis (L.) Scop.* (DS).

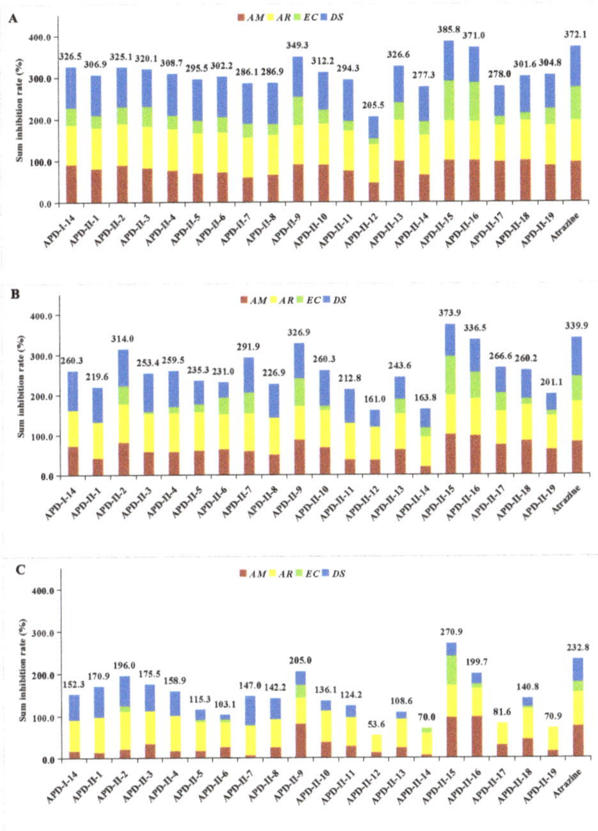

Figure 5. Effects (% inhibition) of compounds **APD-I-14**, **APD-II-1** to **APD-II-19** on loss of plant weight at a different dosage under pre-emergence conditions after treatment at 14 days; (**A**) at a dosage of 750 g ha^{-1}; (**B**) at a dosage of 375 g ha^{-1}; (**C**) at a dosage of 187.5 g ha^{-1}. *Abutilon theophrasti Medicus* (AM), *Amaranthus retroflexus L.* (AR), *Echinochloa crus-galli* (EC) *Digitaria sanguinalis (L.) Scop.* (DS).

Figure 6. Photographs illustrating loss of plant weight at a different dosage of compound **APD-II-15** and atrazine under pre-emergence conditions after treatment at 14 days; (**A**) at a dosage of 750 g ha^{-1}; (**B**) at a dosage of 375 g ha^{-1}; (**C**) at a dosage of 187.5 g ha^{-1}. The top row of each pallet is CK, the middle row is the plant treated by AZ, and the bottom row is the plant treated by APD-II-15. In each pallet, from left to right: *Abutilon theophrasti Medicus* (AM), *Amaranthus retroflexus L.* (AR), *Echinochloa crus-galli* (EC), *Digitaria sanguinalis (L.) Scop.* (DS).

It is noteworthy that introducing a hydroxyl at the terminal of the alkyl chain significantly improved the herbicidal activity (for example, **APD-II-15** (R_2 = 2-hydroxyethyl, sum inhibition = 270.9%) > **APD-II-1** (R_2 = Et, sum inhibition = 170.9%) and **APD-II-16** (R_2 = 1-hydroxypropan-2-yl, sum inhibition = 199.7%) > **APD-II-7** (R_2 = i-propyl, sum inhibition = 147%)), implying that the introduction of hydrophilic groups to the target compound is conducive to improve herbicidal activity. Considering that **APD-II-15** possess the highest sum inhibitory activity against the weeds tested, it was confirmed as a potential herbicide lead compound for further investigation.

3.3. Herbicidal Spectrum and Crop Safety of Compound APD-II-15

Since compound **APD-II-15** stood out as the most potential herbicide lead compound, its herbicidal spectrum was investigated under pre-emergence conditions. As shown in Figure 7, as a whole, **APD-II-15** displays strong control with inhibition >60% against 11 of the 16 weeds tested at a dosage of 187.5 g ha^{-1} under pre-emergence conditions, while atrazine displays strong control with inhibition >60% against 9 of the 16 tested weeds under the same conditions. This indicated that compound **APD-II-15** had a broader spectrum of weed control than the commercial herbicide atrazine. It is noteworthy that atrazine has no control against *Eragrostis curvula*, whereas compound **APD-II-15** displayed strong control with >70% inhibition to this weed, indicating that atrazine could be used in combination with compound **APD-II-15** to broaden the herbicidal spectrum. To evaluate whether compound **APD-II-15** has the potential to be developed as an herbicide candidate, the test of crop selectivity was performed at a dosage of 375 g ha^{-1} under pre-emergence conditions. The results shown in Table 1 revealed that wheat, soybean, millet and sorghum possessed a high tolerance toward **APD-II-15**, but **APD-II-15** was not selective for maize (10.7% injury), cotton (37.6% injury) and peanut (26.9% injury). These promising results indicated that compound **APD-II-15** can be served as a potential herbicide candidate for weed control in wheat, soybean, millet and sorghum fields.

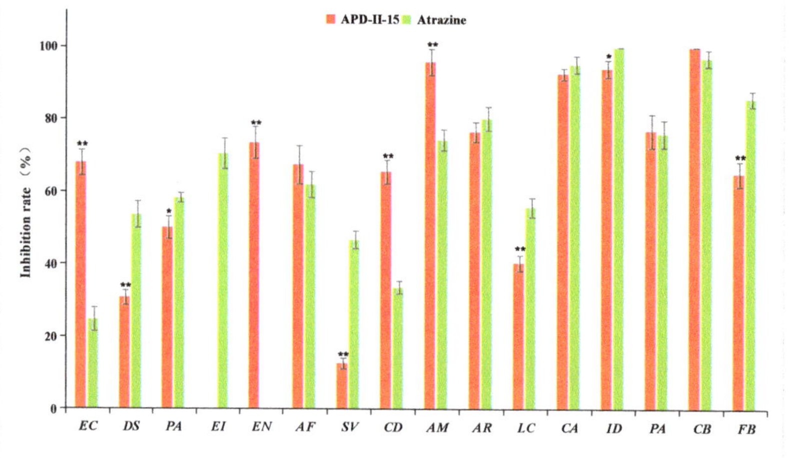

Figure 7. Effects (% inhibition) of the target compound **APD-II-15** and atrazine on loss of plant weight at a dosage of 187.5 g ha^{-1} after treatment at 14 days under the pre-emergence conditions. *Abutilon theophrasti Medicus* (AM), *Amaranthus retroflexus L.* (AR), *Echinochloa crus-galli* (EC) *Digitaria sanguinalis (L.) Scop.* (DS), *Poa annua L.* (PA), *Eleusine indica* (EI), *Eragrostis curvula (Schrad.) Nees* (EN), *Avena fatua L.* (AF), *Setaria viridis* (L.) *Beauv.* (SB), *Cyperus difformis L.* (CD), *Leptochloa chinensis (L.) Nees* (LC), *Chenopodium album L.* (CA), *Ixeris denticulata* (ID), *Plantago asiatica L.* (PA), *Capsella bursa-pastoris (Linn.) Medic* (CB), *Flaveria bidentis (L.) Kuntze* (FB); Asterisk (*) means significant differences between different treatments. *, significant at the $p < 0.05$ level, and **, significant at the $p < 0.01$ level compared with atrazine.

Table 1. Pre-emergence crop selectivity of **APD-II-15** at the dosage of 375 g ha^{-1} (Injury inhibition) [a].

Comp.	% Injury						
	Maize	Wheat	Soybean	Cotton	Millet	Peanut	Sorghum
APD-II-15	10.7 ± 1.3 **	0	0	37.6 ± 3.1 **	0	26.9 ± 1.4 **	0
Atrazine	0	24.8 ± 2.6	2.8 ± 1.1	0	16.7 ± 3.4	0	0

[a] Each value represents the mean ± SD of three experiments. Asterisk (*) means significant differences between different treatments. **, significant at the $p < 0.01$ level compared with atrazine.

3.4. Evaluation of Molecular Mode of Action of the Target Compounds

In order to explore the molecular mode of action of the target compound, **APD-II-15** was selected to study the herbicidal mechanism with *Arabidopsis thaliana* as a model plant. As shown in Figure 8, **APD-II-15** showed a certain inhibition against the root and seedling of *A. thaliana*, and the *A. thaliana* treated with **APD-II-15** developed a symptom of yellowed leaves (Figure 8), similar to that of atrazine. However, from this phenotype, it is difficult to confirm the molecular mode of action of **APD-II-15**.

Figure 8. Photographs illustrating seedling phenotype of *Arabidopsis thaliana* treated with **APD-II-15** and atrazine.

To understand the molecular mode of action, the RNA sequencing was subsequently performed to identify the differential expression genes (DEGs) of *A. thaliana* treated with compound **APD-II-15**. In total, 16362 DEGs with padj <0.05 | log2FoldChange | >0.0 were identified in **APD-II-15** vs CK. The heatmap and volcano plots for those 16,362 DEGs are illustrated in Figure 9. It was found that there were 7977 up-regulated genes and 8385 downregulated genes. To further understand the functions of those DEGs, GO annotations and KEGG annotations are given in Figure 10. As shown in Figure 10A, GO annotations were performed on the 16,362 DEGs, which were classified into the molecular function, cell component and biological process. Most enriched down-regulated GO terms were related to polysaccharide metabolic processes and cytoskeleton part; meanwhile, KEGG pathway enrichment analysis showed that most DEGs were significantly enriched in the carbon metabolism and glycolysis/gluconeogenesis. These analysis results illustrated that a growth inhibition of weeds by **APD-II-15** might result from disruptions of the formation of the cytoskeleton and carbon metabolism. Further research is still ongoing to clarify the herbicidal mode of action of the target compounds.

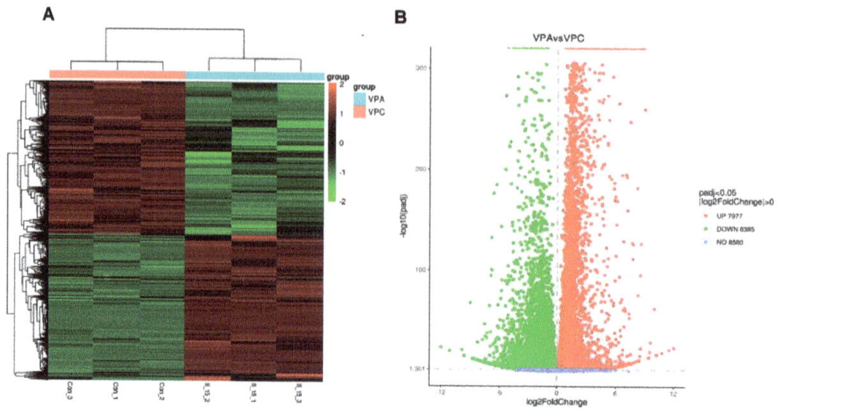

Figure 9. RNA sequencing analysis of *A. thaliana* treated by **APD-II-15**. (**A**) Heatmap of differential expression genes; (**B**) volcano plot of differential expression genes.

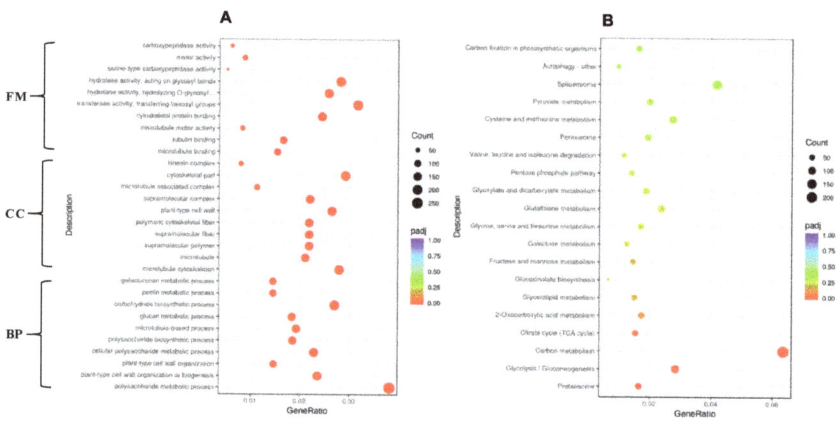

Figure 10. RNA sequencing analysis of *A. thaliana* treated by **APD-II-15**. (**A**) GO analysis of differential expression genes (MF: molecular function; CC: cell component; BP: biological process); (**B**) KEGG analysis of differential expression genes.

4. Conclusions

In summary, a novel 3-(1-aminoethylidene)-6-methyl-2*H*-pyran-2,4(3*H*)-dione structure has been rationally designed by molecular hybridization between the natural enamino diketone skeleton and the reported herbicide lead compound I; thirty-three **APD** derivatives were prepared in moderate to good yield. After the optimization of the R_1 and R_2 group, compound **APD-II-15** was confirmed as a potential herbicide lead compound owing to its good herbicidal activity, broad herbicidal spectrum and good crop safety. The investigation of the phenotypes of *A. thaliana* has difficulty in identifying the molecular mode of action of compound **APD-II-15**; thus, RNA sequencing was performed to understand the herbicidal mechanism. The RNA sequencing analysis results suggested that a growth inhibition of weeds by **APD-II-15** might result from the disruptions of the formation of the cytoskeleton and carbon metabolism. The present work demonstrates that compound **APD-II-15** can be served as a potential lead structure for further optimization. Further studies on the structural optimization of compound **APD-II-15** and on the clarifying of the herbicidal mode of action of target compounds are ongoing in our laboratory.

Supplementary Materials: The following supporting information can be downloaded at: https://www.mdpi.com/article/10.3390/agronomy13010202/s1.

Author Contributions: Conceptualization, K.L.; methodology, K.L. and K.C.; validation, C.-C.W., N.L.; formal analysis, P.L. and X.-W.H.; investigation, C.-C.W. and N.L.; data curation, X.-K.W., S.-B.W. and X.-W.H.; writing—original draft preparation, K.L.; writing—review and editing, K.L. and K.C.; project administration, K.L.; funding acquisition, K.L., X.-K.W. and L.-S.J. All authors have read and agreed to the published version of the manuscript.

Funding: This project was financially supported by the National Natural Science Foundation of China (No. 31701827), the China Postdoctoral Science Foundation (No. 2020M671984), the Guangyue Young Scholar Innovation Team of Liaocheng University (No. LCUGYTD2022-04), the National Key Research and Development Program of China (No. SQ2020YFF0422322) and the Natural Science Foundation of Shandong Province (No. ZR202102180037).

Institutional Review Board Statement: Not applicable.

Informed Consent Statement: Not applicable.

Data Availability Statement: Not applicable.

Conflicts of Interest: The authors declare no conflict of interest.

References

1. Oerke, E.C. Crop losses to pests. *J. Agric. Sci.* **2005**, *144*, 31–34. [CrossRef]
2. Heap, I. Global perspective of herbicide-resistant weeds. *Pest Manag. Sci.* **2014**, *70*, 1306–1315. [CrossRef] [PubMed]
3. Shaner, D.L. Lessons learned from the history of herbicide resistance. *Weed Sci.* **2014**, *62*, 427–431. [CrossRef]
4. Busi, R.; Goggin, D.E.; Heap, I.M.; Horak, M.G.; Jugulam, M.; Masters, R.A.; Napier, R.M.; Riar, D.S.; Satchivi, N.M.; Torra, J.; et al. Weed resistance to synthetic auxin herbicides. *Pest Manag. Sci.* **2018**, *74*, 2265–2276. [CrossRef]
5. Peterson, M.A.; Collavo, A.; Ovejero, R.; Shivraind, V.; Walshe, M.J. The challenge of herbicide resistance around the world: A current summary. *Pest Manag. Sci.* **2018**, *74*, 2246–2259. [CrossRef]
6. Beckiea, H.J.; Harker, K.N. Our top 10 herbicide-resistant weed management practices. *Pest Manag. Sci.* **2017**, *73*, 1045–1052. [CrossRef]
7. Riemens, M.; Sønderskov, M.; Moonen, A.C.; Storkey, J.; Kudsk, P. An integrated weed management framework: A pan-European perspective. *Eur. J. Agron.* **2022**, *133*, 126443. [CrossRef]
8. Tataridas, A.; Kanatas, P.; Chatzigeorgiou, A.; Zannopoulos, S.; Travlos, I. Sustainable crop and weed management in the Era of the EU green deal: A survival guide. *Agronomy* **2022**, *12*, 589. [CrossRef]
9. Duke, S.; Dayan, F.; Romagni, J.; Rimando, A.M. Natural products as sources of herbicides: Current status and future trends. *Weed Res.* **2000**, *40*, 99–111. [CrossRef]
10. Ahrens, H.; Lange, G.; Müller, T.; Rosinger, C.; Wilms, L.; van Almsick, A. 4-Hydroxyphenylpyruvate dioxygenase inhibitors in combination with safeners: Solutions for modern and sustainable agriculture. *Angew. Chem. Int. Ed.* **2013**, *52*, 9388–9398. [CrossRef]
11. Sparks, T.C.; Duke, S.O. Structure simplification of natural products as a lead generation approach in agrochemical discovery. *J. Agric. Food Chem.* **2021**, *69*, 8324–8346. [CrossRef] [PubMed]
12. Sparks, T.C.; Hahn, D.R.; Garizi, N.V. Natural products, their derivatives, mimics and synthetic equivalents: Role in agrochemical discovery. *Pest Manag. Sci.* **2017**, *73*, 700–715. [CrossRef] [PubMed]
13. Wang, D.W.; Lin, H.Y.; Cao, R.J.; Yang, S.G.; Chen, Q.; Hao, G.F.; Yang, W.C.; Yang, G.F. Synthesis and herbicidal evaluation of triketone-containing quinazoline-2,4-diones. *J. Agric. Food Chem.* **2014**, *62*, 11786–11796. [CrossRef] [PubMed]
14. Wang, D.W.; Lin, H.Y.; Cao, R.J.; Ming, Z.Z.; Chen, T.; Hao, G.F.; Yang, W.C.; Yang, G.F. Design, synthesis and herbicidal activity of novel quinazoline-2,4-diones as 4-hydroxyphenylpyruvate dioxygenase inhibitors. *Pest Manag. Sci.* **2015**, *71*, 1122–1132. [CrossRef] [PubMed]
15. Gross, E.M.; Wolk, C.P.; Jüttner, F. Fischerellin, a new allelochemical from the freshwater cyanobacterium Fischerella musicola. *J. Phycol.* **1991**, *27*, 686–692. [CrossRef]
16. Ding, L.; Dahse, H.-M.; Hertweck, C. Cytotoxic alkaloids from Fusarium incarnatum associated with the mangrove tree aegiceras corniculatum. *J. Nat. Prod.* **2012**, *75*, 617–621. [CrossRef]
17. Avdović, E.H.; Stojković, D.L.; Jevtić, V.V.; Milenković, D.; Marković, Z.S.; Vuković, N.; Potočňák, I.; Radojević, I.D.; Čomić, L.R.; Trifunović, S.R. Preparation and antimicrobial activity of a new palladium(II) complexes with a coumarin-derived ligands. crystal structures of the 3-(1-(o-toluidino)ethylidene)-chroman-2,4-dione and 3-(1-(m-toluidino) ethylidene)-chroman-2,4-dione. *Inorg. Chim. Acta.* **2019**, *484*, 52–59. [CrossRef]
18. Mladenović, M.; Vuković, N.; Nićiforović, N.; Sukdolak, S.; Solujić, S. Synthesis and molecular descriptor characterization of novel 4-hydroxy-chromene-2-one derivatives as antimicrobial agents. *Molecules* **2009**, *14*, 1495–1512. [CrossRef]

19. Baldwin, A.G.; Bevan, J.; Brough, D.; Ledder, R.; Freeman, S. Synthesis and antibacterial activities of enamine derivatives of dehydroacetic acid. *Med. Chem. Res.* **2018**, *27*, 884–889. [CrossRef]
20. Garmaise, D.L.; Chu, D.T.W.; Bernstein, E.; Inaba, M. Synthesis and antibacterial activity of 2′-substituted chelocardin analogues. *J. Med. Chem.* **1979**, *22*, 559–564. [CrossRef]
21. Bangalore, P.K.; Vagolu, S.K.; Bollikanda, R.K.; Veeragoni, D.K.; Choudante, P.C.; Misra, S.; Sriram, D.; Sridhar, B.; Kantevari, S. Usnic acid enaminone-coupled 1,2,3-triazoles as antibacterial and antitubercular agents. *J. Nat. Prod.* **2020**, *83*, 26–35. [CrossRef] [PubMed]
22. Canela, M.D.; Pérez-Pérez, M.J.; Noppen, S.; Sáez-Calvo, G.; Díaz, J.F.; Camarasa, M.J.; Liekens, S.; Priego, E.M. Novel colchicine-site binders with a cyclohexanedione scaffold identified through a ligand-based virtual screening approach. *J. Med. Chem.* **2014**, *57*, 3924–3938. [CrossRef] [PubMed]
23. Ilić, D.R.; Jevtić, V.V.; Radić, G.P.; Arsikin, K.; Ristić, B.; Harhaji-Trajković, L.; Vuković, N.; Sukdolak, S.; Klisurić, O.; Trajković, V.; et al. Synthesis, characterization and cytotoxicity of a new palladium(II) complex with a coumarine-derived ligand. *Eur. J. Med. Chem.* **2014**, *74*, 502–508. [CrossRef] [PubMed]
24. Budzisz, E.; Keppler, B.K.; Giester, G.; Wozniczka, M.; Kufelnicki, A.; Nawrot, B. Synthesis, crystal structure and biological characterization of a novel Palladium(II) complex with a coumarin-derived ligand. *Eur. J. Inorg. Chem.* **2004**, *22*, 4412–4419. [CrossRef]
25. Budzisz, E.; Małecka, M.; Lorenz, I.P.; Mayer, P.; Kwiecien, R.A.; Paneth, P.; Krajewska, U.; Rozalski, M. Synthesis, cytotoxic effect, and structure—Activity relationship of Pd(II) complexes with coumarin derivatives. *Inorg. Chem.* **2006**, *45*, 9688–9695. [CrossRef] [PubMed]
26. Budzisz, E.; Brzezinska, E.; Krajewska, U.; Rozalski, M. Cytotoxic effects, alkylating properties and molecular modelling of coumarin derivatives and their phosphonic analogues. *Eur. J. Med. Chem.* **2003**, *38*, 597–603. [CrossRef] [PubMed]
27. Dias, L.C.; Demuner, A.J.; Valente, V.V.M.; Barbosa, L.C.A.; Martins, F.T.; Doriguetto, A.C.; Ellena, J. Preparation of achiral and chiral (E)-enaminopyran-2,4-diones and their phytotoxic activity. *J. Agric. Food Chem.* **2009**, *57*, 1399–1405. [CrossRef]
28. Serban, A.; Watson, K.G.; Bird, G.J.; Farquharson, G.J.; Cross, L.E. Preparation of 2-[1-(ethoxyimino)propyl]-5-indanylcyclohexene-1,3-diones and analogs as herbicides and plant growth regulators. US 4952722A, 28 August 1990.
29. Liu, Y.X.; Zhao, H.P.; Wang, Z.W.; Li, Y.H.; Song, H.B.; Riches, H.; Beattie, D.; Gu, Y.C.; Wang, Q.M. The discovery of 3-(1-aminoethylidene)quinoline-2,4(1*H*,3*H*)-dione derivatives as novel PSII electron transport inhibitors. *Mol. Divers.* **2013**, *17*, 701–710. [CrossRef]
30. Batran, R.Z.; Khedr, M.A.; Abdel Latif, N.A.; Abd El Aty, A.A.; Shehata, A.N. Synthesis, homology modeling, molecular docking, dynamics, and antifungal screening of new 4-hydroxycoumarin derivatives as potential chitinase inhibitors. *J. Mol. Struc.* **2019**, *1180*, 260–271. [CrossRef]
31. Lv, P.; Chen, Y.L.; Zhao, Z.; Shi, T.Z.; Wu, X.W.; Xue, J.Y.; Li, Q.X.; Hua, R.M. Design, synthesis, and antifungal activities of 3-acyl thiotetronic acid derivatives: New fatty acid synthase inhibitors. *J. Agric. Food Chem.* **2018**, *66*, 1023–1032. [CrossRef]
32. Neumann, D.M.; Cammarata, A.; Backes, G.; Palmer, G.E.; Jursic, B.S. Synthesis and antifungal activity of substituted 2,4,6-pyrimidinetrione carbaldehyde hydrazones. *Bioorg. Med. Chem.* **2014**, *22*, 813–826. [CrossRef] [PubMed]
33. Guo, H.Y.; Jin, C.M.; Zhang, H.M.; Jin, C.-M.; Shen, Q.K.; Quan, Z.S. Synthesis and biological evaluation of (+)-usnic acid derivatives as potential anti-toxoplasma gondii agents. *J. Agric. Food Chem.* **2019**, *67*, 9630–9642. [CrossRef] [PubMed]
34. Lei, K.; Li, P.; Yang, X.F.; Wang, S.B.; Wang, X.K.; Hua, X.W.; Sun, B.; Ji, L.S.; Xu, X.H. Design and Synthesis of novel 4-hydroxyl-3-(2-phenoxyacetyl)-pyran-2-one derivatives for use as herbicides and evaluation of their mode of action. *J. Agric. Food Chem.* **2019**, *67*, 10489–10497. [CrossRef] [PubMed]
35. Wang, C.C.; Chen, K.; Li, N.; Fu, S.Y.; Li, P.; Ji, L.S.; Liu, G.Y.; Wang, X.K.; Lei, K. Design, synthesis, mode of action and herbicidal evaluation of quinazolin-4(3*H*)-one derivatives based on aryloxyphenoxypropionate motif. *Agronomy* **2022**, *12*, 1840. [CrossRef]
36. Lei, K.; Li, P.; Zhou, X.Y.; Wang, S.B.; Wang, X.K.; Ji, L.S.; Liu, R.M.; Xu, X.H. Design, synthesis of 5-acylbarbituric acid derivatives and evaluation of herbicidal activity and molecular mode of action. *Chin. J. Org. Chem.* **2020**, *40*, 2788–2797. [CrossRef]
37. Lei, K.; Liu, Y.; Wang, S.B.; Sun, B.; Hua, X.W.; Xu, X.H. Synthesis and herbicidal activity of 3-acetyl-4-hydroxy-2,1-benzothiazine derivatives. *Chem. Res. Chin. Univ.* **2019**, *35*, 609–615. [CrossRef]
38. Yu, G.; Wang, L.G.; Han, Y.; He, Q.Y. ClusterProfiler: An R package for comparing biological themes among gene clusters. *OMICS* **2012**, *16*, 284–287. [CrossRef]

insects

MDPI

Article

Synthesis of Piperine-Based Ester Derivatives with Diverse Aromatic Rings and Their Agricultural Bioactivities against *Tetranychus cinnabarinus* Boisduval, *Aphis citricola* Van der Goot, and *Eriosoma lanigerum* Hausmann

Tianze Li [1], Min Lv [1,*], Houpeng Wen [1], Yanyan Wang [1], Sunita Thapa [1], Shaoyong Zhang [2] and Hui Xu [1,*]

[1] College of Plant Protection, Northwest A&F University, Xianyang 712100, China
[2] Key Laboratory of Vector Biology and Pathogen Control of Zhejiang Province, College of Life Science, Huzhou University, Huzhou 313000, China
* Correspondence: lvmin@nwsuaf.edu.cn (M.L.); orgxuhui@nwsuaf.edu.cn (H.X.)

Simple Summary: Piperine, as a plant-derived natural product, has diverse bioactivities in the agricultural industry. Although the chemical insecticides/acaricides are now used to control pests, the increasingly serious resistance has raised an arduous challenge for their effective pest management. In this paper, by using piperine as a lead compound, a series of its novel ester derivatives were obtained by structural modification with different aromatic rings. Evaluation of the activities was conducted against three crop-threatening pests, *Tetranychus cinnabarinus* Boisduval (Acari: Tetranychidae), *Aphis citricola* Van der Goot (Homoptera: Aphididae), and *Eriosoma lanigerum* Hausmann (Hemiptera: Aphididae). Some compounds exhibited good to excellent acaricidal and aphicidal activities. These results further indicate that the naturally occurring compound piperine can be used as a lead compound to develop potential agrochemicals for pest management.

Abstract: Exploration of plant secondary metabolites or by using them as leads for development of new pesticides has become one of the focal research topics nowadays. Herein, a series of new ester derivatives of piperine were prepared via the Vilsmeier–Haack–Arnold (VHA) reaction, and their structures were characterized by infrared spectroscopy (IR), melting point (mp), proton nuclear magnetic resonance spectroscopy (^{1}H NMR), and carbon nuclear magnetic resonance spectroscopy (^{13}C NMR). Notably, the steric configurations of compounds **6** and **7** were confirmed by single-crystal analysis. Against *T. cinnabarinus*, compounds **9** and **11** exhibited 47.6- and 45.4-fold more pronounced acaricidal activity than piperine. In particular, compounds **9** and **11** also showed 2.6-fold control efficiency on the fifth day of piperine. In addition, compound **6** (>10–fold higher than piperine) displayed the most potent aphicidal activity against *A. citricola*. Furthermore, some derivatives showed good aphicidal activities against *E. lanigerum*. Moreover, the effects of compounds on the cuticles of *T. cinnabarinus* were investigated by the scanning electron microscope (SEM) imaging method. This study will pave the way for future high value added application of piperine and its derivatives as botanical pesticides.

Keywords: piperine; ester; *Tetranychus cinnabarinus* Boisduval; *Aphis citricola* Van der Goot; *Eriosoma lanigerum* Hausmann; structural modification; insecticidal activities; SEM; VHA reaction

1. Introduction

In the 21st century, new technologies have actively impacted many aspects of human life, including health management as well as food quality and safety. However, because of the rapid growth of the population and the expansion of urbanization leading to a decrease in the area of agricultural acreage, it is hard to produce enough food for everyone in the world. Thus, scientific research should focus on deficient food production and improve the yield and quality. Nevertheless, the limitation is that pests on crops cannot be controlled effectively [1,2]. The carmine spider mite, *Tetranychus cinnabarinus* Boisduval (Acari: Tetranychidae), is one of the most dominating agricultural pests, which can cause major destruction for more than 100 species of plants [3,4]. The aphids, *Aphis citricola* Van der Goot (Homoptera: Aphididae) and *Eriosoma lanigerum* Hausmann (Hemiptera: Aphididae), can spread plant viruses on a large scale, generate an infection of fungi, and affect the process of photosynthesis by their hazardous way of feeding [5,6]. For example, because of their specific mouthpart structure, the Hemiptera pests can act as vectors for most plant viruses, such as *Barley yellow dwarf virus* (BYDV), *Potato leafroll virus* (PLRV), *Turnip yellows virus* (TuYV), *Cereal yellow dwarf virus* (CYDV), and *Pea enation mosaic virus 1* (PEMV1) [7,8]. The abovementioned acarids and aphids have become the most destructive pests in agriculture production and cause hundreds of millions of USD in economic losses annually [9]. In addition, the unconscionable abuse of chemical industrial pesticides has aroused increasingly serious resistance, not only because of environmental pollution but also toxicities to natural enemies [10–12]. Therefore, the development of sustainable and eco-friendly pesticides is extremely urgent. Fortunately, plant-derived natural products have been used to discover green pesticidal molecules, which could delay the development of resistance and mitigate the pollution of synthetic insecticides/acaricides/aphicides [13].

Piperine (1, Figure 1) is an alkaloid isolated from *Piper nigrum* Linn. This natural compound has attracted attention because of its numerous bioactivities including anti-cancer [14–16], anti-inflammation [17], anti-Alzheimer's disease [18], anti-malarial [19], anti-diabetic [20], anti-oxidant [21], and anti-mycobacterial activities [22]. Moreover, piperine also has agricultural benefits such as anti-fungal [23] and insecticidal activities [24–27]. On the other hand, the ester group introduced on natural products could result in potent derivatives possessing insecticidal [28–33], anti-fungal [34,35], and herbicidal activities [36]. Recently, the Vilsmeier–Haack–Arnold (VHA) reaction has become a general method for the formylation of various active aromatic hydrocarbons [37]. In this paper, a series of new piperine ester derivatives with diverse aromatic rings were designed and synthesized by application of the VHA reaction (Figure 1). Their bioactivities were evaluated against *T. cinnabarinus*, *A. citricola*, and *E. lanigerum*. Additionally, the scanning electron microscope (SEM) imaging method was used to investigate the effects of piperine ester derivatives on the cuticle of *T. cinnabarinus*.

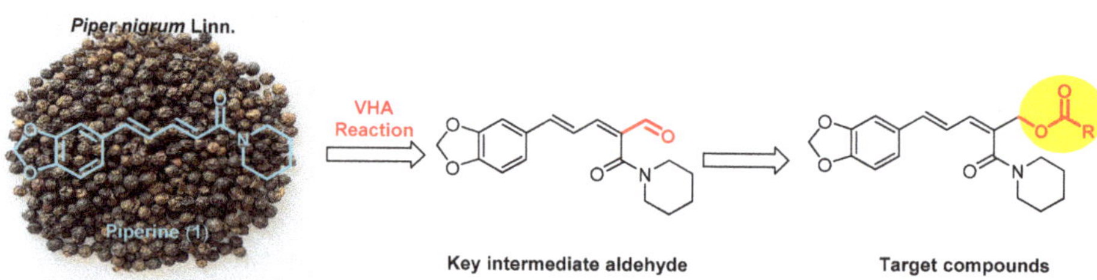

Figure 1. Design of target compounds from piperine (**1**) via the VHA reaction.

2. Materials and Methods

2.1. Insects

The spider mites (*T. cinnabarinus*) were reared by using cowpea seedlings as the material; under the conditions of no acaricides, 26 ± 1 °C, 70–75% RH (relative humidity), and 14 h/10 h (light/dark) photoperiod in the laboratory. The aphids (*A. citricola*, and *E. lanigerum*) were collected from apple trees in experimental fields at the Plant Protection College of the Northwest A&F University (Yangling, Shaanxi province, China).

2.2. Reagents and Instruments

All chemical reagents were purchased and utilized without further purification. Solvents were used directly or treated with standard methods before use. Analytical thin-layer chromatography (TLC) and preparative thin-layer chromatography (PTLC) were performed with silica gel plates using silica gel 60 GF_{254} (Qingdao Haiyang Chemical Co., Ltd., Qingdao, China). Silica-gel column chromatography was performed with silica gel 200–300 mesh (Qingdao Haiyang Chemical Co., Ltd., Qingdao, China). The melting point (mp) was determined using the XT-415 digital melting point apparatus (Beijing Tech Instrument, Ltd., Beijing, China) and was uncorrected. Nuclear magnetic resonance spectra (^{1}H NMR, ^{13}C NMR) were obtained using Avance III 400/500 MHz equipment (Bruker, Karlsruhe, Germany) and using tetramethylsilane (TMS) as the internal standard. X-ray crystallography was performed on SMART APEX II equipment (Bruker, Karlsruhe, Germany). Infrared (IR) spectra were obtained from a TENSOR27 instrument (Bruker, Karlsruhe, Germany). The scanning electron microscope images were shot using an S-3400N scanning electron microscope (Hitachi, Tokyo, Japan). Piperine (**1**) was purchased from Baoji Haoxiang Biotechnology Co. Ltd. (Shaanxi province, China).

2.3. General Procedure for the Synthesis of Aldehyde 2

To a stirred solution of phosphorus oxychloride ($POCl_3$, 20 mmol) in N,N-dimethylformamide (DMF, 20 mmol) at 0 °C for 0.5 h, a solution of piperine (**1**, 4 mmol) in DMF (2 mL) was added dropwise. After the addition, the mixture was allowed to stir at 60 °C. When the reaction was complete and checked by TLC analysis, the mixture was poured into ice water (200 mL). Subsequently, 1 M aq. NaOH solution was used to neutralize the pH value to 8–9 and the precipitated product was collected and washed with ice water and MeOH to afford aldehyde **2**.

Data for **2**: Yield: 30%; yellow solid; mp 184–186 °C; IR cm^{-1} (KBr): 3052, 2937, 2858, 2733, 1678, 1607, 1445, 1256, 1188, 812, 707; ^{1}H NMR (500 MHz, $CDCl_3$) δ: 9.49 (s, 1H, –*C*HO), 7.15 (d, *J* = 11.0 Hz, 1H), 6.97–7.03 (m, 3H), 6.80–6.89 (m, 2H), 6.01 (s, 2H, –O*C*H_2O–), 3.74 (s, 2H, –N*C*H_2–), 3.26 (t, *J* = 6.0 Hz, 2H, –N*C*H_2–), 1.66 (s, 4H), 1.54 (s, 2H); ^{13}C NMR (125 MHz, $CDCl_3$) δ: 189.7, 163.8, 149.6, 149.4, 148.4, 144.5, 137.4, 129.9, 124.4, 121.5, 108.6, 106.3, 101.6, 47.9, 42.5, 26.6, 25.8, 24.4. HRMS [ESI]: calcd for $C_{18}H_{19}NO_4Na$ ([M + Na]+), 336.1206; found, 336.1206.

2.4. General Procedure for the Synthesis of Compound **3**

A mixture of aldehyde **2** (1 mmol) and sodium borohydride ($NaBH_4$, 1.5 mmol) in MeOH (5 mL) was stirred at 0 °C for 0.5 h. Then, it was quenched using 0.2 M aq. HCl solution and extracted with ethyl acetate (EtOAc, 20 mL × 3). Then, the combined organic phase was washed with brine (30 mL × 3), dried over anhydrous Na_2SO_4, concentrated in vacuo, and purified by silica-gel column chromatography to give compound **3**.

Data for **3**: Yield: 99%; yellow solid; mp 123–125 °C; IR cm^{-1} (KBr): 3270, 2939, 2866, 1618, 1441, 1249, 1190, 811, 719; ^{1}H NMR (500 MHz, $CDCl_3$) δ: 6.87 (d, *J* = 0.5 Hz, 1H), 6.77–6.79 (m, 1H), 6.74 (d, *J* = 8.0 Hz, 1H), 6.48–6.55 (m, 2H), 6.32–6.34 (m, 1H), 5.95 (s, 2H, –OCH_2O–), 4.30 (s, 2H, –CH_2OH), 3.67–3.69 (m, 2H, –NCH_2–), 3.45 (t, *J* = 5.5 Hz, 2H, –NCH_2–), 1.63–1.64 (m, 4H), 1.52 (t, *J* = 5.5 Hz, 2H); ^{13}C NMR (125 MHz, $CDCl_3$) δ: 168.8, 148.1, 147.6, 136.8, 134.9, 131.2, 127.4, 122.2, 121.8, 108.4, 105.5, 101.2, 63.9, 47.8, 42.4, 26.6, 25.8, 24.4. HRMS [ESI]: calcd for $C_{18}H_{21}NO_4Na$ ([M + Na]$^+$), 338.1363; found, 338.1354.

2.5. General Procedure for the Synthesis of Target Compounds **4–12**

A mixture of compound **3** (0.2 mmol), aromatic carboxylic acids (RCO_2H, 0.4 mmol), 1-(3-dimethylaminopropyl)-3-ethylcarbodiimide hydrochloride (EDCI, 0.4 mmol), and 4-dimethylaminopyridine (DMAP, 0.04 mmol) in dry dichloromethane (CH_2Cl_2, 5 mL) was stirred at 0 °C to room temperature. When the reaction was complete the solution was checked by TLC analysis and the mixture was concentrated and purified by PTLC to afford compounds **4–12** in 44–57% yields. The characterization data for **4** and **5** are presented here, whereas **6–12** are characterized in the Supporting Information.

Data for **4**: Yield: 53%; yellow solid; mp 106–107 °C; IR cm^{-1} (KBr): 3429, 3051, 2932, 2353, 1723, 1588, 1484, 1447, 1241, 1183, 1029, 958, 764, 598; ^{1}H NMR (400 MHz, $CDCl_3$) δ: 7.58 (d,

J = 0.8 Hz, 1H), 7.19 (d, J = 2.8 Hz, 1H), 6.90 (d, J = 1.2 Hz, 1H), 6.82–6.84 (m, 1H), 6.77 (d, J = 8.0 Hz, 1H), 6.51–6.60 (m, 3H), 6.46 (d, J = 10.4 Hz, 1H), 5.97 (s, 2H), 5.04 (s, 2H), 3.71 (s, 2H), 3.46 (t, J = 5.2 Hz, 2H), 1.64 (s, 4H), 1.48 (s, 2H); ^{13}C NMR (100 MHz, $CDCl_3$) δ: 167.3, 158.2, 148.2, 148.0, 146.5, 144.3, 136.7, 131.7, 130.9, 130.5, 122.2, 121.8, 118.2, 111.9, 108.5, 105.5, 101.2, 66.5, 47.8, 42.5, 26.5, 25.8, 24.5.

Data for **5**: Yield: 53%; yellow oil; IR cm^{-1} (KBr): 3399, 3307, 3075, 2373, 1723, 1579, 1491, 1450, 1368, 1284, 1121, 1026, 931, 741, 701, 605; ^{1}H NMR (400 MHz, $CDCl_3$) δ: 9.21 (d, J = 1.6 Hz, 1H), 8.79 (dd, J_1 = 4.8 Hz, J_2 = 1.6 Hz, 1H), 8.32 (dt, J_1 = 8.0 Hz, J_2 = 2.0 Hz, 1H), 7.43 (dd, J_1 = 8.0 Hz, J_2 = 5.2 Hz, 1H), 6.91 (d, J = 1.2 Hz, 1H), 6.83–6.85 (m, 1H), 6.78 (d, J = 8.0 Hz, 1H), 6.53–6.66 (m, 2H), 6.50 (d, J = 10.4 Hz, 1H), 5.97 (s, 2H), 5.11 (s, 2H), 3.72 (s, 2H), 3.47 (t, J = 5.2 Hz, 1H), 1.64 (s, 4H), 1.46 (s, 2H); ^{13}C NMR (100 MHz, $CDCl_3$) δ: 167.2, 164.8, 153.5, 150.7, 148.2, 148.1, 137.2, 137.0, 132.0, 130.7, 130.2, 125.9, 123.4, 122.3, 121.7, 108.5, 105.5, 101.3, 67.2, 47.7, 42.4, 26.6, 25.8, 24.4.

2.6. Biological Assay

2.6.1. Acaricidal Activity against *T. cinnabarinus*

The acaricidal activities of compounds **1–12** against *T. cinnabarinus* were tested by using the slide-dipping method [38,39]. The commercial acaricidal pesticide spirodiclofen was used as a positive control. The solutions of spirodiclofen and piperine and its derivatives were prepared in aq. Tween-80 (0.1 g/L) at 0.5 mg/mL. A total of 35–45 healthy, active, and size-uniform female adults of *T. cinnabarinus* were selected and affixed dorsally in two lines to a strip of double-coated masking tape on a microscope slide (triplicate for each compound). Then the slides were dipped into the corresponding solution for 5 s and taken out. Additionally, the slides treated with 0.1 g/L of aq.Tween-80 were used as a blank control group (CK). Thereafter, these slides with mites were moisturized and kept at 26 ± 1 °C and 70–75% RH (relative humidity) with a 14 h/10 h (light/dark) photoperiod. The results were checked by a binocular dissecting microscope and the number of dead mites was recorded at 48 and 72 h after treatment. Finally, the corrected mortality rate was calculated as the formula:

$$\text{Corrected mortality rate (\%)} = \frac{(T - C)}{(100\% - C)} 100$$

where C is the mortality rate of CK and T is the mortality rate of the treated *T. cinnabarinus*. Furthermore, linear regressions of the 72 h mortality rates (%) versus five concentrations of some potent compounds and spirodiclofen were performed and their LC_{50} values were calculated.

2.6.2. Control Efficiency of Some Compounds against *T. cinnabarinus* in the Greenhouse

The solutions of compounds **1**, **9**, **11**, and spirodiclofen were prepared in aq. Tween-80 (1 g/L) at 0.3 mg/mL. Each cowpea seedling (three seedlings per replicate and three replicates per treatment) was infected with 50 healthy and size-uniform female adults of *T. cinnabarinus* in the greenhouse. An airbrush was used to spray 10 mL of the corresponding solution for each treatment. The blank control group was tested with aq. Tween-80 (1 g/L) only. The experimental conditions and evaluation method were as mentioned above [38,39].

2.6.3. Aphicidal Activity against *A. citricola*

The aphicidal activity against *A. citricola* was assessed by the topical application method. The solutions of compounds **1–12** were prepared at 1 mg/mL in acetone (CK was only treated with acetone), and methomyl (0.1 mg/mL) was used as a positive control. Then, approximately 90 apterous adult aphids were selected for each compound (30 aphids per dish). Subsequently, 0.04 μL corresponding solution was dropped on the pronotum of each aphid. The experiment was carried out at 25 ± 1 °C, 50 ± 7% RH, and on a 14 h/10 h (light/dark) photoperiod. Their 24 and 48 h corrected mortality rate values were calculated as above. Finally, the 48 h LD_{50} values of some potent compounds were calculated [40,41].

2.6.4. Aphicidal Activity against *E. lanigerum*

The solutions of compounds **1–12** were prepared at 0.2 mg/mL in 0.1 g/L aq. Tween-80 (0.1 g/L aq. Tween-80 as CK). Two commercial pesticides, thiamethoxam and chlorpyrifos, were used as positive controls. Some 2-year-old apple twigs were cut into pieces of 3–5 cm in length with one or two aphid groups (about 60 aphids) on each twig. Then, the 2 mL corresponding solution was sprayed on the twigs with an average of three replicates and the twigs were moisturized with cotton on both ends and incubated at 25 ± 1 °C, 60–80% RH, and with a 14 h/10 h (light/dark) photoperiod. The number of dead aphids was recorded at 48 and 72 h after treatment and the corrected mortality rate values were calculated as mentioned above [42].

2.6.5. The Effects of Piperine Derivatives **9** and **11** on the Features of Mite Cuticles

After the mites were treated with compounds **9** (at 0.298 mg/mL) and **11** (at 0.313 mg/mL) for 24, 48, and 72 h and the corresponding mites were collected (0.1 g/L aq. Tween-80 as CK). They were treated with 4% aq. glutaraldehyde at 4 °C for 6 h, then the samples were washed 3 times with 0.1 M aq. phosphate-buffered saline (PBS) and a series of ethanol solutions (10%, 30%, 50%, 70%, 80%, 90%, 95%, and absolute ethanol). Subsequently, a series of *tert*–butanol solutions (25%, 50%, 75%, and 100% *tert*–butanol, *v/v* = ethanol/*tert*–butanol) were used to rinse the mites. After the samples were dried and gold sprayed, they were observed by SEM [43].

2.7. *Statistical Analysis*

All the recorded data were statistically analyzed using the software of IBM SPSS Statistics 20.0. The probit analysis was used to estimate the LC_{50} and LD_{50} values. The mortality rates were subjected to one-way ANOVA and Duncan's test ($p < 0.05$).

3. Results and Discussion

3.1. *Synthesis*

As shown in Figure 2, the formyl group was firstly introduced at the C–2 position of piperine (**1**) by the VHA reaction to afford the aldehyde **2** [44]. Then, in the presence of $NaBH_4$ and MeOH, compound **2** was reduced to give compound **3** [45]. Finally, in the presence of EDCI and DMAP, target compounds **4–12** were obtained by reaction of compound **3** with aromatic acids RCO_2H in 44–57% yields [46,47]. The chemical structures of compounds **2–12** were all determined by mp, IR, ^{1}H NMR, and ^{13}C NMR. Fortunately, the steric configurations of compounds **6** (CCDC: 2215963) and **7** (CCDC: 2215965) were confirmed by single-crystal X-ray diffraction (Figure 3). Obviously, the amide groups of compounds **6** and **7** were all in *Z* configuration. That is, the formylation reaction of compound **1** (its amide group in *E* configuration) with $POCl_3$ and DMF was at its C-2 position and the amide group of compound **2** was in *Z* configuration. When compound **1** reacted with $POCl_3$ and DMF, if the amide group of compound **1** was in *Z* configuration, there was possibly a larger space at the C-2 position for the VHA reaction.

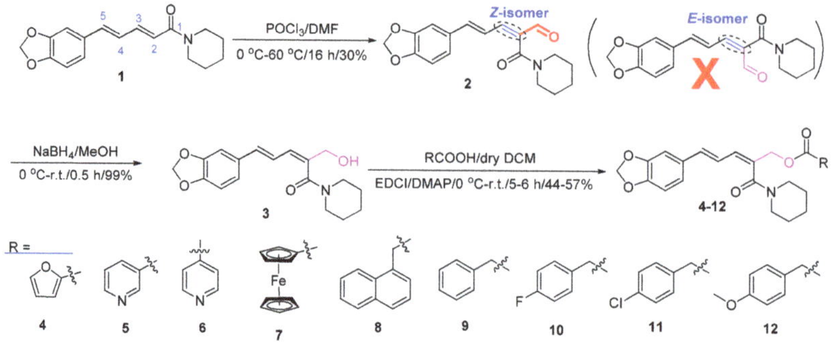

Figure 2. Synthetic route for the preparation of piperine ester derivatives (**2–12**).

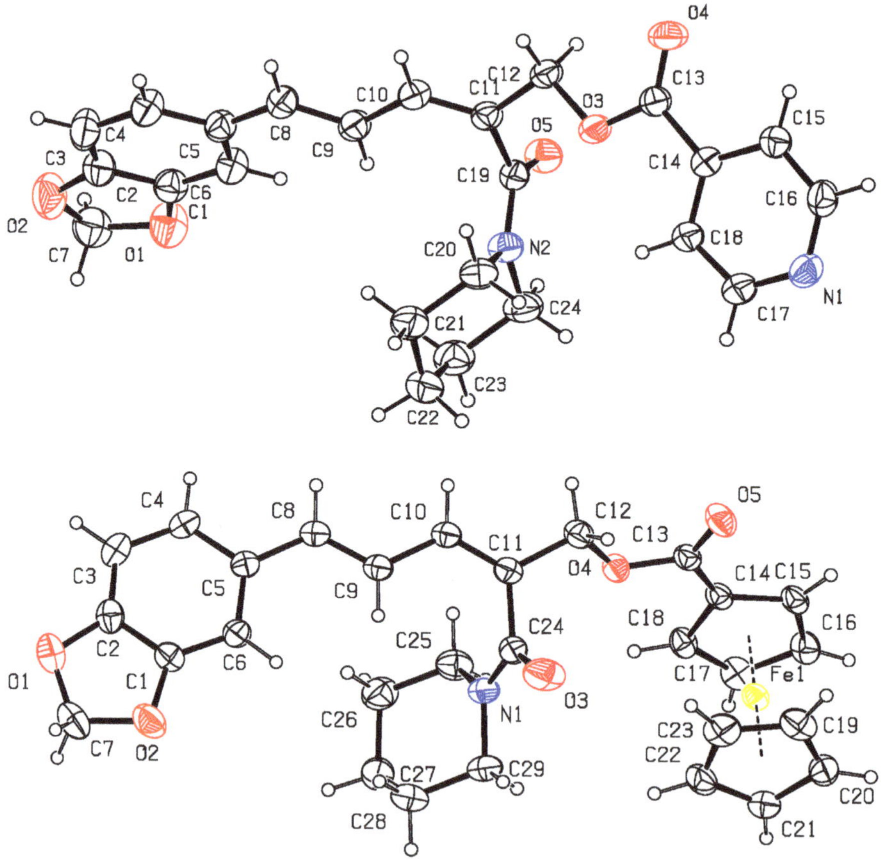

Figure 3. X–ray crystal structures of **6** (**top**) and **7** (**bottom**).

3.2. Pesticidal Activities

3.2.1. Acaricidal Activity against *T. cinnabarinus*

The 48 and 72 h acaricidal results of compounds **1–12** at 0.5 mg/mL against the female adults of *T. cinnabarinus* are described in Table 1. The 72 h corrected mortality rates (CMRs) of all derivatives against *T. cinnabarinus* ranged from 30.4% to 59.4%. In particular, the

72 h CMRs of compounds **9** (R as $PhCH_2$) and **11** (R as *p*–$ClPhCH_2$) were 59.4% and 56.3%, respectively; whereas the 72 h CMR of compound **1** was only 25.2%. Compounds **9** and **11** exhibited significant toxicities. It was revealed that R as benzyl or *p*–Clbenzyl was necessary for improving the acaricidal activity. In addition, the LC_{50} values of compounds **9** and **11** against *T. cinnabarinus* were 0.298 and 0.313 mg/mL, respectively (Table 2). Nevertheless, the LC_{50} value of compound **1** was 14.198 mg/mL. Particularly, compounds **9** and **11** were 45.4– and 47.6–fold stronger than piperine. Notably, their acaricidal activities were more potent than those of 7-oxycarbonylandrographolide, cholesterol, and osthole ester/amide derivatives [38,41,45,46]. When compared with our previous results [48], it suggested that introduction of the CH_2 between the phenyl/4-Fphenyl/4-Clphenyl and the carbonyl groups was very important for the acaricidal activity. Later, to obtain good acaricidal agents, R as the different substituted benzyl should be considered. The cowpea seedling leaves turned from green to light yellow and the symptoms of densely packed white spots (sucked by *T. cinnabarinus*) were presented to the piperine-treated and the blank control groups (Figure 4). However, the leaves of the **9**- and **11**-treated groups appeared with sporadic white spots. These phenomena were the same as our previous reports [47]. The control effects on the fifth day of compounds **9** and **11** were 56.3% and 54.9%, respectively, which were 2.6-fold that of **1** (21.5%). The photographs for evaluation of the control effects of compounds **1**, **9**, **11**, and spirodiclofens are shown in the supporting information (Figures S1–S5).

Table 1. Acaricidal activity of compounds **1–12** against female adults of *T. cinnabarinus* at 0.5 mg/mL.

Compound	Corrected Mortality Rate (Mean ± SE, %) [a]	
	48 h	72 h
1	8.32 ± 2.3	25.2 ± 3.1 i [b]
2	16.1 ± 3.9	30.9 ± 0.6 hi
3	13.7 ± 2.2	33.0 ± 3.2 gh
4	18.2 ± 2.8	41.7 ± 1.2 ef
5	13.5 ± 2.7	30.4 ± 3.8 hi
6	15.3 ± 3.2	45.0 ± 0.8 def
7	22.1 ± 3.1	46.3 ± 0.2 de
8	18.7 ± 2.8	39.1 ± 0.7 fg
9	39.1 ± 3.2	59.4 ± 0.9 b
10	32.0 ± 3.6	50.8 ± 2.4 cd
11	25.0 ± 1.1	56.3 ± 2.6 bc
12	27.4 ± 2.0	41.7 ± 2.5 ef
spirodiclofen	50.4 ± 1.8	88.6 ± 0.9 a

[a] Average of three replicates. [b] Multiple range tests using Duncan's test ($p < 0.05$). The same letters denote treatments that are not significantly different from each other.

Table 2. Toxicity regression analysis of compounds **1**, **9**, and **11** against *T. cinnabarinus* at 72 h.

Compound	LC_{50} (mg/mL)	Confidence Interval 95% (mg/mL)	Regression Equation [a]	*r*
1	14.198	9.545–25.947	Y = –0.735 + 0.638X	0.990
9	0.298	0.243–0.368	Y = 0.656 + 1.247X	0.996
11	0.313	0.252–0.395	Y = 0.614 + 1.216X	0.997
spirodiclofen	0.115	0.093–0.141	Y = 1.147 + 1.220X	0.990

[a] Regression analysis by IBM SPSS Statistics 20.0, $p < 0.05$.

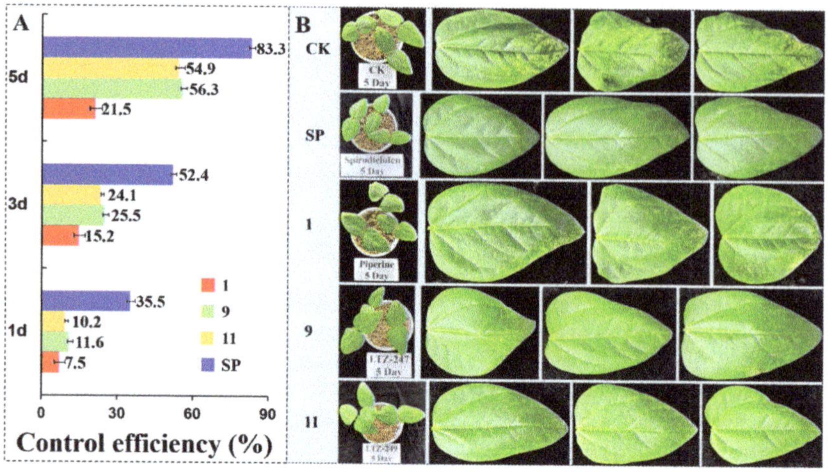

Figure 4. The control efficiency (**A**) and symptoms (**B**) of cowpea leaves infected with *T. cinnabarinus* on the 5th day after treatment with SP (spirodiclofen), **1** (piperine), **9** (LTZ–247), and **11** (LTZ–249) (CK: aq. Tween–80 (1 g/L)).

3.2.2. Aphicidal Activity against *A. citricola*

The 24 and 48 h aphicidal results of compounds **1–12** against apterous adults of *A. citricola* at 0.04 μg/nymph are depicted in Table 3. The CMRs at 48 h of compounds **2** (R as CHO) and **3** (R as CH_2OH) were 18.2% and 27.3%, respectively, whereas the CMR at 48 h of compound **1** was 25.0%. It suggested that introduction of the formyl or hydroxymethyl group at the C-2 position of compound **1** was unnecessary for improving the aphicidal activity. However, introduction of the ester group at the C-2 position of compound **1** could lead to potent derivatives. For example, CMRs at 48 h of compounds **4–12** were 28.4%–55.7%, in particular compound **6** (R as pyridin-4-yl) showed the most potent aphicidal activity with a CMR at 48 h of 55.7%. The CMRs at 48 h of compounds **4** (R as furan-2-yl) and **5** (R as pyridin-3-yl) were only 29.5% and 28.4%, respectively. It indicated that introduction of the pyridin-4-ylcarbonyloxymethylene group at the C-2 position of compound **1** was very important for the aphicidal activity against *A. citricola*. Furthermore, the LD_{50} value of compound **6** was 0.030 μg/nymph, that is, the aphicidal activity of compound **6** was >10–fold of that of compound **1** (LD_{50}: 0.308 μg/nymph) (Table 4). To our delight, the aphicidal activity of compound **6** was as promising as those of 7-oxycarbonylandrographolide and osthole ester/amide derivatives [38,46].

3.2.3. Aphicidal Activity against *E. lanigerum*

The results of the aphicidal activity at 48 and 72 h of compounds **1–12** against *E. lanigerum* at 200 μg/mL are shown in Table 5. The CMRs at 72 h of compounds **2–12** ranged from 22.7% to 38.1%. In particular, the CMR at 72 h of compound **4** (R as furan-2-yl) was 38.1%, which was 1.6-fold that of compound **1** (72 h CMR: 24.5%). Whereas CMRs at 72 h of compounds **5** (R as pyridin-3-yl) and **6** (R as pyridin-4-yl) were 33.4% and 24.8%, respectively. The CMRs at 72 h of compounds **8–12** (R as benzyl and naphthalen-1-ylmethylene) were 22.7%–32.9%. Obviously, it suggested that introduction of the furan-2-ylcarbonyloxymethylene group at the C-2 position of compound **1** was very important for aphicidal activity against *E. lanigerum*. Based on our previous results [48], it demonstrated that introduction of the CH_2 between the (un)substituted phenyl and the carbonyl groups was not necessary for aphicidal activity against *E. lanigerum*.

Table 3. Aphicidal activity of compounds **1–12** against apterous adults of *A. citricola* at 0.04 μg/nymph.

Compound	Corrected Mortality Rate (Mean ± SE, %) [a]	
	24 h	48 h
1	7.8 ± 1.1	25.0 ± 1.9 e [b]
2	6.7 ± 1.9	18.2 ± 1.9 f
3	8.9 ± 1.1	27.3 ± 3.0 e
4	16.7 ± 1.9	29.5 ± 2.2 de
5	15.6 ± 1.1	28.4 ± 1.9 de
6	35.6 ± 1.1	55.7 ± 1.9 b
7	15.6 ± 2.2	35.2 ± 3.4 cd
8	20.0 ± 1.9	29.5 ± 3.0 de
9	11.1 ± 1.1	35.2 ± 1.9 cd
10	16.7 ± 1.9	37.5 ± 3.0 c
11	14.4 ± 1.1	39.8 ± 1.1 c
12	30.0 ± 1.9	42.0 ± 1.9 c
methomyl [c]	50.0 ± 1.9	88.6 ± 1.1 a

[a] Average of three replicates. [b] Multiple range tests using Duncan's test ($p < 0.05$). The same letters denote treatments that are not significantly different from each other. [c] 0.004 μg/nymph.

Table 4. LD_{50} values of compounds **1** and **6** against *A. citricola* at 48 h.

Compound	LD_{50} (μg/nymph)	Confidence interval 95% (μg/nymph)	Regression Equation [a]	*r*
1	0.308	0.217–0.491	Y = 0.419 + 0.821X	0.997
6	0.030	0.024–0.037	Y = 2.366 + 1.549X	0.982

[a] Regression analysis by IBM SPSS Statistics 20.0, $p < 0.05$.

Table 5. Aphicidal activity of compounds **1–12** against *E. lanigerum* at 200 μg/mL.

Compound	Corrected Mortality Rate (Mean ± SE, %) [a]	
	48 h	72 h
1	16.7 ± 1.4	24.5 ± 1.0 e [b]
2	11.0 ± 1.8	22.9 ± 2.3 e
3	10.7 ± 2.0	29.9 ± 0.5 cd
4	23.9 ± 2.2	38.1 ± 0.6 b
5	11.1 ± 3.3	33.4 ± 2.1 c
6	12.3 ± 1.8	24.8 ± 1.8 e
7	13.2 ± 2.1	22.8 ± 0.7 e
8	15.2 ± 1.9	32.4 ± 2.6 c
9	17.7 ± 1.4	32.9 ± 0.5 c
10	17.3 ± 2.5	25.9 ± 2.2 de
11	13.7 ± 2.8	29.7 ± 1.6 cd
12	12.7 ± 1.1	22.7 ± 0.6 e
thiamethoxam	50.2 ± 0.2	90.9 ± 1.3 a
chlorpyrifos	48.8 ± 0.8	91.3 ± 0.6 a

[a] Average of three replicates. [b] Multiple range tests using Duncan's test ($p < 0.05$). The same letters denote treatments that are not significantly different from each other.

3.2.4. The Effects on the Features of Mite Cuticles by SEM

To investigate the preliminary mechanism of potent compounds **9** and **11** against *T. cinnabarinus*, we conducted an experiment to observe the cuticle symptoms of compound **9**- and **11**-treated *T. cinnabarinus* by using scanning electron microscope (SEM) imaging method (Figure 5). These distinct phenomena were obtained by comparison with the blank control and the treated groups. The arrangement of cuticles in the blank control group (Figure 5A–C) was well regulated and aligned. However, after 24 h, the crests of the mite cuticles of the compound **9**- and **11**-treated groups (Figure 5D,G) began to fracture and appeared irregular. After 48 h (Figure 5E,H), the sectional cuticles of the **9**-treated group were assembled into masses; additionally, the crests of **11**-treated mites partially disappeared. After 72 h, the crests of mite cuticles of the **9**- and **11**-treated groups (Figure 5F,I) all appeared broken and rough and their surfaces were accompanied by the characteristics of collapsed and lacunal symptoms [48,49].

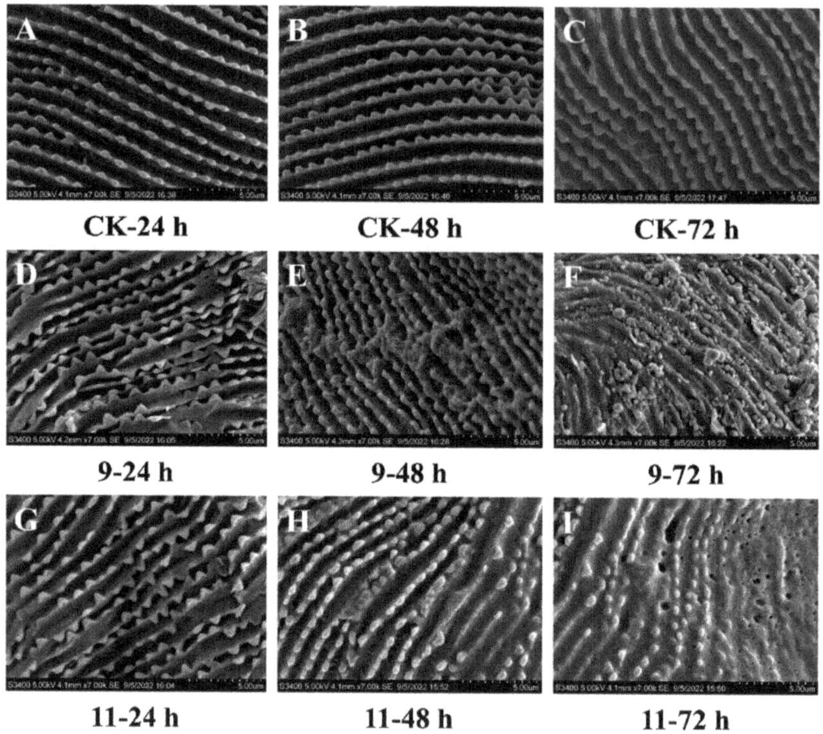

Figure 5. The scanning electron microscope images of *T. cinnabarinus* treated with compounds **9** (**D–F**), **11** (**G–I**), and aq. Tween-80 (0.1 g/L) (**A–C**) after 24, 48, and 72 h, respectively. Bars (**A–I**): 5.00 μm.

4. Conclusions

In this study, by using piperine as a lead compound, we synthesized a series of new ester derivatives of piperine by applying the Vilsmeier–Haack–Arnold reaction under mild conditions. The key steric configurations of compounds **6** and **7** were confirmed by single-crystal X–ray diffraction. Their agricultural bioactivities were evaluated against three crop-threatening pests, *T. cinnabarinus*, *A. citricola*, and *E. lanigerum*. Compounds **9** and **11** showed >45-fold higher acaricidal activity than piperine and exhibited almost 2.6-fold the control efficiency of piperine against *T. cinnabarinus* in the greenhouse. In addition, compound **6** (R as pyridin-4-yl: had >10-fold higher aphicidal activity than that of piperine) displayed the most potent aphicidal activity against *A. citricola*. Compound **4** (R as furan-2-yl) showed 1.6-fold the aphicidal activity of piperine against *E. lanigerum*. The effects of the compounds on the cuticles of *T. cinnabarinus* were observed by the SEM imaging method. These results will pave the way for future applications of piperine and its derivatives as botanical pesticides.

Supplementary Materials: The following supporting information can be downloaded at: https://www.mdpi.com/article/10.3390/insects14010040/s1, Figures S1–S5: The photographs of compounds **1**, **9**, **11**, spirodiclofen, and the blank control groups for control effect evaluation. Structural characterization data (IR, ^{1}HNMR, ^{13}CNMR and melting points) of all compounds. ^{1}H NMR and ^{13}C NMR spectra of all compounds (PDF).

Author Contributions: T.L., H.W., M.L., H.X., Y.W., S.T., S.Z. who have made substantial contributions to present work are named in the manuscript. H.X. and M.L. designed the work, analyzed the data, and wrote and revised the paper. All authors have read and agreed to the published version of the manuscript.

Funding: This research was funded by the National Natural Science Foundation of China (Project No. 31872013).

Data Availability Statement: The dataset utilized in this study is available upon request.

Acknowledgments: M.L. wants to thank the National Natural Science Foundation of China (No. 31872013).

Conflicts of Interest: The authors declare no conflict of interest.

References

1. Gu, B.J.; Zhang, X.L.; Bai, X.M.; Fu, B.J.; Chen, D.L. Four steps to food security for swelling cities. *Nature* **2019**, *566*, 31–33. [CrossRef] [PubMed]
2. Hilliou, F.; Chertemps, T.; Maibeche, M.; Le Goff, G. Resistance in the genus Spodoptera: Key insect detoxification genes. *Insects* **2021**, *12*, 544. [CrossRef] [PubMed]
3. Huang, X.; Lv, M.; Ma, Q.; Zhang, Y.; Xu, H. High value–added application of natural products in crop protection: Semisynthesis and acaricidal activity of limonoid–type derivatives and investigation of their biocompatible O/W nanoemulsions as agronanopesticide candidates. *J. Agric. Food Chem.* **2021**, *69*, 14488–14500. [CrossRef] [PubMed]
4. Zhou, H.; Zhang, Y.Q.; Lai, T.; Liu, X.J.; Guo, F.Y.; Guo, T.; Ding, W. Acaricidal mechanism of scopoletin against *Tetranychus cinnabarinus*. *Front. Physiol.* **2019**, *10*, 164. [CrossRef]
5. Scott, I.M.; McDowell, T.; Renaud, J.B.; Krolikowski, S.W.; Chen, L.; Dhaubhadel, S. Investigation of metabolic resistance to soybean aphid (*Aphis glycines* Matsumura) feeding in soybean cultivars. *Insects* **2022**, *13*, 356. [CrossRef]
6. Barczak, T.; Bennewicz, J.; Korczynski, M.; Blazejewicz-Zawadzinska, M.; Piekarska-Boniecka, H. Aphid assemblages associated with urban park plant communities. *Insects* **2021**, *12*, 173. [CrossRef]
7. O'Hara, F.M.; Davis, J.A.; Swale, D.R. Profile of commercialized aphicides on the survivorship and feeding behavior of the cotton aphid, *Aphis gossypii*. *Pestic. Biochem. Physiol.* **2022**, *186*, 105174. [CrossRef]
8. Smirnova, E.; Firth, A.E.; Miller, W.A.; Scheidecker, D.; Brault, V.; Reinbold, C.; Rakotondrafara, A.M.; Chung, B.Y.W.; Ziegler–Graff, V. Discovery of a small non–AUG–initiated ORF in poleroviruses and luteoviruses that is required for long–distance movement. *PLoS Pathog.* **2015**, *11*, e1004868. [CrossRef]
9. Zhang, C.; Liu, R.; He, J.; Ma, Z.; Zhang, X. Chemical compositions of *Ligusticum chuanxiong* oil and lemongrass oil and their joint action against *Aphis citricola* Van Der Goot (Hemiptera: Aphididae). *Molecules* **2016**, *21*, 1359. [CrossRef]
10. Liu, X.; Wu, D.; Zhang, Y.; Zhou, H.; Lai, T.; Ding, W. RNA–seq analysis reveals candidate targets for curcumin against *Tetranychus cinnabarinus*. *Biomed. Res. Int.* **2016**, *2016*, 2796260.
11. Kumar, A.; Gill, J.P.S.; Bedi, J.S.; Kumar, A. Pesticide residues in Indian raw honeys, an indicator of environmental pollution. *Environ. Sci. Pollut. Res. Int.* **2018**, *25*, 34005–34016. [CrossRef] [PubMed]
12. Liu, J.; Jiang, Z.; Feng, K.; Lu, W.; Wen, X.; Sun, J.; Li, J.; Liu, J.; He, L. Transcriptome analysis revealed that multiple genes were related to the cyflumetofen resistance of *Tetranychus cinnabarinus* (Boisduval). *Pestic. Biochem. Physiol.* **2021**, *173*, 104799. [CrossRef] [PubMed]
13. Granados–Echegoyen, C.; Loera–Alvarado, G.; Miranda-Salcedo, M.A.; Hernández–Cruz, J.; Luna–Cruz, A.; Loera–Alvarado, E. Field efficacy of synthetic and botanical-derived insecticides against *Melanaphis sacchari*[1], and non-target and beneficial species associated with cultivated sorghum. *Southwest. Entomol.* **2021**, *46*, 33–46. [CrossRef]
14. de Almeida, G.C.; Oliveira, L.F.S.; Predes, D.; Fokoue, H.H.; Kuster, R.M.; Oliveira, F.L.; Mendes, F.A.; Abreu, J.G. Piperine suppresses the Wnt/β–catenin pathway and has anti–cancer effects on colorectal cancer cells. *Sci. Rep.* **2020**, *10*, 11681. [CrossRef] [PubMed]
15. Gunasekaran, V.; Elangovan, K.; Niranjali Devaraj, S. Targeting hepatocellular carcinoma with piperine by radical–mediated mitochondrial pathway of apoptosis: An in vitro and in vivo study. *Food Chem. Toxicol.* **2017**, *105*, 106–118. [CrossRef] [PubMed]
16. Tawani, A.; Amanullah, A.; Mishra, A.; Kumar, A. Evidences for piperine inhibiting cancer by targeting human G–quadruplex DNA sequences. *Sci. Rep.* **2016**, *6*, 39239. [CrossRef]
17. Viswanadha, V.P.; Dhivya, V.; Beeraka, N.M.; Huang, C.Y.; Gavryushova, L.V.; Minyaeva, N.N.; Chubarev, V.N.; Mikhaleva, L.M.; Tarasov, V.V.; Aliev, G. The protective effect of piperine against isoproterenol–induced inflammation in experimental models of myocardial toxicity. *Eur. J. Pharmacol.* **2020**, *885*, 173524. [CrossRef]

18. Nazifi, M.; Oryan, S.; Esfahani, D.E.; Ashrafpoor, M. The functional effects of piperine and piperine plus donepezil on hippocampal synaptic plasticity impairment in rat model of Alzheimer's disease. *Life Sci.* **2021**, *265*, 118802. [CrossRef]
19. Wansri, R.; Lin, A.C.K.; Pengon, J.; Kamchonwongpaisan, S.; Srimongkolpithak, N.; Rattanajak, R.; Wilasluck, P.; Deetanya, P.; Wangkanont, K.; Hengphasatporn, K.; et al. Semi–synthesis of *N*–aryl amide analogs of piperine from *Piper nigrum* and evaluation of their antitrypanosomal, antimalarial, and anti–SARS-CoV-2 main protease activities. *Molecules* **2022**, *27*, 2841. [CrossRef]
20. Arcaro, C.A.; Gutierres, V.O.; Assis, R.P.; Moreira, T.F.; Costa, P.I.; Baviera, A.M.; Brunetti, I.L. Piperine, a natural bioenhancer, nullifies the antidiabetic and antioxidant activities of curcumin in streptozotocin–diabetic rats. *PLoS ONE* **2014**, *9*, e113993. [CrossRef]
21. Hu, X.; Wu, D.; Tang, L.; Zhang, J.; Zeng, Z.; Geng, F.; Li, H. Binding mechanism and antioxidant activity of piperine to hemoglobin. *Food Chem.* **2022**, *394*, 133558. [CrossRef]
22. Philipova, I.; Valcheva, V.; Mihaylova, R.; Mateeva, M.; Doytchinova, I.; Stavrakov, G. Synthetic piperine amide analogs with antimycobacterial activity. *Chem. Biol. Drug Des.* **2018**, *91*, 763–768. [CrossRef] [PubMed]
23. Wang, J.; Wang, W.; Xiong, H.; Song, D.; Cao, X. Natural phenolic derivatives based on piperine scaffold as potential antifungal agents. *BMC Chem.* **2020**, *14*, 24. [CrossRef] [PubMed]
24. Qu, H.; Yu, X.; Zhi, X.Y.; Lv, M.; Xu, H. Natural-product-based insecticidal agents 14. semisynthesis and insecticidal activity of new piperine-based hydrazone derivatives against *Mythimna separata* Walker in vivo. *Bioorg. Med. Chem. Lett.* **2013**, *23*, 5552–5557. [CrossRef] [PubMed]
25. Qu, H.; Lv, M.; Xu, H. Piperine: Bioactivities and structural modifications. *Mini-Rev. Med. Chem.* **2015**, *15*, 145–156. [CrossRef]
26. Qu, H.; Lv, M.; Yu, X.; Lian, X.; Xu, H. Discovery of some piperine-based phenylsulfonylhydrazone derivatives as potent botanically narcotic agents. *Sci. Rep.* **2015**, *5*, 13077. [CrossRef]
27. Nesterkina, M.; Bernier, U.R.; Tabanca, N.; Kravchenko, I. Repellent activity of monoterpenoid esters with neurotransmitter amino acids against yellow fever mosquito, *Aedes aegypti*. *Open Chem.* **2018**, *16*, 95–98. [CrossRef]
28. Yu, X.; Che, Z.; Xu, H. Recent advances in the chemistry and biology of podophyllotoxins. *Chem.-Eur. J.* **2017**, *23*, 4467–4526. [CrossRef]
29. Sun, G.S.; Xu, X.; Jin, S.H.; Lin, L.; Zhang, J.J. Ovicidal and insecticidal activities of pyriproxyfen derivatives with an oxime ester group. *Molecules* **2017**, *22*, 958. [CrossRef]
30. Huang, X.B.; Li, T.; Shan, X.; Lu, R.; Hao, M.; Lv, M.; Sun, Z.; Xu, H. High value-added use of citrus industrial wastes in agriculture: Semisynthesis and anti-tobacco mosaic virus/insecticidal activities of ester derivatives of limonin modified in the B ring. *J. Agric. Food Chem.* **2020**, *68*, 12241–12251. [CrossRef]
31. Tharamak, S.; Yooboon, T.; Pengsook, A.; Ratwatthananon, A.; Kumrungsee, N.; Bullangpoti, V.; Pluempanupat, W. Synthesis of thymyl esters and their insecticidal activity against *Spodoptera litura* (Lepidoptera: Noctuidae). *Pest Manag. Sci.* **2020**, *76*, 928–935. [CrossRef] [PubMed]
32. Li, Q.; Huang, X.; Li, S.; Ma, J.; Lv, M.; Xu, H. Semisynthesis of esters of fraxinellone C4/10-oxime and their pesticidal activities. *J. Agric. Food Chem.* **2016**, *64*, 5472–5478. [CrossRef] [PubMed]
33. Zhang, Y.Y.; Xu, H. Recent progress in the chemistry and biology of limonoids. *RSC Adv.* **2017**, *7*, 35191–35220. [CrossRef]
34. Wang, K.; Jiang, S.; Yang, Y.; Fan, L.; Su, F.; Ye, M. Synthesis and antifungal activity of carvacrol and thymol esters with heteroaromatic carboxylic acids. *Nat. Prod. Res.* **2019**, *33*, 1924–1930. [CrossRef]
35. Huang, M.; Duan, W.G.; Lin, G.S.; Li, K.; Hu, Q. Synthesis and antifungal activity of novel 3–caren–5–one oxime esters. *Molecules* **2017**, *22*, 1538. [CrossRef]
36. Xu, X.; Cai, X.; Wang, B.; Min, W.; Wang, Q.; Lai, C.; Hu, H.; Xu, D. Synthesis and herbicidal activities of novel substituted acetophenone oxime esters of pyrithiobac. *Chem. Sel.* **2020**, *5*, 69–74. [CrossRef]
37. Herbivo, C.; Comel, A.; Kirsch, G.; Raposo, M.M.M. Synthesis of 5–aryl–5′–formyl–2,2′–bithiophenes as new precursors for nonlinear optical (NLO) materials. *Tetrahedron* **2009**, *65*, 2079–2086. [CrossRef]
38. Hao, M.; Sun, Z.; Xu, J.; Lv, M.; Xu, H. Semisynthesis and pesticidal activities of derivatives of the diterpenoid andrographolide and investigation on the stress response of *Aphis citricola* Van der Goot (Homoptera: Aphididae). *J. Agric. Food Chem.* **2020**, *68*, 4131–4143. [CrossRef]
39. Li, S.; Lv, M.; Li, T.; Hao, M.; Xu, H. Spirodiclofen ether derivatives: Semisynthesis, structural elucidation, and pesticidal activities against *Tetranychus cinnabarinus* Boisduval, *Aphis citricola* Van der Goot and *Mythimna separata* Walker. *Pest Manag. Sci.* **2021**, *77*, 2395–2402. [CrossRef]
40. Xu, H.; Zhang, K.; Lv, M.; Hao, M. Construction of cholesterol oxime ether derivatives containing isoxazoline/isoxazole fragments and their agricultural bioactive properties/control efficiency. *J. Agric. Food Chem.* **2021**, *69*, 8098–8109. [CrossRef]
41. Li, T.; Lv, M.; Xu, H. Construction of new oxime esters of cholesterol containing piperic acid-like fragments as insecticidal agents against *Aphis citricola* Van der Goot (Homoptera: Aphididae) and *Plutella xylostella* Linnaeus (Lepidoptera: Plutellidae), *Bioorganic Med. Chem. Lett.* **2022**, *62*, 128634. [CrossRef] [PubMed]
42. Ma, S.; Jiang, W.; Li, Q.; Li, T.; Wu, W.; Bai, H.; Shi, B. Design, synthesis, and study of the insecticidal activity of novel steroidal 1,3,4–oxadiazoles. *J. Agric. Food Chem.* **2021**, *69*, 11572–11581. [CrossRef] [PubMed]
43. Chandran, R.; Williams, L.; Hung, A.; Nowlin, K.; LaJeunesse, D. SEM characterization of anatomical variation in chitin organization in insect and arthropod cuticles. *Micron* **2016**, *82*, 74–85. [CrossRef]
44. Netz, N.; Opatz, T. A modular formal total synthesis of (±)–cycloclavine. *J. Org. Chem.* **2016**, *81*, 1723–1730. [CrossRef] [PubMed]

45. Shan, X.; Lv, M.; Wang, J.; Qin, Y.; Xu, H. Acaricidal and insecticidal efficacy of new esters derivatives of a natural coumarin osthole. *Ind. Crops Prod.* **2022**, *182*, 114855. [CrossRef]
46. Hao, M.; Lv, M.; Zhou, L.; Li, H.; Xu, J.; Xu, H. Construction, pesticidal activities, control effects, and detoxification enzyme activities of osthole ester/amide derivatives. *J. Agric. Food Chem.* **2022**, *70*, 9337–9345. [CrossRef] [PubMed]
47. Li, S.; Lv, M.; Sun, Z.; Hao, M.; Xu, H. Optimization of osthole in the lactone ring: Structural elucidation, pesticidal activities, and control efficiency of osthole ester derivatives. *J. Agric. Food Chem.* **2021**, *69*, 6465–6474. [CrossRef] [PubMed]
48. Li, T.; Lv, M.; Wen, H.; Wang, J.; Wang, Z.; Xu, J.; Fang, S.; Xu, H. High value-added application of natural plant products in crop protection: Construction and pesticidal activities of piperine-type ester derivatives and their toxicology study. *J. Agric. Food Chem.* **2022**. [CrossRef]
49. Balabanidou, V.; Grigoraki, L.; Vontas, J. Insect cuticle: A critical determinant of insecticide resistance. *Curr. Opin. Insect Sci.* **2018**, *27*, 68–74. [CrossRef]

agronomy

MDPI

Article

Antifungal Effect of Plant Extracts on the Growth of the Cereal Pathogen *Fusarium* spp.—An In Vitro Study

Weronika Kursa [1], Agnieszka Jamiołkowska [1,*], Jakub Wyrostek [2] and Radosław Kowalski [2]

[1] Department of Plant Protection, University of Life Sciences in Lublin, 20-950 Lublin, Poland
[2] Department of Analysis and Food Quality Assessment, University of Life Sciences in Lublin, 20-950 Lublin, Poland
* Correspondence: aguto@wp.pl

Abstract: The aim of the study was a laboratory evaluation of the antifungal effect of leaf extracts from yarrow (*Achillea millefolium* L.), tansy (*Tanacetum vulgare* L.), sage (*Salvia officinalis* L.) and wormwood (*Artemisia absinthium* L.) on fungi of the genus *Fusarium*, major cereal pathogens. The study used 5%, 10%, and 20% concentrations of plant extracts, evaluating their effect on the linear growth of *Fusarium avenaceum*, *F. culmorum*, *F. graminearum*, *F. sporotrichioides* and the percentage of their growth inhibition compared to control. The study also included the assessment of the content of selected biologically active compounds in plant extracts and their impact on the development of the aforementioned pathogenic fungi. The total content of polyphenols and flavonoids in the extracts was assessed by spectrophotometry, and antioxidant activity was determined using the synthetic 2,2-diphenyl-1-picrylhydrazyl (DPPH) radical. Plant extracts from sage were characterized by the highest polyphenol contents (81.95 mg/mL) and flavonoids (21.12 mg/mL) compared to other plant extracts, and also showed the highest antioxidant activity (102.44 mM Trolox). Wormwood extract contained the lowest amount of phenolic compounds (flavonoids—5.30 mg/mL, polyphenols—43.83 mg/mL). Plant extracts inhibited the mycelia growth of fungal pathogen depending upon the fungus species, type of extract and its concentration. The extracts of sage (S) and tansy (T) plants at a concentration of 20% demonstrated strong inhibitory effect against the tested fungi (the highest inhibition coefficient for S20: 83.53%; T20: 72.58%), while 10% and 5% extracts of these plants were less effective in inhibiting the growth of *Fusarium* (highest inhibition coefficient for S10: 71.33%; S5: 54.14%; T10: 56.67%; T5: 38.64%). Yarrow (Y) and wormwood (W) extracts showed low fungistatic effect. Their 20% concentration inhibited the development of mycelia growth of fungi at the level of 63.82% (W20) and 67.57% (Y20). The 5% and 10% concentrations of these plant extracts had the weakest effect on the tested fungi (Y5: 34.09; W5: 42.06%; Y10: 45.01%; W10: 57.44%), even stimulating the *Fusarium* growth compared to the control (Y5: −23.7%). Based on the study, it was found that each species of fungus reacted differently to the addition of the extract to the culture medium and its concentration, however *F. avenaceum* and *F. culmorum* were the most sensitive fungi, while the least sensitive was *F. graminearum*. The results of the research are the preliminary phase for further field tests to determine the fungistatic effect of plant extracts in field conditions, their phytotoxicity and biological stability, as well as the possibility of producing a biopreparation to protect plants against fusariosis.

Keywords: plant extracts; *Fusarium*; antifungal activity

1. Introduction

Crop plants are constantly attacked by pathogens both during pre- and post-harvest stages, often causing economically important yield losses. The main culprits of plant diseases are pathogenic fungi, which cause reduced yields, and their lower nutritional and organoleptic value [1]. In some cases, fungi are also indirectly responsible for the occurrence of allergic disorders and poisoning among consumers. Fungi of the genus *Fusarium* deserve special attention in this regard [2]. They infect cereal plants cultivated in different climatic zones [3–7], mainly causing head blight [8]. Grain is therefore the main source of inoculum of fungi causing cereal diseases during the growing season [7,9,10]. *Fusarium* spp. can survive in the soil in the form of saprotrophic mycelium, and some species also in the form of chlamydospores [8,10]. Fungi of the genus *Fusarium* are the cause of pre-emergence and post-emergence blight of cereal seedlings. From a toxicological and economic point of view, however, the most dangerous disease caused by *Fusarium* spp. is head blight (FHB), which is accompanied by contamination of grain with mycotoxins (fumonisins, trichothecenes, and zearalenone) [4,11,12]. The ability of these fungi to produce mycotoxins is a very important factor determining the harmfulness of *Fusarium* spp. and reducing cereal grain quality [2,13]. In most regions of the world, the main causes of head blight are *F. graminearum* Schwabe, *F. culmorum* (Wm.G. Sm.) Sacc., *F. avenaceum* (Fr.) Sacc., *F. graminearum* Schwabe and *F. sporotrichioides* Sherb. [14–16]. The risk of contamination of agricultural products with mycotoxins is not limited to raw materials. Consumption of contaminated feed by livestock can lead to contamination of meat, milk, eggs and related products [17,18].

In this context, the use of synthetic fungicides is still the most effective method of protecting cereals against pathogenic fungi. However, their application causes long-term persistence of active ingredients of pesticides in food and the environment [19]. Researchers are searching for new solutions to safely protect crops against pathogens. One such effort is the search for new biologically active compounds, whose purpose is to limit the development of pathogenic fungi, while inhibiting the production of mycotoxins and having a low negative impact on the environment [20,21]. Plants are the source of many biologically active compounds, especially herbal plants, including thyme, oregano, garlic, sage [22,23]. The plants' tremendous biosynthetic capacities make it possible to use them to produce biological preparations and apply them as alternatives to synthetic chemicals [22,23]. Natural bioactive compounds called natural fungicides are non-specific, and their effect on pathogens is comprehensive [24]. The effect of natural fungicides based on plant extracts depends mainly on the content of phenols, terpenes and alkaloids [25]. Phenols have anti-radical and antioxidant properties, and selected groups of phenolic compounds, such as phenolic acids, flavonoids and tannins have an additional direct anti-fungal effect [26–28]. The mechanisms of action of these compounds are still poorly understood [28,29]. It is assumed that the antifungal properties of phenolics are attributed to their lipophilicity and/or the occurrence of the hydroxyl groups in their structure. Due to their binding properties to adhesions and proteins, they are qualified to disrupt membranes, inactivate enzymes and complex metal ions, thereby exhibiting toxic effects upon fungi. In particular, the lipophilicity of phenolics facilitates penetration of the cytoplasmic membrane, whereas hydroxyl groups are involved in the uncoupling of oxidative phosphorylation [25]. The fungistatic effect is also exhibited by alkaloids, which are usually found in poisonous plants. These compounds are effective against bacteria, parasites and fungi [28]. They penetrate the cell wall and/or DNA of the fungus [25]. Natural bioactive compounds used in plant protection not only limit the growth of fungi (fungistatic effect) but also engage plant defense responses [20,30]. Direct action of these compounds is based on inhibition of fungal sporulation, germination of spores, and reduction of hyphae growth [31]. Natural secondary metabolites produced by plants under the influence of elicitors also have a protective role in relation to pathogens [30].

The aim of the study was a laboratory evaluation of the fungistatic effect of plant extracts from tansy, yarrow, common sage and wormwood on selected *Fusarium* spp., important in agricultural phytopathology. One of the research stages was the assessment of the content of selected biologically active compounds in alcohol plant extracts and the analysis of their effect on the growth of polyphagic pathogenic fungi. The fungistatic activity of the extracts was evaluated based on the analysis of the fungal growth inhibition coefficient. The results of the conducted research are fundamental in deciding on further work addressed to the development of commercial preparations and their formulation in order to effectively protect cereals against fusarioses.

2. Materials and Methods

2.1. Chemical Characterization of Plant Extracts

2.1.1. Plant Material

The research plant material consisted leaves of yarrow (*Achillea millefolium* L.), tansy (*Tanacetum vulgare* L.), sage (*Salvia officinalis* L.) and wormwood (*Artemisia absinthium* L.) collected prior to flowering of plants. The material was collected in the natural environment (Podlasie Province, Poland), dried and powdered. Raw materials were stored in sealed packages, protected from light, moisture and the influence of foreign odors. The plants used in the research are plants commonly found in the natural state of a temperate climate and are easy to obtain herbal material.

2.1.2. Extract Preparation

To prepare the extracts, 300 g of each powdered herb was suspended in 3000 mL of 70%. Extraction was carried out under a reflux condenser at the boiling point of ethanol for 6 h. The resulting extract was filtered through filter paper and concentrated to 300 mL (extract 1:1) using a rotary evaporator (Heidolph Instruments, Schwabach, Germany). The test extracts were stored in a refrigerator (4 °C).

2.1.3. Total Polyphenol Analysis

The concentration of total polyphenols and total flavonoids was determined in the extracts. The concentration of polyphenols (calculated as gallic acid) was determined using the spectrophotometric method (λ = 765 nm) with the Folin–Ciocalteau reagent, according to the modified method of Singelton and Rossi [32]. The results were calculated from the equation of the calibration curve prepared for gallic acid standards in the concentration range of 10–60 mg/L (10, 20, 30, 40, 50, 60 mg/L). Each sample, depending on the initial concentration, was diluted according to the range on the standard curve. All analyses were performed in triplicate.

2.1.4. Flavonoid Analysis

The content of flavonoids (in terms of epicatechin) was determined by spectrophotometry according to a modified method described by Karadeniz et al. [33]. The results were calculated based on the calibration curve prepared for epicatechin standards in the concentration range of 10–400 mg/L (10, 50, 100, 150, 200, 250, 300 and 400 mg/L). Each sample, depending on the starting concentration, was diluted according to the range of the standard curve. All analyses were performed in triplicate.

2.1.5. Assessment of Extract Antioxidant Activities

The antioxidant activity of plant extracts was determined using the modified method of Brand–Willams et al. [34] using the synthetic radical DPPH (1,1-diphenyl-2-picrylhydrazyl, Sigma) converted to mM Trolox [35]. The ability of the tested antioxidant to counteract the oxidation reaction was calculated from the following formula: % inhibition = 100 $(A_0 - A_m)/A_0$, where A_m is the mean absorbance of the test solution containing the antioxidant, and A_0 is the absorbance of the DPPH radical solution.

2.2. Biological Assay

2.2.1. Isolation of Fungal Cultures

Fungal cultures were obtained in 2020–2021 as a result of mycological analyses of winter wheat grains (*Triticum aestivum* L.) of the cultivars 'Hondia' (DANKO Plant Breeding Company, Poland), 'Euforia' (Plant Breeding Company in Strzelce, Poland), 'Linus' (RAGT Semences Group, France). The following species were studied: *Fusarium avenaceum* (Fr.) Sacc. (strain P27), *F. culmorum* (Wm.G. Sm.) Sacc. (strain Fc37), *F. graminearum* Schwabe (strain Fg54), *F. sporotrichioides* Sherb. (strain P41). The fungal inoculum was derived from 10-day-old single-spore colonies grown on Potato Dextrose Agar (PDA Difco, Becton, Dickinson & C., France) stored in the fungal collection of the Department of Plant Protection,

University of Life Sciences in Lublin. Confirmation of the affiliation of the fungal strain to the species was made on the basis of microscopic analysis of each isolate/strain (structure and size of spores, colony colour) using appropriate mycological keys.

2.2.2. In Vitro Evaluation of Antifungal Potential of Plant Extracts

The study evaluated the effect of 5%, 10%, and 20% extracts of yarrow (Y), tansy (T), sage (S), and wormwood (W) on the linear growth of the fungi tested. For this purpose, the method of poisoned media was used [31]. The control consisted of fungal colonies growing on PDA medium with the addition of 5%, 10%, and 20% residue after evaporation of the extraction solvent (70% ethanol; the total volume of 1000 mL was evaporated to 100 mL in a rotary evaporator under the same conditions as for the preparation of plant extracts). The experiment was performed in 5 replicates. Experimental combinations were incubated for 10 days at 25 °C. The diameter of fungal colonies (mm) was measured after 2, 4, 6, 8 and 10 days. The measure of the antifungal effect was the inhibition of mycelial growth on the culture medium with extract addition compared to control. The fungistatic activity of plant extracts was calculated on the basis of the growth inhibition/stimulation coefficient of fungal colonies calculated using the Abbott formula: $I = [(C - T)/C] \times 100\%$, where: I—inhibition index of fungus linear growth (%), C—diameter of fungus colony in the control sample, E—diameter of fungus colony in the experimental sample containing the test substance in agar [31].

2.3. Statistical Analysis

Data were analyzed by analysis of variance (Duncan's test) at the $p \leq 0.05$ significance level using the Statistica program 12.6 (StatSoft Polska, Kraków, Poland).

3. Results

3.1. Content of Polyphenols, Flavonoids and Antioxidant Activity of Extracts

The chemical composition of yarrow, tansy, common sage and wormwood extracts was determined in the study. The average contents of flavonoids in the extracts are presented in Table 1. The concentration of flavonoids in the tested extracts ranged from 7.82 to 21.12 mg/mL and differed significantly ($p \leq 0.05$) depending on the type of extract. The highest content of flavonoids was recorded in sage extracts (21.12 mg/mL), and the lowest in wormwood extracts (5.30 mg/mL). The highest content of polyphenols was also found in the extracts from sage (81.95 mg/mL) and tansy (77.12 mg/mL), while significantly ($p \leq 0.05$) lower in the wormwood extract (43.83 mg/mL). The lowest antioxidant activity was recorded for yarrow extract (84.40 mM Trolox) and the highest for sage (102.44 mM Trolox) (Table 1).

Table 1. Concentration of flavonoids (epicatechin equivalent mg/mL), polyphenols (gallic acid equivalent mg/mL) and antioxidant activity in the basal extract (100%).

Plant Extract	Flavonoids (mg/mL) ± SD	Polyphenols (mg/mL) ± SD	Antioxidant Activity, Free Radical-Scavenging Ability	
			% Inhibition ± SD	mM Trolox ± SD
yarrow (Y)	5.05 ^{c} ± 0.130	59.56 ^{b} ± 4.080	56.13 ^{a} ± 0.386	84.40 ^{a} ± 0.608
tansy (T)	7.82 ^{b} ± 0.069	77.12 ^{a} ± 3.075	57.79 ^{a} ± 6.383	87.02 ^{a} ± 10.065
sage (S)	21.12 ^{a} ± 0.904	81.95 ^{a} ± 8.117	67.58 ^{a} ± 2.855	102.44 ^{a} ± 4.502
wormwood (W)	5.30 ^{c} ± 0.218	43.83 ^{c} ± 3.280	62.92 ^{a} ± 5.857	95.10 ^{a} ± 9.235

SD—Standard Deviation; a, b, c—values in the rows marked with the same letter do not differ significantly at a significance level of $p \leq 0.05$.

3.2. Inhibition of Fungal Growth

Laboratory tests of biotic activity allowed to determine the direct effect of plant extracts on the growth dynamics of the fungi. Each fungal species reacted differently to the addition of plant extracts in the substrate, their concentration and time of exposure. The strongest antifungal effect was recorded for extracts from sage and tansy (Figures 1 and 2, Tables 2–5).

The strength of the fungistatic effect of plant extracts increased with increasing concentrations. Extracts at a concentration of 20% showed the strongest effect.

Table 2. Inhibition (%) of the growth of *Fusarium avenaceum* (mm) after application of plant extracts.

Experimental Combination		Number of Days ± SD				
		2nd	4th	6th	8th	10th
Y	5% concentration	34.09 ^{ef} ± 3.94	16.43 ^{e} ± 10.68	7.96 ^{e} ± 8.33	−9.43 ^{e} ± 6.38	−23.7 ^{e} ± 3.33
T		38.64 ^{de} ± 13.64	23.19 ^{e} ± 12.88	11.07 ^{e} ± 8.45	−10.03 ^{e} ± 8.42	−19.4 ^{e} ± 3.03
S		29.55 ^{ef} ± 14.19	54.11 ^{bc} ± 14.22	41.87 ^{e} ± 9.51	27.05 ^{c} ± 9.86	13.6 ^{cd} ± 10.49
W		19.32 ^{f} ± 5.21	19.32 ^{e} ± 7.29	9.34 ^{e} ± 9.07	−9.73 ^{e} ± 7.76	−21.6 ^{e} ± 6.84
Y	10% concentration	42.55 ^{cde} ± 12.77	20.93 ^{e} ± 13.32	10.83 ^{e} ± 9.38	8.26 ^{d} ± 7.34	2.77 ^{d} ± 6.43
T		51.06 ^{bcd} ± 3.69	29.07 ^{de} ± 2.01	17.5 ^{e} ± 2.50	15.29 ^{cd} ± 1.06	7.56 ^{d} ± 3.05
S		35.11 ^{de} ± 14.39	40.12 ^{cd} ± 13.09	31.25 ^{d} ± 8.20	22.94 ^{c} ± 8.75	19.65 ^{c} ± 7.87
W		38.30 ^{de} ± 3.69	20.35 ^{e} ± 5.04	10.42 ^{e} ± 3.15	9.48 ^{d} ± 2.65	3.78 ^{d} ± 4.43
Y	20% concentration	67.57 ^{a} ± 0.00	59.35 ^{ab} ± 3.71	50.3 ^{bc} ± 10.80	47.05 ^{b} ± 8.74	36.25 ^{b} ± 7.24
T		67.57 ^{a} ± 5.41	65.42 ^{ab} ± 3.53	56.81 ^{b} ± 7.17	53.61 ^{b} ± 6.20	44.81 ^{b} ± 6.53
S		63.96 ^{ab} ± 3.12	73.83 ^{a} ± 1.62	76.33 ^{a} ± 1.02	72.43 ^{a} ± 3.99	61.30 ^{a} ± 8.90
W		56.76 ^{abc} ± 5.41	52.80 ^{bc} ± 3.53	51.19 ^{bc} ± 1.54	47.26 ^{b} ± 1.00	39.51 ^{b} ±0.61

Y—yarrow; T—tansy; S—sage, W—wormwood extract; SD—Standard Deviation; a, b, c … —values in the rows marked with the same letter do not differ significantly at a significance level of $p \leq 0.05$.

Table 3. Inhibition (%) of the growth of *Fusarium culmorum* (mm) after application of plant extracts.

Experimental Combination		Number of Days ± SD				
		2nd	4th	6th	8th	10th
Y	5% concentration	26.98 ^{ef} ± 2.75	2.14 ^{f} ± 18.33	10.2 ^{h} ± 9.33	16.48 ^{d} ± 5.56	6.48 ^{e} ± 2.85
T		34.92 ^{de} ± 15.31	11.54 ^{f} ± 23.54	27.86 ^{g} ± 15.30	28.52 ^{c} ± 14.92	16.30 ^{d} ± 11.98
S		57.14 ^{b} ± 4.76	46.58 ^{cde} ± 1.96	47.01 ^{def} ± 2.69	45.19 ^{b} ± 1.95	29.07 ^{c} ± 4.72
W		42.06 ^{cd} ± 4.96	35.47 ^{e} ± 7.06	38.31 ^{fg} ± 2.83	38.52 ^{bc} ± 2.10	23.33 ^{cd} ± 1.11
Y	10% concentration	15.46 ^{f} ± 3.57	42.67 ^{de} ± 1.53	45.61 ^{ef} ± 2.50	30.93 ^{c} ± 3.16	16.67 ^{d} ± 2.22
T		50.52 ^{bc} ± 12.37	59.67 ^{bcd} ± 13.50	50.76 ^{cde} ± 5.15	37.96 ^{bc} ± 10.28	23.89 ^{cd} ± 7.78
S		54.64 ^{bc} ± 9.45	71.33 ^{ab} ± 4.16	67.56 ^{ab} ± 7.45	45.00 ^{b} ± 5.47	26.85 ^{c} ± 5.89
W		24.74 ^{ef} ± 9.94	52.33 $^{b–e}$ ± 6.35	57.44 ^{bcd} ± 2.82	46.11 ^{b} ± 5.64	30.74 ^{c} ± 3.78
Y	20% concentration	48.39 ^{bcd} ± 7.39	59.41 ^{bcd} ± 4.67	56.53 $^{b–e}$ ± 1.80	38.89 ^{bc} ± 5.47	24.07 ^{cd} ± 3.90
T		72.58 ^{a} ± 2.79	71.47 ^{ab} ± 6.85	67.54 ^{ab} ± 4.58	56.48 ^{a} ± 2.85	41.85 ^{b} ± 4.98
S		74.19 ^{a} ± 2.79	83.53 ^{a} ± 1.02	75.00 ^{a} ± 1.71	65.37 ^{a} ± 2.74	52.59 ^{a} ± 1.95
W		61.29 ^{ab} ± 4.84	63.82 ^{bc} ± 6.18	59.7 ^{bc} ± 5.92	44.63 ^{b} ± 5.04	28.89 ^{c} ± 4.19

Y—yarrow; T—tansy; S—sage, W—wormwood extract; SD—Standard Deviation; a, b, c … —values in the rows marked with the same letter do not differ significantly at a significance level of $p \leq 0.05$.

Table 4. Inhibition (%) of the growth of *Fusarium graminearum* (mm) after application of plant extracts.

Experimental Combination		Number of Days ± SD				
		2nd	4th	6th	8th	10th
Y	5% concentration	28.57 ^{a} ± 0.00	13.01 ^{g} ± 15.42	8.05 ^{d} ± 4.09	−9.09 ^{c} ± 5.21	−22.41 ^{c} ± 3.33
T		28.57 ^{a} ± 0.00	36.99 $^{b–e}$ ± 4.75	22.88 ^{bcd} ± 31.66	4.17 ^{bc} ± 44.25	−2.41 ^{bc} ± 56.23
S		16.67 ^{abc} ± 4.12	23.97 ^{efg} ± 5.44	22.03 ^{bcd} ± 7.77	4.17 ^{bc} ± 11.50	−20.00 ^{c} ± 23.29
W		21.40 ^{ab} ± 7.14	21.23 ^{fg} ± 5.17	11.00 ^{cd} ± 5.83	−9.84 ^{c} ± 4.59	−27.93 ^{c} ± 6.57
Y	10% concentration	16.67 ^{abc} ± 8.33	41.24 $^{a–d}$ ± 3.09	33.10 ^{ab} ± 6.37	30.52 ^{ab} ± 6.54	22.32 ^{ab} ± 7.09
T		4.17 ^{c} ± 12.50	53.61 ^{a} ± 0.00	46.13 ^{a} ± 1.06	39.78 ^{a} ± 2.06	35.94 ^{a} ± 1.02
S		16.67 ^{abc} ± 8.33	42.27 $^{a–d}$ ± 12.50	42.96 ^{ab} ± 9.21	41.96 ^{a} ± 10.04	38.39 ^{a} ± 10.46
W		11.08 ^{bc} ± 4.81	51.03 ^{ab} ± 3.89	48.93 ^{a} ± 11.20	43.87 ^{a} ± 13.32	35.27 ^{a} ± 21.29

Table 4. *Cont.*

Experimental Combination		Number of Days ± SD				
		2nd	4th	6th	8th	10th
Y	20% concentration	25.71 ab ± 9.90	32.00 de ± 4.00	37.75 ab ± 1.84	30.60 ab ± 0.55	29.95 a ± 2.74
T		28.57 a ± 4.95	48.00 abc ± 5.29	49.40 a ± 3.19	42.90 a ± 4.47	39.06 a ± 2.07
S		22.86 ab ± 13.09	54.00 a ± 7.21	42.17 ab ± 3.19	39.75 a ± 5.78	36.98 a ± 3.16
W		14.31 bc ± 4.95	34.68 c–f ± 13.32	30.12 abc ± 4.34	19.55 ab ± 5.27	17.96 ab ± 1.35

Y—yarrow; T—tansy; S—sage, W—wormwood extract; SD—Standard Deviation; a, b, c . . . —values in the rows marked with the same letter do not differ significantly at a significance level of $p \leq 0.05$.

Table 5. Inhibition (%) of the growth of *Fusarium sporotrichioides* (mm) after application of plant extracts.

Experimental Combination		Number of Days ± SD				
		2nd	4th	6th	8th	10th
Y	5% concentration	3.64 e ± 15.75	13.67 f ± 6.01	21.09 f ± 2.73	9.63 d ± 3.57	0.93 d ± 1.60
T		23.64 cde ± 11.89	24.82 def ± 3.47	28.18 ef ± 0.36	27.59 bc ± 16.52	5.37 cd ± 4.85
S		37.27 bcd ± 4.72	37.41 cde ± 9.22	41.96 cd ± 7.12	25.93 bc ± 6.51	6.11 cd ± 3.89
W		25.45 cd ± 7.87	32.73 cde ± 7.19	37.58 de ± 5.06	26.85 bc ± 5.04	7.96 cd ± 2.80
Y	10% concentration	2.63 e ± 12.06	23.47 ef ± 3.68	29.01 ef ± 2.88	14.44 cd ± 5.36	4.44 cd ± 2.94
T		31.58 cd ± 15.79	42.86 bc ± 10.75	43.41 cd ± 5.80	29.81 bc ± 5.34	12.96 bc ± 3.21
S		43.86 abc ± 6.08	46.94 abc ± 6.12	48.68 bc ± 4.40	30.93 abc ± 4.10	7.04 cd ± 5.16
W		17.54 de ± 20.44	38.44 cd ± 12.17	44.02 cd ± 9.09	29.26 bc ± 13.22	13.52 bc ± 8.36
Y	20% concentration	60.28 a ± 2.46	55.21 ab ± 6.44	46.36 cd ± 9.68	35.37 ab ± 10.40	20.19 ab ± 13.38
T		64.54 a ± 6.50	54.57 ab ± 13.65	49.80 abc ± 6.14	29.26 bc ± 12.35	20.00 ab ± 4.34
S		63.12 a ± 8.86	59.31 a ± 8.09	58.10 ab ± 2.65	39.07 ab ± 6.24	14.26 bc ± 6.86
W		56.03 ab ± 2.46	58.99 a ± 3.94	58.50 a ± 3.56	46.11 a ± 2.94	26.11 a ± 5.47

Y—yarrow; T—tansy; S—sage, W—wormwood extract; SD—Standard Deviation; a, b, c . . . —values in the rows marked with the same letter do not differ significantly at a significance level of $p \leq 0.05$.

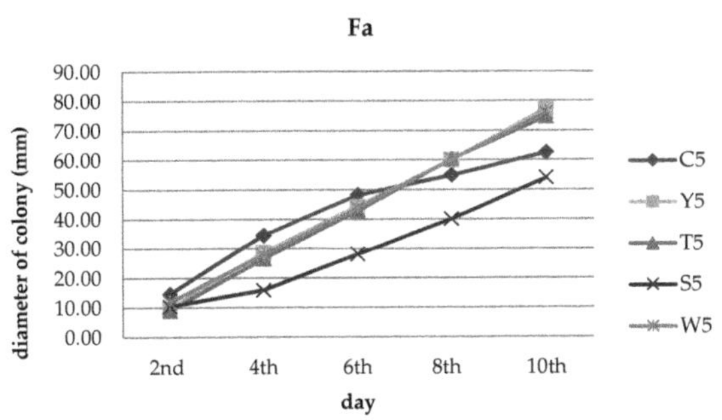

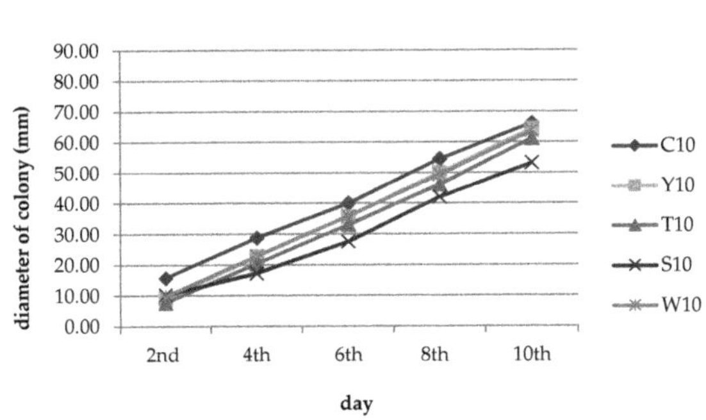

Figure 1. *Cont.*

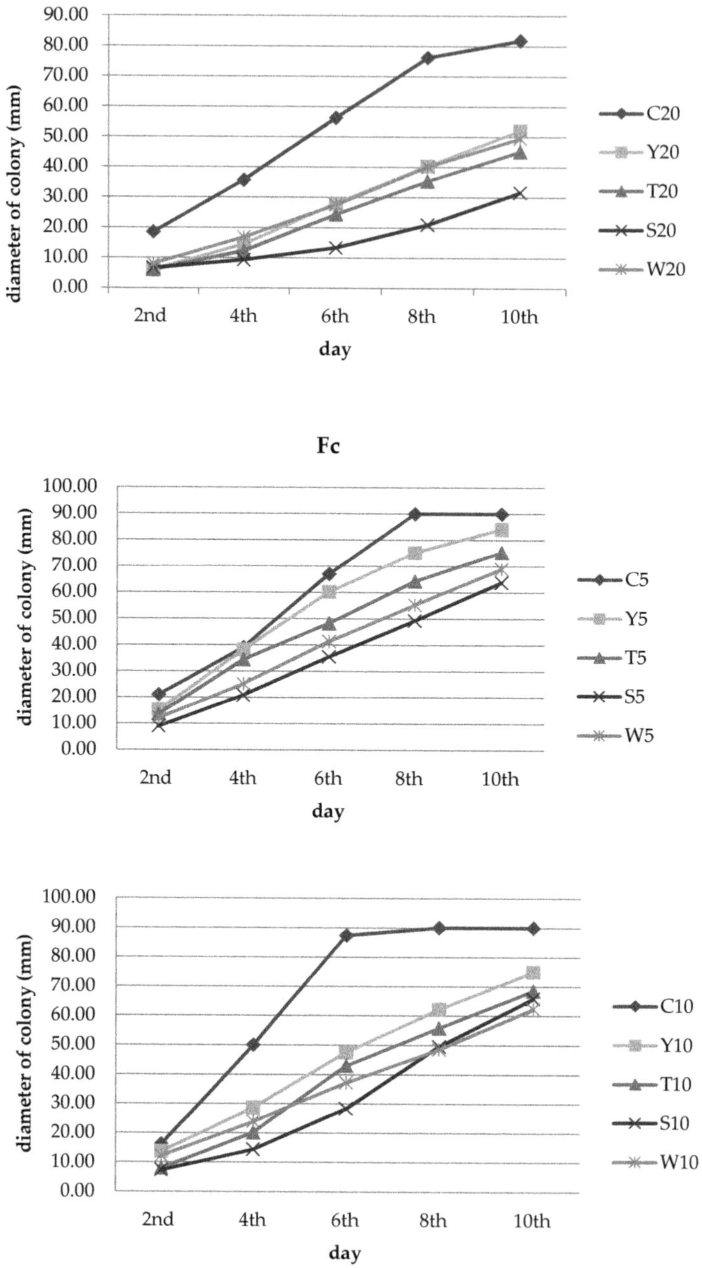

Figure 1. *Cont.*

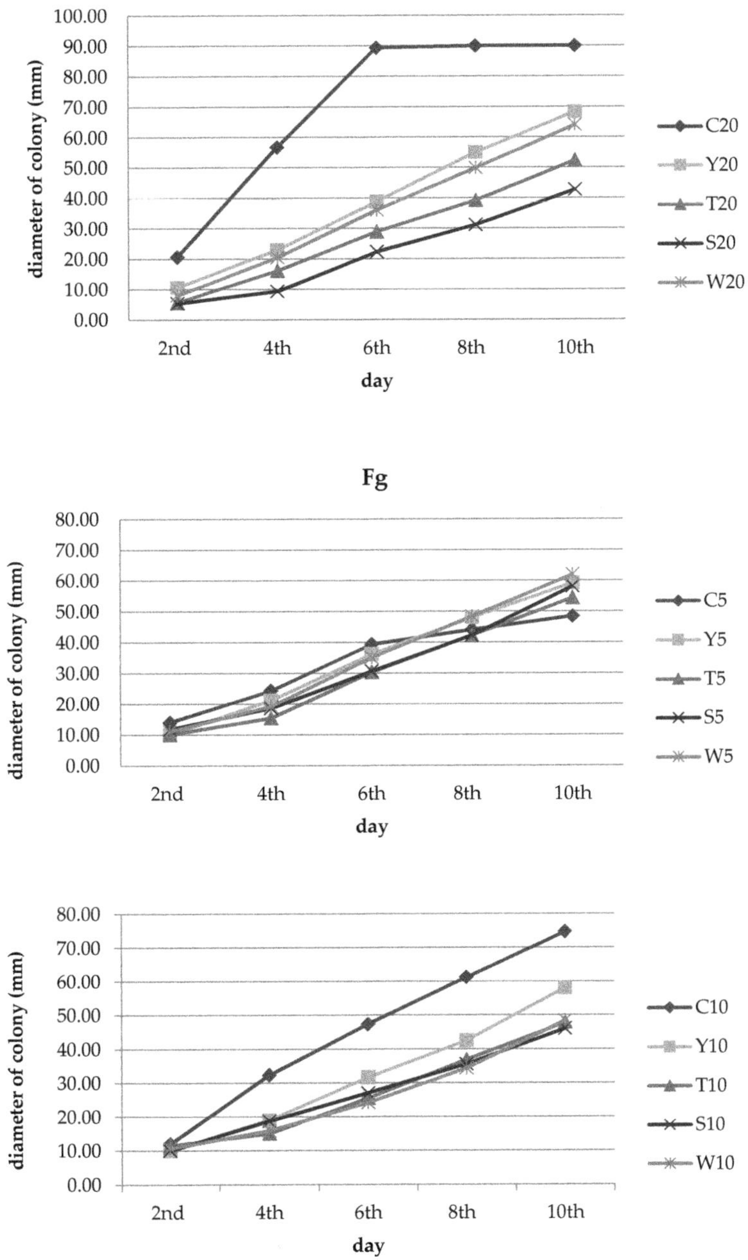

Figure 1. *Cont.*

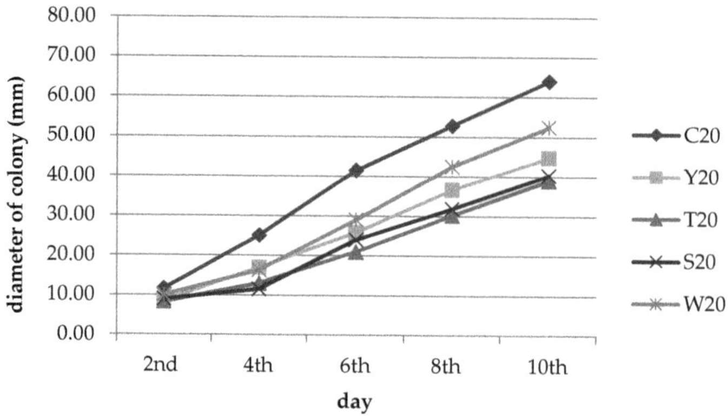

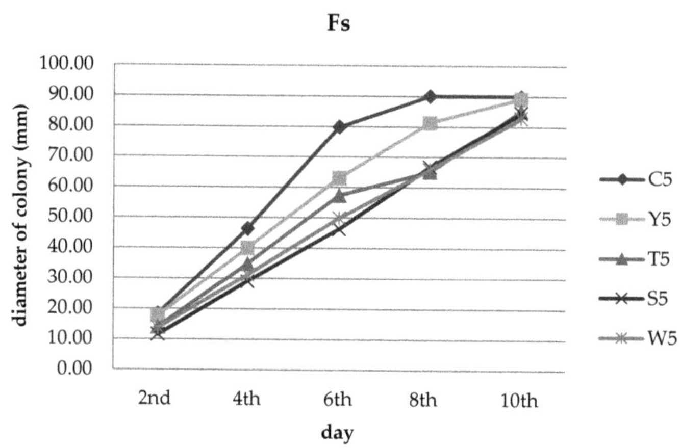

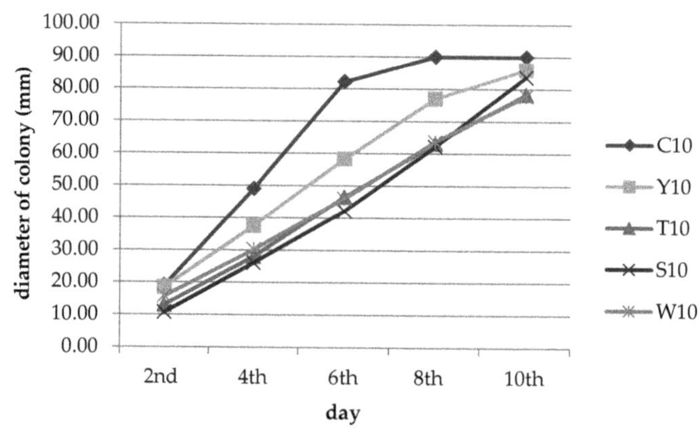

Figure 1. *Cont.*

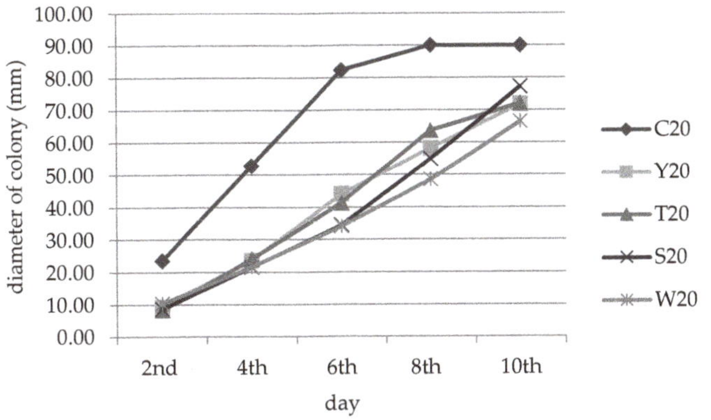

Figure 1. Diameter of colony (mm) of fungi growing on potato-dextrose agar (PDA) with different plant extract concentrations; Fa—*F. avenaceum*, Fc—*F. culmorum*, Fg—*F. graminearum*, Fs—*F. sporotrichioides*; C5, C10, C20—control with 5%, 10%, 20% solvent residue; Y5, Y10, Y20—5%, 10%, 20% yarrow extract concentration, T5, T10, T20—5%, 10%, 20% tansy extract concentration; S5, S10, S20—5%, 10%, 20% sage extract concentration; W5, W10, W20—5%, 10%, 20% wormwood extract concentration.

Sage extracts (S) significantly ($p \leq 0.05$) inhibited fungal growth for the duration of the experiment, and the best results were obtained for 20% extract. The best antifungal effects were recorded against *F. culmorum* (S20: 52.59–83.53%), *F. avenaceum* (S20: 61.30–76.33%) and *F. sporotrichioides* (S20: 14.26–63.12%) throughout the experiment (Figure 1, Tables 2, 3 and 5). Of the fungi tested, *F. graminearum* was the least sensitive to the 20% concentration of this extract in the culture medium (S20: 22.86–54.00%) (Figure 1, Table 4). The weakest effect of the sage extract was recorded for the concentration of 5%, especially against *F. graminearum* (S5: 4.17–23.97%), where the fungistatic effect of the extract decreased already on day 4 of the experiment, and a faster experimental colony growth was observed on days 8 and 10 day compared to control (Figure 1, Table 4). However, the antifungal effect of sage extract against *F. graminearum* was not significantly ($p \leq 0.05$) different compared to other plant extracts (Tables 2–5).

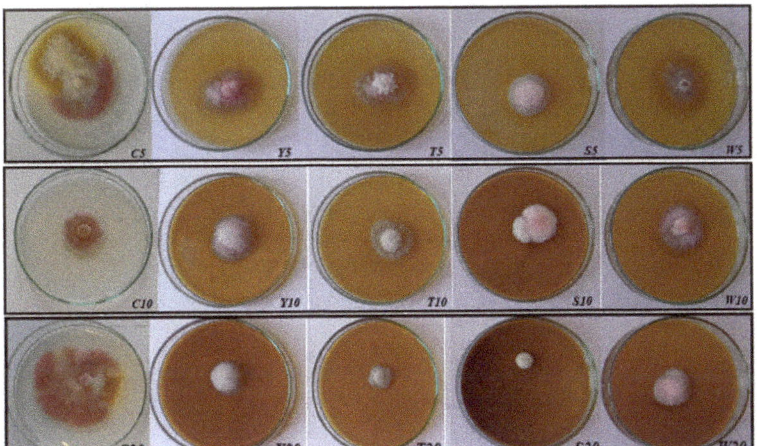

Figure 2. *Cont.*

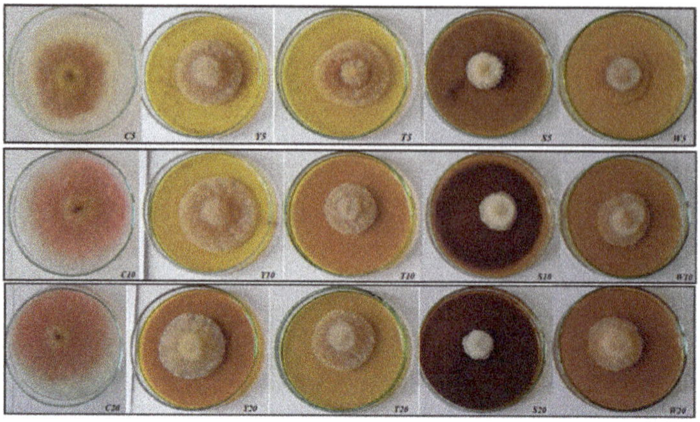

Fusarium culmorum

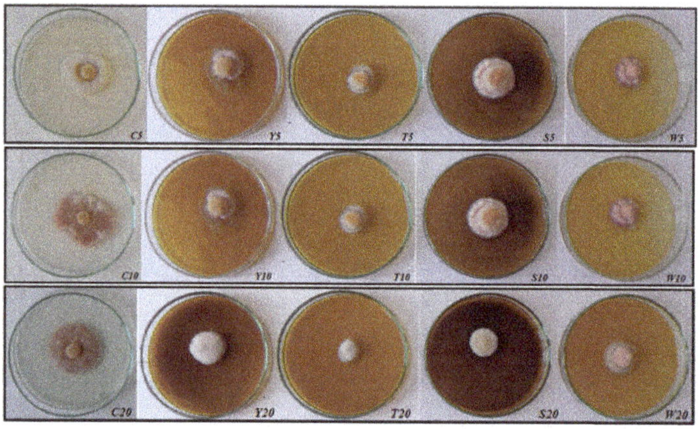

Fusarium graminearum

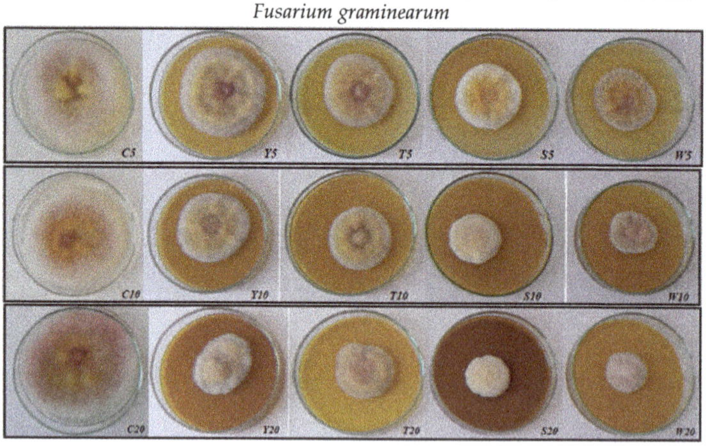

Fusarium sporotrichioides

Figure 2. Linear growth of fungal colonies on PDA with the addition of plant extracts; C5, C10, C20—control with 5%, 10%, 20% solvent residue; Y5, Y10, Y20—5%, 10%, 20% yarrow extract, T5, T10, T20—5%, 10%, 20% tansy extract; S5, S10, S20—5%, 10%, 20% sage extract; W5, W10, W20—5%, 10%, 20% wormwood extract; 6th day of growth.

Interesting results were also noted after applying 20% concentration of tansy extract (T). The fungi *F. culmorum* (T20: 41.85–72.58%), *F. avenaceum* (T20: 44.81–67.57%), and *F. sporotrichioides* (T20: 20.00–64.54%) were particularly sensitive to the effect of tansy, but the fungistatic effect of sage was not significantly ($p \leq 0.05$) different compared to the other plant extracts at this concentration for individual fungal species (exception: *F. culmorum* Y20) (Tables 2–5). Tansy extract exerted the strongest effect in the first days of the experiment (days 2, 4), while its effectiveness decreased in the following days. The weakest effect of tansy extract was recorded for a concentration of 5% and against *F. sporotrichioides* (T5: 5.37–28.18%) (Figure 1, Table 5).

Yarrow (Y) and wormwood (W) extracts showed slightly lower fungistatic activity, the weakest at a concentration of 5%, which slightly inhibited the surface growth of *F. sporotrichioides* (Y5: 0.93–21.09%; W5: 7.96–37.58%), and they even stimulated colony growth of *F. graminearum* and *F. avenaceum* on day 10 of the experiment (Tables 2, 4 and 5). The 20% concentration of yarrow extract was most effective in limiting the growth of *F. sporotrichioides* (Y20: 20.19–60.28%), and least effective against *F. graminearum* (Y20: 25.71–37.75%). Interesting results were also recorded for wormwood extracts, especially with regard to *F. culmorum* colonies, whose surface growth at the beginning of the experiment (2–4 days) was inhibited at the level of 61.29–63.82%. *F. graminearum* strain turned out to be particularly resistant to wormwood extract, and its 20% concentration inhibited the growth of the fungus only at the level of 14.31–34.68% (Tables 3 and 4). The study found that each fungal species reacted differently to the addition of the extract to the culture medium and its concentration, however, *F. avenaceum* and *F. culmorum* were the most sensitive fungi, while the least sensitive was *F. graminearum*.

The conducted experiment also focused on changes in the morphology of the fungi under the influence of plant extracts (Figure 2, Table 6). The most common were changes in mycelial structure, and coloration of the obverse and reverse of the colony. With increasing extract concentration in the medium, the changes in the colony structure intensified. Fungal colonies with plant extracts in the medium were more compact and of elevated growth, while in the control mycelium the mycelium was looser and less compact. The reverse side of the test fungi was lighter or colorless compared to the control colonies (Table 6). The fungistatic effect of plant extracts demonstrated in the study on *Fusarium* spp., i.e., species important in agricultural phytopathology, warrants further research in this area. There is a need to conduct field experiments in this area, which should evaluate the effects of individual extracts on phytotoxicity and plant health during vegetation, as well as to test their effectiveness for grain treatment.

Table 6. Selected features of fungal morphology under the influence of plant extracts (6th day of the experiment).

Fungus Species	Experimental Combination	Mycelium Surface and Structure	Obverse	Reverse
F. avenaceum	C5, C10, C20	fluffy, slightly elevated	pink-white; pink	maroon
	Y5, Y10, Y20	substrate, slightly compact, elevated in the center	white-purple	purple, light purple
	T5, T10, T20	slightly compact; elevated	white	slightly pink; colorless
	S5, S10, S20	fluffy, slightly elevated	pink-white; white-cream	slightly pink; colorless
	W5, W10, W20	flat, centrally slightly elevated	white-gray-pink	pink-white
F. culmorum	C5, C10, C20	fluffy, regular, even growth	white-pink	purple
	Y5, Y10, Y20	fluffy, slightly elevated in the center	purple	maroon
	T5, T10, T20	fluffy, slightly raised	purple; pink-white	purple; maroon
	S5, S10, S20	compact; elevated	white-pink	light pink
	W5, W10, W20	fluffy, slightly elevated in the center	gray-pink-white	maroon
F. graminearum	C5, C10, C20	regular growth, elevated in the center	white-pink; pink	maroon
	Y5, Y10, Y20	irregular, substrate, elevated in the center	pink-purple-white	brown and maroon
	T5, T10, T20	regular growth, elevated	white-pink	pink
	S5, S10, S20	regular growth, elevated	white-pink-yellow	light brown
	W5, W10, W20	regular growth, elevated in the center	white-purple	gray-purple; light pink

Table 6. *Cont.*

Fungus Species	Experimental Combination	Mycelium Surface and Structure	Obverse	Reverse
F. sporotrichioides	C5, C10, C20	fluffy, regular, even growth	pink-white-yellow	maroon
	Y5, Y10, Y20	fluffy, slightly raised	pink-white-yellow	maroon
	T5, T10, T20	fluffy, slightly raised	pink-white-yellow	maroon
	S5, S10, S20	compact; elevated	pink-white-yellow	maroon
	W5, W10, W20	fluffy, elevated and compact	pink-white-yellow; white-yellow	maroon

C5, C10, C20—control with 5%, 10%, 20% solvent residue; Y5, Y10, Y20—5%, 10%, 20% yarrow extract, T5, T10, T20—5%, 10%, 20% tansy extract; S5, S10, S20—5%, 10%, 20% sage extract; W5, W10, W20—5%, 10%, 20% wormwood extract.

4. Discussion

The conducted laboratory experiments allowed for determining the direct effect of plant extracts on the growth dynamics of fungi of the genus *Fusarium*, important in the pathology of agricultural plants. Studies of the fungistatic activity of plant extracts are conducted all over the world and concern various pathogens. The most important in this regard are herbal plants, which show strong biocidal properties due to the content of various biologically active compounds [36–43]. Currently, it is well recognized that hundreds of biological active chemical compounds are present in plants, working in synergism, and conferring a broad variety of bioactivities. Phenolic compounds are the most common group of plant components with strong antimicrobial, including antifungal properties. The present study has shown that the sage (*Salvia officinalis*) extract demonstrated best fungistatic properties. It is a medicinal plant with antioxidant, antimicrobial and anti-inflammatory properties. Its diverse biological activity is mainly conditioned by polyphenolic compounds [44]. The main components of the secondary metabolites of *Salvia* spp. are terpenoids and flavonoids. The aerial parts of these plants contain mainly flavonoids, triterpenoids and monoterpenes, especially concentrated in flowers and leaves [45–47]. Numerous studies also confirmed the high biocidal activity of sage secondary metabolites [46,47]. However, differences regarding this effect were identified depending on the dosage plant extracts applied. It was reported in different studies that biologically active compounds of *Salvia* spp. exerted antifungal effect on *Fusarium* species, such as *F. tricintum*, *F. sporotrichioides* and *F. oxysporum* f. sp. *radicis-lycopersici* [48,49]. Studies have also shown the antimycotoxigenic properties of major monoterpene constituent 1,8-cineole from *S. officinalis* in relation to ochratoxin A (OTA), produced by *Aspergillus carbonarius* [47]. Secondary metabolites of sage, especially phenolic acids, flavonoids and terpenes, also show high antioxidant activity [50–54]. Rowshan and Najafian [50] showed in their study the highest content of such polyphenols as rosemarinic acid, catechin, vanillin, chlorogenic acid, quercetin and p-coumaric acid. The antioxidant capacity of the tested extracts was correlated with the total phenolic and flavonoid content [55], and these reports were confirmed in the present study. Sage, due to the highest phenolic content among the tested plant extracts, was also the most effective against pathogenic fungi, inhibiting their growth even by 83.53% (*F. culmorum*—20% concentration).

High antifungal activity was also observed for tansy extracts, especially for the concentration of 20%. Tansy is characterized by a high content of secondary metabolites [56,57], thus it exerts a strong antibacterial and antifungal effect [58,59]. Tansy plants contain many phenolic compounds (flavonoids and phenolic acids). Among flavonoids, there are mainly luteolin, apigenin and quercetin glycosides, while phenolic acids are predominantly represented by chlorogenic, caffeic and dicaffeoylquinic acids [57,60]. *T. vulgare* extracts also show high antioxidant activity, strongly correlated with the content of polyphenols, especially flavonoids [61]. The high antioxidant potential (FRAP and DPPH) of tansy extract was shown by Bączek et al. [60], and the present study confirmed the aforementioned reports. However, the world literature lacks detailed data on the effect of tansy extracts on phytopathogens, especially fungi of the genus *Fusarium*. Korpinen et al. [56] showed a strong inhibitory effect of tansy extracts on *Penicillium venetum* and *Aspergillus niger*. However, Wens and Geuens [62] showed that the addition of 500 μL of *T. vulgare* extract to the culture medium inhibited the growth of *F. oxysporum* but only at the level of 32% compared to control. Different results were obtained in the current study, indicating a strong antifungal effect of tansy extract, especially during the first days of the experiment (2nd day), mainly against *F. culmorum* (72.5%), *F. avenaceum*, (67.5%) and *F. sporotrichioides* (64.5%) at 20% concentration of the extract in the medium. Furthermore, it should be noted that one plant from a given genus has a specific chemical composition and its significant medicinal value does not mean that all the other plants of the same genus had the same properties. Some studies even reported that the same species, but of different origin showed considerable differences in chemical composition, and thus, different bioactive efficacy and potency [43,63].

Yarrow is also a rich source of flavonoids. *Achillea millefolium* is a medicinal plant containing over 100 different biologically active compounds with antimicrobial properties [64]. Flavonoids, apigenin and quercetin, phenolic acid and caffeoylquinic acid have been described as the main phenolic compounds present in yarrow [65]. The chemical composition of a plant can vary significantly due to various factors, such as the geographic origin of the plant, the time of harvest and the analyzed plant parts, extraction and analysis methods [43,66]. This was confirmed by Georgieva et al. [67], indicating that the total content of polyphenols in *A. millefolium* water extracts (2.74–7.92 mg GAE/100 g fw) and the total content of flavonoids (from 0.05% to 0.07%) changed over time. The highest content of these compounds was determined at the beginning and the lowest at the end of growth. The high content of flavonoids, polyphenols and the high antioxidant potential of water extracts in our study was determined at a level similar to that presented by other researchers [68,69]. Fierascu et al. [69] showed that yarrow extract strongly inhibited the growth of *Aspergillus niger* (70.19%) and *Penicillium hirsutum* (47.40%) [69], as well as the oil had fungistatic effect against *Rhizopus stolonifer* (65.7%) and *Verticillium dahliae* (65.3%) [63]. High content of secondary metabolites determines the strength of the extract's antifungal effect, which was also confirmed by our research. The strongest antifungal effect of the extract was recorded for a concentration of 20% at the beginning of the experiment, for all *Fusarium* species tested, especially against *F. sporotrichioides* (60.2%). The extracts showed the weakest effect against *F. graminearum*, 5% concentration of the extract in the culture medium even stimulated mycelial growth in subsequent days.

Of the plant extracts tested, woodworm extract showed the weakest antifungal properties against *Fusarium* spp. Kordali et al. [70] confirmed the fungistatic properties of wormwood from the Turkish population against 34 fungal species, including *F. solani* and *F. oxysporum*. Although many authors confirmed the fungistatic effects of wormwood extracts and oils [70–73], the strength of their biocidal action varied. The high efficacy of *A. absinthium* ethanol extracts has been described in laboratory studies against such phytopathogens as *Penicillium expansum*, *Botrytis cinerea*, *B. allii*, *Monilinia laxa*, *M. fructigena* or *Plasmopara viticola* [72]. The main antifungal compounds contained in wormwood extracts include flavonoids, thiophene and terpenoid [74,75]. Canadanovic-Brunet et al. [27] showed that the biocidal activity of extracts was elevated with increasing content of phenolic compounds and flavonoids, indicating that they contributed to antiradical and antioxidant activity.

Plant extracts from sage, yarrow, tansy and wormwood, in addition to inhibiting the growth of *Fusarium* spp. fungi, caused changes in the colour and structure of aerial mycelium. Jamiołkowska and Kowalski [31] reported that the addition of grapefruit extract (Biosept 33 SL) to the culture medium strongly inhibited the growth and formation of morphological elements of *F. avenaceum*, *F. equiseti*, *F. culmorum* and *B. cinerea*, thereby limiting fungal sporulation. The authors also found structural changes in the hyphae, such as cytoplasm dehydration and hyphae deformation. The inhibitory effect of biopreparations on the growth of phytopathogenic fungi in vitro was studied by Orlikowski et al. [76,77]. These authors showed that the compounds present in the grapefruit extract (Biosept 33 SL preparation) inhibited not only the germination of *F. oxysporum* and *B. cinerea* spores, but also the growth of germ hyphae by dehydrating the cytoplasm of mycelial cells.

The strength of the biocidal effect of plant extracts varies and depends on the type of extract and the microorganism it acts on. The results of the present study and literature data indicate that the chemical composition of plant extracts varies depending on the plant species and the sampling site [78]. While the antibacterial and antioxidant effects of various plant extracts are well understood, the antifungal and antimycotoxigenic have not yet been thoroughly studied. The present study complements the data on the bioprotective effect of plant extracts against selected phytopathogens of the genus *Fusarium*, important in agricultural phytopathology. However, the results obtained require confirmation of the antifungal effect of the extracts under field conditions, and a properly prepared mixture of plant extracts may increase the strength of fungistatic effect and provide the basis for the development of a new biofungicide of plant origin and become a safer alternative for sustainable and organic crop production.

5. Conclusions

The use of natural compounds for pathogen control is very attractive, and the availability of biotechnological methods open new avenues for plant protection approaches. The results of the conducted experiments indicated a positive effect of plant extracts from sage, tansy, yarrow and wormwood on the growth of fungi of the genus *Fusarium* spp. The extracts inhibited the growth of *Fusarium* spp. fungi significantly better at a higher concentration (20%) compared to lower concentrations. Sage and tansy extracts showed the strongest antifungal activity against *F. culmorum*, *F. avenaceum* and *F. sporotrichioides*, *F. graminearum*, and wormwood extracts exerted the weakest effect. It was also demonstrated that the greater fungistatic effect of plant extracts depended on the higher content of secondary metabolites (polyphenols and flavonoids) and their high antioxidant activity. The conducted research is the basis for further research to determine the fungistatic effect of plant extracts in field conditions, their phytotoxicity and biological stability. Plant extracts obtained in a conventional way may, however, be unstable and difficult to obtain on an industrial scale, therefore, efforts should be made to develop such biotechnological methods that ensure stable biologically active compounds, the production of which will be economically justified. The results of field research will allow for the development of the proper composition of the plant component (mixture of extracts), which will be the basis for the production of a natural preparation for cereal protection against fusariosis as well as seed treatment.

Author Contributions: Conceptualization, A.J. and W.K.; methodology, A.J., J.W. and R.K.; software, W.K., A.J. and J.W.; validation, A.J. and R.K.; formal analysis, W.K., J.W. and A.J.; investigation, W.K. and A.J.; resources, W.K. and A.J.; data curation, A.J. and J.W.; writing—original draft preparation, W.K. and A.J.; writing—review and editing, A.J. and W.K.; visualization, A.J.; supervision, A.J.; project administration, A.J.; funding acquisition, A.J., W.K. and R.K. All authors have read and agreed to the published version of the manuscript.

Funding: This research was supported by projects no.SD/38/RiO/2022, no OKK/s/44/2022 provided by University of Life Sciences in Lublin, Poland.

Conflicts of Interest: The authors declare no conflict of interest.

References

1. Agrios, G.N. Losses caused by plant diseases. In *Plant Pathology*; Elsevier: Oxford, UK, 2004; pp. 29–45.
2. Mielniczuk, E.; Skwaryło-Bednarz, B. Fusarium Head Blight, Mycotoxins and Strategies for Their Reduction. *Agronomy* **2020**, *10*, 509. [CrossRef]
3. Bottallico, A.; Perrone, G. Toxigenic Fusarium species and mycotoxins associated with head blight in small-grain cereals in Europe. *J. Plant Pathol.* **2002**, *108*, 611–624. [CrossRef]
4. Logrieco, A.; Bottalico, A.; Mule, G.; Morietti, A.; Perrone, G. Epidemiology of toxigenic fungi and their associated mycotoxins for some miediterranean crops. *Eur. J. Plant Pathol.* **2003**, *109*, 645–667. [CrossRef]
5. Xu, X.M.; Nicholson, P.; Thomsett, M.A.; Simpson, D.; Cooke, B.M.; Doohan, F.M.; Edwards, S.G. Relationship between the fungal complex causing Fusarium head blight of wheat and environmental conditions. *Phytopathology* **2008**, *98*, 69–78. [CrossRef]

6. Backhouse, D.; Burgess, L.W.; Summerell, B.A. BiogeoFigurey of *Fusarium*. In *Fusarium Nelson Memorial Symposium*; Summerell, B.A., Leslie, J.F., Backhouse, D., Bryden, W.L., Burgess, L.W., Eds.; APS Press: St Paul, MN, USA, 2012; pp. 122–137.
7. Tekauz, A.; Mitchell Fetch, J.W.; Rossnagel, B.G.; Savard, M.E. Progress in assessing the impact of Fusarium head blight on oat in western Canada and screening of *Avena germplasm* for resistance. *Cereal Res. Commun.* **2008**, *36*, 49–56. [CrossRef]
8. Hietaniemi, V.; Rämö, S.; Yli-Mattila, T.; Jestoi, M.; Peltonen, S.; Kartio, M.; Sieviläinen, E.; Koivisto, T.; Parikka, P. Updated survey of Fusarium species and toxins in Finnish cereal grains. *Food Addit. Contam. Part A* **2016**, *33*, 831–848. [CrossRef]
9. Tekle, S.; Skinnes, H.; Bjørnstad, A. The germination problem of oat seed lots affected by Fusarium head blight. *Eur. J. Plant Pathol.* **2013**, *135*, 147–158. [CrossRef]
10. Köhl, J.; de Haas, B.H.; Kastelein, P.; Burgers, S.L.G.E.; Waalwijk, C. Population dynamics of *Fusarium* spp. and *Microdochium nivale* in crops and crop residues of winter wheat. *Phytopathology* **2007**, *97*, 971–978. [CrossRef]
11. Desjardins, A.E. *Fusarium-Mycotoxins Chemistry Genetics and Biology*; APS Press: St. Paul, MN, USA, 2006.
12. Stenglein, S.A.; Dinolfo, M.I.; Barros, G.; Bongiorno, F.; Chulze, S.N.; Moreno, M.V. *Fusarium poae* pathogenicity and mycotoxin accumulation on selected wheat and barley genotypes at a single location in Argentina. *Plant Dis.* **2014**, *98*, 1733–1738. [CrossRef]
13. Goliński, P.; Waśkiewicz, A.; Wiśniewska, H.; Kiecana, I.; Mielniczuk, E.; Gromadzka, M.; Kostecki, M.; Bocianowski, J.; Rymaniak, E. Reaction of winter wheat (*Triticum aestivum* L.) cultivars to infection with *Fusarium* spp. mycotoxins contamination in grain and chaff. *Food Addit. Contam. Part A* **2010**, *27*, 1015–1024. [CrossRef]
14. Yli-Mattila, T. Ecology and evolution of toxigenic *Fusarium* species in cereals in Northern Europe and Asia. *J. Plant Pathol.* **2010**, *92*, 7–18. Available online: https://www.jstor.org/stable/41998764 (accessed on 1 January 2020).
15. Beccari, G.; Covarelli, L.; Nicholson, P. Infection processes and soft wheat response to root rot and crown rot caused by *Fusarium culmorum*. *Plant Pathol.* **2011**, *60*, 671–684. [CrossRef]
16. Hellin, P.; Dedeurwaerder, G.; Duvivier, M.; Scauflaire, J.; Huybrechts, B.; Callebaut, A.; Munaut, F.; Legrève, A. Relationship between *Fusarium* spp. diversity and mycotoxin contents of mature grains in southern Belgium. *Food Addit. Contam. Part A* **2016**, *33*, 1228–1240. [CrossRef] [PubMed]
17. Stanciu, O.; Juan, C.; Miere, D.; Loghin, F.; Mañes, J. Presence of Enniatins and Beauvericin in Romanian Wheat Samples: From Raw Material to Products for Direct Human Consumption. *Toxins* **2017**, *9*, 189. [CrossRef] [PubMed]
18. Gruber-Dorninger, C.; Jenkins, T.; Schatzmayr, G. Global Mycotoxin Occurrence in Feed: A Ten-Year Survey. *Toxins* **2019**, *11*, 375. [CrossRef] [PubMed]
19. Directive 2009/128/EC of the European Parliament and of the Council of 21 October 2009 establishing a framework for Community action to achieve the sustainable use of pesticides. *Off. J. Eur. Union L* **2009**, *309*, 71–86.
20. Orsoni, N.; Degola, F.; Nerva, L.; Bisceglie, F.; Spadola, G.; Chitarra, W.; Terzi, V.; Delbono, S.; Ghizzoni, R.; Morcia, C.; et al. Double Gamers—Can Modified Natural Regulators of Higher Plants Act as Antagonists against Phytopathogens? The Case of Jasmonic Acid Derivatives. *Int. J. Mol. Sci.* **2020**, *21*, 8681. [CrossRef]
21. Jamiołkowska, A.; Skwaryło-Bednarz, B.; Mielniczuk, E.; Bisceglie, F.; Pelosi, G.; Degola, F.; Gałązka, A.; Gręda, E. Effect of Thiosemicarbazone Derivatives and *Fusarium culmorum* (Wm.G. Sm.) Sacc. Infection of Winter Wheat Seedlings on Their Health Status and Soil Biological Activity. *Agronomy* **2022**, *12*, 116. [CrossRef]
22. Jamiołkowska, A. Preparaty biotechniczne i biologiczne w ochronie papryki słodkiej (*Capsicum annuum* L.) przed grzybami chorobotwórczymi i indukowaniu reakcji obronnych roślin. *Rozpr. Nauk. UP W Lub.* **2013**, *379*, 117.
23. Kawka, M.; Pilarek, M.; Sykłowska-Baranek, K.; Pietrosiuk, A. Roślinne metabolity jako kluczowy bioprodukt biotechnologii roślin. *Biul. Wydz. Farm. WUM* **2017**, *8*, 68–79. [CrossRef]
24. Isman, M.B. Botanical insecticides, deterrents, and repellents in modern agriculture and an increasingly regulated world. *Annu. Rev. Entomol.* **2006**, *51*, 45–66. [CrossRef] [PubMed]
25. Acheuk, F.; Basiouni, S.; Shehata, A.A.; Dick, K.; Hajri, H.; Lasram, S.; Yilmaz, M.; Emekci, M.; Tsiamis, G.; Spona-Friedl, M.; et al. Status and Prospects of Botanical Biopesticides in Europe and Mediterranean Countries. *Biomolecules* **2022**, *12*, 311. [CrossRef] [PubMed]
26. Martins, N.; Barros, L.; Henriques, M.; Silva, S.; Ferreira, I.C.F.R. Activity of phenolic compounds from plant origin against *Candida* species. *Ind. Crop. Prod.* **2015**, *74*, 648–670. [CrossRef]
27. Canadanovic-Brunet, J.M.; Djilas, S.M.; Cetkovic, G.S.; Tumbas, V.T. Free-radical scavenging activity of wormwood (*Artemisia absinthium* L) extracts. *J. Sci. Food Agric.* **2005**, *85*, 265–272. [CrossRef]
28. Šernaitė, L. Plant extracts: Antimicrobial and antifungal activity and appliance in plant protection. *Sodininkystė Ir Daržininkystė* **2017**, *36*, 58–68.
29. Shabana, Y.M.; Abdalla, M.E.; Shanin, A.A.; El-Sawy, M.M.; Draz, I.S.; Youssif, A.W. Efficacy of plant extracts in controlling wheat leaf rust disease caused by *Puccinia triticina*. *Egypt. J. Basic Appl. Sci.* **2017**, *4*, 67–73. [CrossRef]
30. Jamiołkowska, A. Natural compounds as elicitors of plant resistance against diseases and new biocontrol strategies. *Agronomy* **2020**, *10*, 173. [CrossRef]
31. Jamiołkowska, A.; Kowalski, R. Laboratory effect of *Silphium perfoliatum* L. on the growth of tested fungi. *Acta Sci. Pol. Hortorum Cultus* **2012**, *11*, 43–55.
32. Singleton, V.L.; Rossi, J.A. Colorimetry of total phenolics with phosphomolybdic-phosphotungstic acid reagents. *Am. J. Enol. Viticult.* **1965**, *16*, 144–158.

33. Karadeniz, F.; Burdurlu, H.S.; Koca, N.; Soyer, Y. Antioxidant activity of selected fruits and vegetables grown in Turkey. *Turk. J. Agric. For.* **2005**, *29*, 297–303.
34. Brand-Williams, W.; Cuvelier, M.E.; Berset, C. Use of a Free Radical Method to Evaluate Antioxidant Activity. *Lebensm.-Wiss. Und-Technol.* **1995**, *28*, 25–30. [CrossRef]
35. Wyrostek, J.; Kowalski, R. Effect of ultrasound and fragmentation of the raw material on the extraction of phenolic compounds from peppermint leaves and black tea. *Przemysł Chem.* **2022**, *101*, 928–933.
36. Alsirrag, M.; Ali, R. Determination of antimicrobial and antioxidants activity of *Salvia przewalskii* seed oil against pathogenic bacteria and fungi. *J. Phys. Conf. Ser.* **2018**, *1032*, 012070. [CrossRef]
37. Dulger, G.; Dulger, B. Antifungal activity of *Salvia verticillata* subsp. *verticillata* against fungal pathogens. *Düzce Üniversitesi Sağlık Bilimleri Enstitüsü Derg.* **2021**, *11*, 305–307. [CrossRef]
38. Medjahed, F.; Merouane, A.; Saadi, A.; Bader, A.; Cioni, P.L.; Flamini, G. Chemical profile and antifungal potential of essential oils from leaves and flowers of *Salvia algeriensis* (Desf.): A comparative study. *Chil. J. Agric. Res.* **2016**, *76*, 195–200. [CrossRef]
39. Abdulrahman, A.; Alkhail, A. Antifungal activity of some extract against some plant pathogenic fungi. *Pak. J. Biol. Sci.* **2005**, *8*, 413–417. [CrossRef]
40. Sebeşan, M.; Cărăban, A. Analysis of the Essential Oils from Thyme (*Thymus vulgaris* L.) and from Peppermint (*Mentha piperita* L.). *Chem. Bull. Politeh. Univ. Timis.* **2008**, *53*, 1–2.
41. Slusarenko, A.J.; Patel, A.; Portz, D. Control of plant diseases by natural products: Allicin from garlic as a case study. *Eur. J. Plant Pathol.* **2008**, *121*, 313–322. [CrossRef]
42. Parveen, S.; Wani, A.H.; Ganie, A.A.; Pala, S.A.; Mir, R.A. Antifungal activity of some plant extracts on some pathogenic fungi. *Arch. Phytopathol. Plant Prot.* **2014**, *47*, 279–284. [CrossRef]
43. Mutlu-Ingok, A.; Devecioglu, D.; Dikmetas, D.N.; Karbancioglu-Guler, F.; Capanoglu, E. Antibacterial, Antifungal, Antimycotoxigenic, and Antioxidant Activities of Essential Oils: An Updated Review. *Molecules* **2020**, *25*, 4711. [CrossRef]
44. Poulios, E.; Giaginis, C.; Vasios, G.K. Current advances on the extraction and identification of bioactive components of sage (*Salvia* spp.). *Curr Pharm Biotechnol.* **2019**, *20*, 845–857. [CrossRef]
45. Topçu, G. Bioactive triterpenoids from *Salvia* species. *J. Nat. Prod.* **2006**, *69*, 482–487. [CrossRef]
46. Redondo-Blanco, S.; Fernandez, J.; Lopez-Ibanez, S.; Miguelez, E.M.; Villar, C.J.; Lombo, F. Plant phytochemicals in food preservation: Antifungal bioactivity: A review. *J. Food Protect.* **2020**, *83*, 163–171. [CrossRef]
47. Dammaka, I.; Hamdi, Z.; Kammun, S.; Euch, E.; Zemni, H.; Mliki, A.; Hassouna, M.; Lasram, S. Evaluation of antifungal and anti-ochratoxigenic activities of Salvia officinalis, Lavandula dentata and Laurus nobilis essential oils and a major monoterpene constituent 1,8-cineole against Aspergillus carbonarius. *Ind. Crop Prod.* **2019**, *128*, 85–93. [CrossRef]
48. Džamič, A.; Sokovič, M.; Ristič, M.; Grujič-Jovanovič, S.; Vukojevič, J.; Marin, P.D. Chemical composition and antifungal activity of *Salvia sclarea* (Lamiaceae) essential oil. *Arch. Biol. Sci.* **2008**, *60*, 233–237. [CrossRef]
49. Yilar, M.; Kadioğlu, I. Antifungal Activities of some Salvia Species Extracts on *Fusarium oxysporum* f. sp. *radicis- lycopersici* (Forl) Mycelium Growth In-Vitro. *Egypt. J. Pest Cont.* **2016**, *26*, 115–118.
50. Rowshan, V.; Najafian, S. Polyphenolic contents and antioxidant activities of aerial parts of *Salvia multicaulis* from the Iran flora. *Nat. Prod. Res.* **2020**, *34*, 2351–2353. [CrossRef]
51. Benedec, D.; Hanganu, D.; Oniga, I.; Tiperciuc, B.; Olah, N.K.; Raita, O.; Bischin, C.; Silaghi-Dumitrescu, R.; Vlase, L. Assessment of rosmarinic acid content in six Lamiaceae species extracts and their antioxidant and antimicrobial potential. *Pak. J. Pharm. Sci.* **2015**, *28*, 2297–2303.
52. Molnar, M.; Jerković, I.; Suknović, D.; Rajs, B.B.; Aladić, K.; Šubarić, D.; Jokić, S. Screening of six medicinal plant extracts obtained by two conventional methods and supercritical CO- extraction targeted on coumarin content, 2, 2-diphenyl-1-picrylhydrazyl radical scavenging capacity and total phenols content. *Molecules* **2017**, *22*, 348. [CrossRef]
53. Neagu, E.; Roman, G.P.; Radu, G.L. Antioxidant capacity of some *Salvia officinalis* concentrated extracts. *Rev. Roum. Chim.* **2011**, *56*, 777–782.
54. Cieśla, Ł.; Staszek, D.; Hajnos, M.; Kowalska, T.; Waksmundzka-Hajnos, M. Development of chromatographic and free radical scavenging activity fingerprints by thin-layer chromatography for selected Salvia species. *Phytochem. Anal.* **2011**, *22*, 59–65. [CrossRef] [PubMed]
55. Safari, M.R.; Azizi, O.; Heidary, S.S.; Kheiripour, N.; Ravan, A.P. Antiglycation and antioxidant activity of four Iranian medical plant extracts. *J. Pharmacopuncture* **2018**, *21*, 82–89. [CrossRef] [PubMed]
56. Korpinen, R.I.; Välimaa, A.L.; Liimatainen, J.; Kunnas, S. Essential Oils and Supercritical CO2 Extracts of Arctic Angelica (*Angelica archangelica* L.), Marsh Labrador Tea (*Rhododendron tomentosum*) and Common Tansy (*Tanacetum vulgare*)—Chemical Compositions and Antimicrobial Activities. *Molecules* **2021**, *26*, 7121. [CrossRef] [PubMed]
57. Nurzyńska-Wierdak, R.; Sałata, A.; Kniaziewicz, M. Tansy (*Tanacetum vulgare* L.)-A Wild-Growing Aromatic Medicinal Plant with a Variable Essential Oil Composition. *Agronomy* **2022**, *12*, 277. [CrossRef]
58. Keskitalo, M.; Pehu, E.; Simon, J.E. Variation in volatile compounds from tansy (*Tanacetum vulgare* L.) related to genetic and morphological differences of genotypes. *Biochem. Syst. Ecol.* **2001**, *29*, 267–285. [CrossRef] [PubMed]
59. Holetz, F.B.; Pessini, G.L.; Sanches, N.R.; Cortez, D.A.G.; Nakamura, C.V.; Dias Filho, B.P. Screening of some plants used in the Brazilian folk medicine for the treatment of infectious diseases. *Mem. Inst. Oswaldo Cruz* **2002**, *97*, 1027–1031. [CrossRef]

60. Bączek, K.; Kosakowska, O.; Przybył, J.; Kuźma, P.; Ejdys, M.; Obiedziński, M.; Węglarz, Z. Intraspecific variability of yarrow (*Achillea millefolium* L.) in respect of developmental and chemical traits. *Herba Pol.* **2015**, *61*, 7–52. [CrossRef]
61. Ivănescu, B.; Tuchiluș, C.; Corciovă, A.; Lungu, C.; Mihai, C.T.; Gheldiu, A.M.; Vlase, L. Antioxidant, antimicrobial and cytotoxic activity of *Tanacetum vulgare*, *Tanacetum corymbosum* and *Tanacetum macrophyllum* extracts. *Farmacia* **2018**, *66*, 282–288.
62. Wens, A.; Geuens, J. In vitro and in vivo antifungal activity of plant extracts against common phytopathogenic fungi. *J. BioSci. Biotechnol.* **2022**, *11*, 15–21. [CrossRef]
63. El-Kalamouni, C.; Venskutonis, P.R.; Zebib, B.; Merah, O.; Raynaud, C.; Talou, T. Antioxidant and Antimicrobial Activities of the Essential Oil of *Achillea millefolium* L. Grown in France. *Medicines* **2017**, *4*, 30. [CrossRef]
64. Ghanati, F.; Bakhtiarian, S.; Parast, B.M.; Behrooz, M.K. Production of new active phytocompounds by Achillea millefolium L. after elicitation with silver nanoparticles and methyl jasmonate. *Biosci. Biotechnol. Res. Asia* **2014**, *11*, 391–399. [CrossRef]
65. Dias, M.I.; Barros, L.; Dueñas, M.; Pereira, E.; Carvalho, A.M.; Alves, R.C.; Oliveira, M.B.; Santos-Buelga, C.; Ferreira, I.C. Chemical composition of wild and commercial Achillea millefolium L. and bioactivity of the methanolic extract, infusion and decoction. *Food Chem.* **2013**, *141*, 4152–4160. [CrossRef] [PubMed]
66. Bimbiraitė, K.; Ragažinskienė, O.; Maruška, A.; Kornyšova, O. Comparison of the chemical composition of four yarrow (*Achilea millefolium* L.) morphotypes. *Biologija* **2008**, *54*, 208–212. [CrossRef]
67. Georgieva, L.; Gadjalova, A.; Mihaylova, D.; Pavlov, A. *Achillea millefolium* L.-phytochemical profile and in vitro antioxidant activity. *Int. Food Res. J.* **2015**, *22*, 1347–1352.
68. Wojdyło, A.; Oszmiański, J.; Czemerys, R. Antioxidant activity and phenolic compounds in 32 selected herbs. *Food Chem.* **2007**, *105*, 940–949. [CrossRef]
69. Fierascu, I.; Ungureanu, C.; Avramescu, S.M.; Fierascu, R.C.; Ortan, A.; Soare, L.C.; Paunescu, A. In vitro antioxidant and antifungal properties of *Achillea millefolium* L. *Rom Biotechnol. Lett.* **2015**, *20*, 10626–10636.
70. Kordali, S.; Kotan, R.; Mavi, A.; Cakir, A.; Ala, A.; Yildirim, A. Determination of the chemical composition and antioxidant activity of the essential oil of *Artemisia dracunculus* and of the antifungal and antibacterial activities of Turkish *Artemisia absinthium*, *A. dracunculus*, *Artemisia santonicum*, and *Artemisia spicigera* essential oils. *J. Agric. Food Chem.* **2005**, *53*, 9452–9458.
71. Umpiérrez, M.L.; Lagreca, M.E.; Cabrera, R.; Grille, G.; Rossini, C. Essential oils from Asteraceae as potential biocontrol tools for tomato pests and diseases. *Phytochem. Rev.* **2012**, *11*, 339–350. [CrossRef]
72. Andreu, V.; Levert, A.; Amiot, A.; Cousin, A.; Aveline, N.; Bertrand, C. Chemical composition and antifungal activity of plant extracts traditionally used in organic and biodynamic farming. *Environ. Sci. Pollut. Res. Int.* **2018**, *25*, 29971–29982. [CrossRef]
73. Msaada, K.; Salem, N.; Bachrouch, O.; Bousselmi, S.; Tammar, S.; Alfaify, A.; Al Sane, K.; Ben Ammar, W.; Azeiz, S.; Brahim, A.H.; et al. Chemical composition and antioxidant and antimicrobial activities of wormwood (Artemisia absinthium L.) essential oils and phenolics. *J. Chem.* **2015**, *2015*, 1–12. [CrossRef]
74. Gonzalez-Coloma, A.; Bailen, M.; Diaz, C.E.; Fraga, B.M.; Martínez-Díaz, R.; Zuniga, G.E.; Contreras, R.A.; Cabrera, R.; Burillo, J. Major components of Spanish cultivated *Artemisia absinthium* populations: Antifeedant, antiparasitic, and antioxidant effects. *Ind. Crop. Prod.* **2012**, *37*, 401–407. [CrossRef]
75. Aberham, A.; Cicek, S.S.; Schneider, P.; Stuppner, H. Analysis of sesquiterpene lactones, lignans, and flavonoids in wormwood (*Artemisia absinthium* L.) using high-performance liquid chromatography (HPLC)− mass spectrometry, reversed phase HPLC, and HPLC− solid phase extraction− nuclear magnetic resonance. *J. Agric. Food Chem.* **2010**, *58*, 10817–10823. [CrossRef] [PubMed]
76. Orlikowski, L.B.; Skrzypczak, C.; Harnaj, I. Biological activity of grapefruit extract in the control of forme speciales of *Fusarium oxysporum*. *J. Plant Prot. Res.* **2001**, *41*, 104–111.
77. Orlikowski, L.B.; Skrzypczak, C.; Jaworska-Marosz, A. Influence of grapefruit extract on the growth and development of *Botrytis* spp. and grey mold development on lily and peony. *Bull. Pol. Acad. Sci. Biol. Sci.* **2001**, *49*, 373–378.
78. Nazzaro, F.; Fratianni, F.; Coppola, R.; Feo, V.D. Essential oils and antifungal activity. *Pharmaceuticals* **2017**, *10*, 86. [CrossRef]

toxins

Review

Recent Advances on Biological Activities and Structural Modifications of Dehydroabietic Acid

Meng Hao [1], Jianwei Xu [1], Houpeng Wen [1], Jiawei Du [1], Shaoyong Zhang [2], Min Lv [1,2,*] and Hui Xu [1,*]

[1] College of Plant Protection, Northwest A&F University, Yangling 712100, China
[2] Key Laboratory of Vector Biology and Pathogen Control of Zhejiang Province, College of Life Science, Huzhou University, Huzhou 313000, China
* Correspondence: lvmin@nwsuaf.edu.cn (M.L.); orgxuhui@nwsuaf.edu.cn (H.X.)

Abstract: AbstractDehydroabietic acid is a tricyclic diterpenoid resin acid isolated from rosin. Dehydroabietic acid and its derivatives showed lots of medical and agricultural bioactivities, such as anticancer, antibacterial, antiviral, antiulcer, insecticidal, and herbicidal activities. This review summarized the research advances on the structural modification and total synthesis of dehydroabietic acid and its derivatives from 2015 to 2021, and analyzed the biotransformation and structure-activity relationships in order to provide a reference for the development and utilization of dehydroabietic acid and its derivatives as drugs and pesticides.

Keywords: dehydroabietic acid; bioactivity; structural modification; total synthesis; structure-activity relationship

Key Contribution: This review presents an overview on the bioactivities, structural modifications, total synthesis, biotransformation, and structure-activity relationships of dehydroabietic acid and its derivatives.

1. Introduction

Terpenoids containing a variety of biological activities, are the promising lead structures for the development of drugs and pesticides [1–6]. Rosin is an important natural renewable resource in pine trees [7,8]. Dehydroabietic acid (**1**, Figure 1), a tricyclic diterpenoid, is a natural resin acid isolated from rosin and shows a wide range of biological activities [9]. The content of dehydroabietic acid in rosin was low, and it was not suitable for direct extraction of dehydroabietic acid. Storage-stable dehydroabietic acid can be purified by disproportionating abietic acid or rosin. In the industry, disproportionated rosin was obtained by disproportionation reaction of rosin as raw material, and then dehydroabietic acid can be obtained by organic amine salt and solvent recrystallization methods. Therefore, the research focus of purification was mainly on the preparation of disproportionated rosin, and the most effective catalyst for the disproportionation process of rosin was the expensive Pd/C catalyst, so it was very important to seek an efficient and economical purification condition [10]. In addition, dehydroabietic acid can also be isolated from two cyanobacteria strains [11].

The molecular skeleton of dehydroabietic acid contains one carboxyl group, one aromatic ring, and two alicyclic rings, with a total of twenty carbons. Dehydroabietic acid and its derivatives showed a variety of biological activities such as antiviral [12], antitumor [13], wound-healing [14], antiulcer [15], gastroprotective [16], anxiolytic [17], herbicidal [18] and antibacterial properties [19]. Furthermore, Xie et al. found that dehydroabietic acid had insecticidal activity against *Peridroma saucia* (Lepidoptera: Noctuidae) [20]. It demonstrated that larch sawfly exposure to dehydroabietic acid resulted in reduced feeding and slowed growth [21]. Dehydroabietic acid acted as an antagonist of insect juvenile hormone, interfering with the endocrine regulation of insects [22], and its derivatives also had an attracting effect on *Spodoptera litura* [23]. Recently, Xin et al. reported that novel multifunctional nanomedicines assembled from chitosan oligosaccharide-melanin complexes and dehydroabietic acid hexamers can achieve efficient and precise treatment of tumors [24]. Huang et al. found that biomass-based carbon dots prepared from dehydroabietic acid by hydrothermal reaction not only can sensitively and selectively detect heavy metal ions, but also can be used for cell imaging with low cytotoxicity [25]. However, while its biological activities were diverse, it had certain toxic and side effects. It reported that resin acids may be toxic to fish [26,27]. Based on the above studies, with the characteristics of good stability (due to the aromatic ring) and wide biological activities, dehydroabietic acid was a candidate of interest in the fields of medicine and agriculture. Moreover, to solve the disadvantages of its toxic and side effects, it was necessary to develop new dehydroabietic acid derivatives with high bioactivity and low toxicity. Therefore, this review summarized the structural modifications and biological activities of dehydroabietic acid and its derivatives from 2015 to 2021.

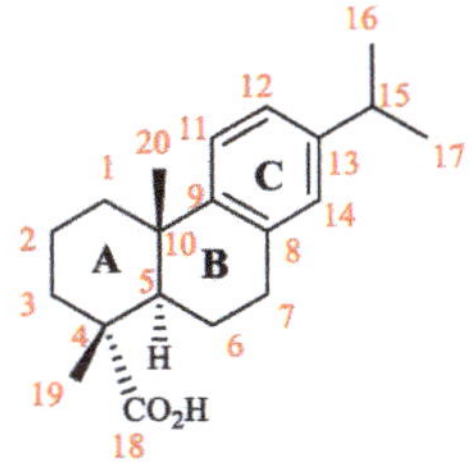

dehydroabietic acid (1)

Figure 1. Chemical structure of dehydroabietic acid (**1**).

2. Bioactivities of Dehydroabietic Acid and Its Derivatives

2.1. Antitumor Activity

Dehydroabietic acid derivatives exhibited anti-cancer activity through various mechanisms including inhibiting tumor cell migration and inducing tumor apoptosis [7]. Lee et al. found that dehydroabietic acid induced apoptosis of human lung cells by inducing the division of caspase-3 and PARP in these cells and interfering with mitochondria [28]. Moreover, Luo et al. suggested that a derivative of dehydroabietic acid (**2**, Figure 2), maybe a therapeutic drug for gastric cancer because it induced damage to cell membranes and organelles, and ultimately led to apoptosis of gastric cancer cells [29]. Through flow cytometry and cell cycle analysis, it was found that compound **3** (Figure 2) can induce HepG2 cell apoptosis and block the HepG2 cell line in the G1 phase to exert anti-cancer effects [30]. Dehydroabietic oxime was a possible method for the treatment of pancreatic cancer and its related inflammation. It can up-regulate the level of p27, and it can also down-regulate the expression of cyclin D1, thereby preventing the growth of pancreatic cancer cells in the G1 phase [31]. Recently, some dehydroabietic acid derivatives were potential drugs targeting the kinase domain of EGFR, which showed anticancer activity against HepG2 cancer cell lines [32].

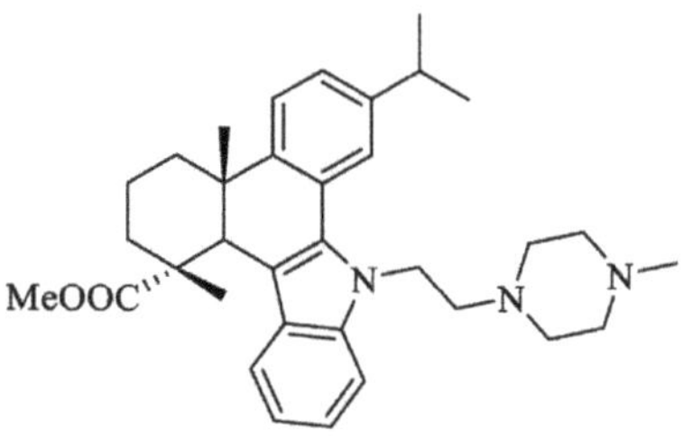

2

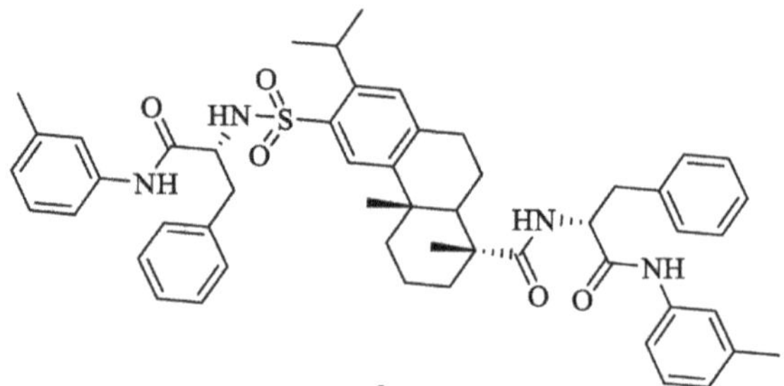

3

Figure 2. Chemical structures of compounds **2** and **3**.

2.2. Anti-Inflammatory Activity

Kim et al. evaluated the anti-inflammatory effects of dehydroabietic acid and found that not only the production of nitric oxide (NO) in the macrophage cell line was reduced under the action of dehydroabietic acid, but also the expression of inflammatory genes was descended. In addition, through the treatment of dehydroabietic acid, both the activity of kinases in the NF-κB cascade and TAK1 (transforming growth factor β-activated kinase 1) in the AP-1 cascade was inhibited [33]. A study conducted by Kang et al. proved that the treatment of dehydroabietic acid was an effective way to ameliorate inflammatory changes related to obesity-related diabetes. It demonstrated that dehydroabietic acid was an activator of PPARα and PPARγ and can inhibit the production of MCP-1, TNF-α, and NO (pro-inflammatory mediators) [34].

2.3. Antibacterial and Antifungal Activities

Experiments by Tretyakova et al. proved that pyrrolidine-containing dehydroabietic acid acetylene derivatives (**4**, Figure 3) showed growth inhibitory effects on the fungi *Candida albicans* and *Cryptococcus neoformans*, and had low hemolytic capacity [35]. In addition, Chen et al. reported that the derivative of dehydroabietic acid (**5**, Figure 3) exhibited excellent antibacterial activity against Gram-positive bacteria such as *Bacillus subtilis* and *Staphylococcus aureus* with the MIC (minimum inhibitory concentration) value of 4 and 2 μg/mL, respectively [36]. Dehydroabietic acid-containing serine derivatives had excellent antibacterial activity against Gram-positive bacteria. For example, the MIC_{90} values of compound **6** (Figure 3) against methicillin resistant *S. aureus*, *Staphylococcus epidermidis*, and *Streptococcus mitis* were 8 μg/mL [19]. Furthermore, through the broth micro-dilution method, it was found that dehydroabietic acid showed antibacterial activity against four kinds of *Streptococcus mutans* (including *S. mutans* ATCC 12175, *S. mutans* NRPC 801, *S. mutans* NRPC 804, and *S. mutans* DMST 18777) [37]. Hassan et al. used dehydroabietic acid derivatives to prepare antibacterial nanocellulose membranes to solve the inevitable toxicity of silver and cationic antimicrobial agents. By simulating the physiological environment of chronic wounds, they found that it prevented the colonization of bacteria on the surface [38]. It was revealed that the dehydroabietic acid analog (**7**, Figure 3) exhibited antimicrobial activity against methicillin resistant *S. aureus* (MIC: 32 μg/mL) [39]. It is urgent to develop new antibacterial agents against methicillin-resistant and methicillin-sensitive *S. aureus* infections (MRSA and MSSA). Compound **8** (Figure 3) had effective inhibitory activity against the above-mentioned bacteria, and its MIC values were from 3.9 to 15.6 μg/mL. Moreover, it had no cytotoxicity and hemolytic activity in mammalian cells [40]. In addition, Liu et al. elucidated that some 12-oxime and *O*-oxime ether derivatives of dehydroabietic acid had strong antistaphylococcal activity. Compound **9** showed high inhibitory activity against *S. aureus* Newman with MIC value of 0.39–0.78 μg/mL, while compounds **10**–**12** had MIC values of 1.25–3.13 μg/mL for multidrug-resistant *S. aureus*. [41]. Additionally, some MIC values of dehydroabietic acid derivatives were listed in Table 1.

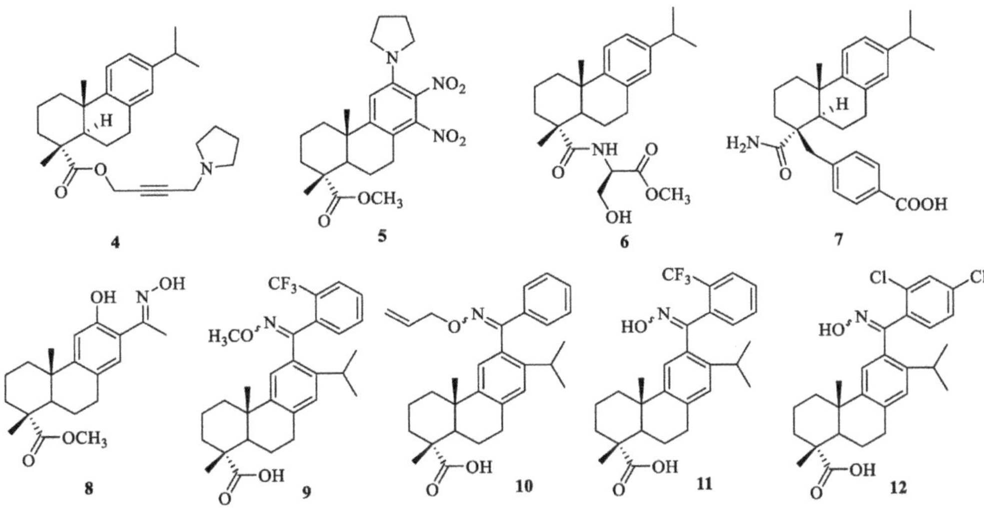

Figure 3. Chemical structures of compounds **4**−**12**.

Table 1. Some MIC values of dehydroabietic acid derivatives against different bacteria.

Compound	Bacteria	MIC Value
5	*B. subtilis*	4 μg/mL [36]
5	*S. aureus*	2 μg/mL [36]
7	methicillin resistant *S. aureus*	32 μg/mL [39]
8	LA-MRSA LGA251(ST425-XI)	3.9 μg/mL [40]
9	*S. aureus* Newman	0.39–0.78 μg/mL [41]
10	NRS-70	1.25–1.56 μg/mL [41]
11	*S. aureus* Newman, NRS-1, NRS-70, NRS-100, NRS-108, and NRS-271	1.56–3.13 μg/mL [41]
12	*S. aureus* Newman, NRS-70, NRS-108, and NRS-271	1.56–3.13 μg/mL [41]

2.4. Insecticidal Activity

Dehydroabietic acid displayed antifeedant activity against Indian meal moth *Plodia interpunctella*, but had no obvious effect on the growth of larvae [22]. Additionally, dehydroabietic acid had certain insecticidal activity against *Aedes aegypti* larvae, with a lethality rate of 65% at 10 ppm [42]. Gao et al. found that dehydroabietic acid amide derivatives containing the thiadiazole fragment (**13a**−**13c**, Figure 4) exhibited remarkable insecticidal activities against diamondback moth (*Plutella xylostella*) with the LC_{50} of 0.222–0.224 μg/mL [43]. Dehydroabietic acid had good antifeedant activity and a certain toxic effect on *Mythimna separata* larvae [44]. Moreover, the mountain pine beetle was often affected by the chemical defense of host oleoresin secretions which contained dehydroabietic acid [45].

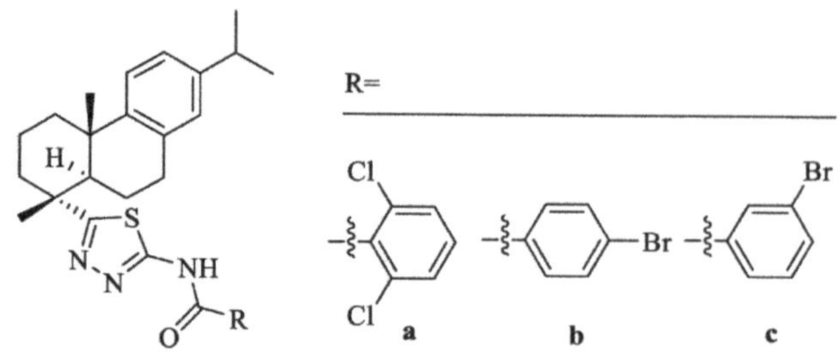

13a-13c

Figure 4. Chemical structures of compounds **13a**–**13c**.

2.5. Antiprotozoal Activity

Dehydroabietic acid not only inhibited the proliferation of the promastigote form, but also induced the production and output of ROS (reactive oxygen species) and downregulated Nrf2 (nuclear factor erythroid 2-related factor 2) to exert its anti-Leishmania activity, mainly due to its antioxidant properties [46]. Pertino et al. reported that dehydroabietic acid had certain antiprotozoal activity against *Trypanosoma cruzi, Leishmaniabraziliensis,* and *Leishmania infantum* [47]. Moreover, the dehydroabietic acid derivative containing amino acid structure was a new and effective antiprotozoal drug with good selectivity to *Leishmania donovani* and *T. cruzi* [48].

2.6. Other Activities

The treatment of dehydroabietic acid alleviated insulin resistance and liver steatosis caused by a high-fat diet in mice as a dual agonist of PPAR-α/γ [49]. Moreover, the anti-obesity effect of dehydroabietic acid may be exerted by improving the levels of plasma glucose and insulin via PPARα/γ-dependent pathways [34,50,51]. Dehydroabietic acid reduced hypertrophy by activating the Keap1/Nrf2-ARE signaling pathway, increasing the expression of the ferroptosis suppressor protein 1 (FSP1) gene, and inhibiting the accumulation of ROS to ameliorate non-alcoholic fatty liver disease [52]. Dehydroabietic acid inhibited angiotensin converting enzyme in human umbilical vein endothelial cells and induced p-Akt, which could reduce the systolic blood pressure of spontaneously hypertensive rats, thereby exerting an anti-hypertensive effect [53]. Dehydroabietic acid as an anti-aging agent could directly bind to SIRT1 protein and activated SIRT1 [54]. Park et al. suggested that dehydroabietic acid protected and regenerated the collagen fibers in the skin irradiated by ultraviolet B [55]. In addition, through *in vivo* and *in vitro* experiments, it was found that the assembled dehydroabietic acid derivative (**14**, Figure 5)/ZnAlTi-LDH (layered double hydroxide) composite material can promote wound healing, kill Gram-negative and positive bacteria infecting the wound, and block ultraviolet rays to protect the skin [56]. Since dehydroabietic acid can promote osteogenic differentiation, it may be a new idea for the treatment of osteoporosis in the elderly [57]. Through *in vitro* experimental studies, Nachar et al. found that dehydroabietic acid may be an effective drug for the treatment of type 2 diabetes. It not only reduced the activity of G6Pase to descend glucose production but also stimulated GS activity to ascend glucose storage [58]. Additionally, a dehydroabietic acid polymer (**15**, Figure 5) can be used as a toner for xerography [59]. BK channels, K^+ channels activated by large-conductance calcium, had vital physiological functions such as the release of neurotransmitters. A thiophene derivative of dehydroabietic acid (**16**, Figure 5) showed BK channel opening activity [60,61]. A dehydroabietic acid derivative (**17**, Figure 5) can be used as fluorescent probes and can sensitively detect Fe^{3+} and Hg^{2+} ions in cells [62]. Moreover, dehydroabietic acid derivatives (**18** and **19**, Figure 5) as natural resources provided a new method for the synthesis of fluorescent materials [63].

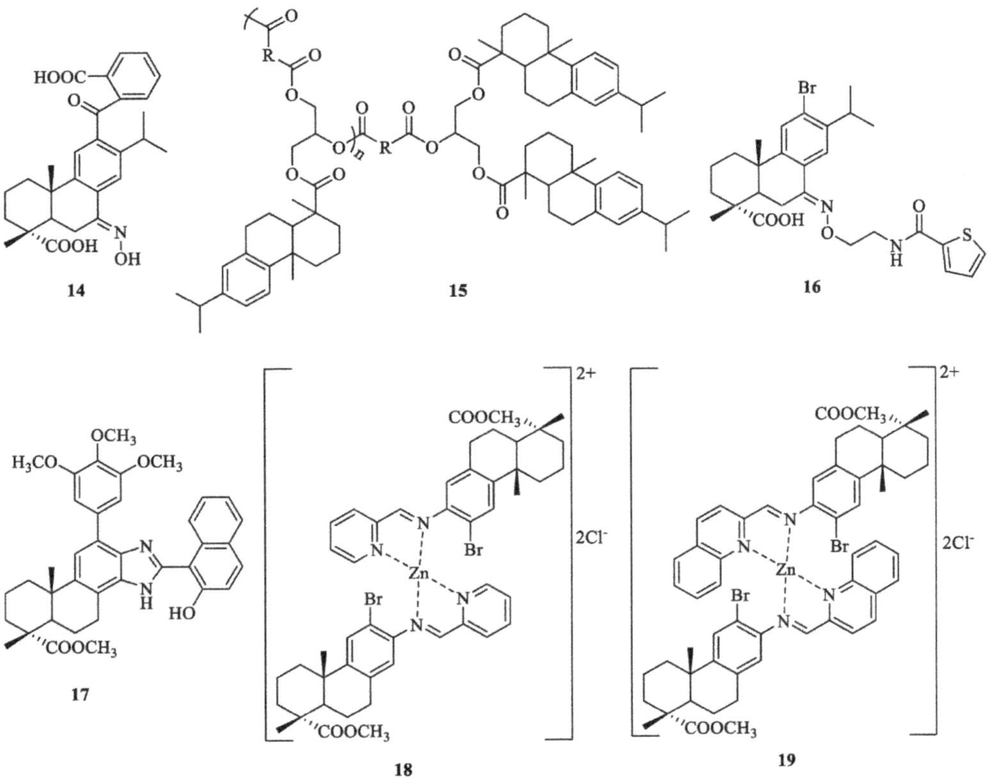

Figure 5. Chemical structures of compounds **14**–**19**.

3. Structural Modification of Dehydroabietic Acid and Its Derivatives

3.1. Structural Modification at C-18 of Dehydroabietic Acid

As depicted in Scheme 1, Huang et al. synthesized a series of chiral dipeptide derivatives (**22(a-h)**–**25(a-h)**) of dehydroabietic acid and tested their anticancer activities by the MTT method against HeLa, NCI-H460, and MGC-803 tumor cell lines. Among them, compound **22f** had the strongest inhibitory effect on the HeLa cancer cell line, with an IC_{50} value of 7.76 ± 0.98 μM, and induced cell apoptosis through the mitochondrial pathway [64].

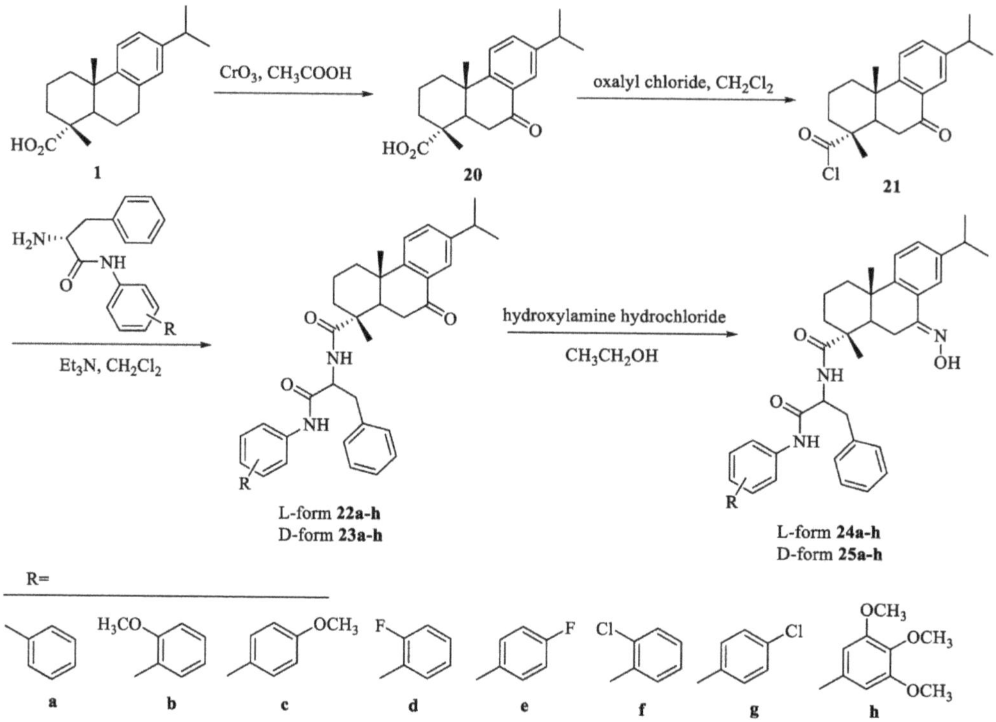

Scheme 1. Synthesis of compounds **22(a-h)**–**25(a-h)**.

Subsequently, compounds containing thiourea and bisphosphonates (**28a**–**28k**) were synthesized as effective antitumor agents (Scheme 2). Compound **28e** showed the best anticancer activity against the SK-OV-3 cell line with an IC_{50} value of 1.79 ± 0.43 μM, which arrested the cell cycle in the G1 phase and induced apoptosis [65]. As shown in Scheme 2, a novel class of dehydroabietic acid acyl-thiourea derivatives (**29(a-o)**–**30(a-o)**) were obtained and their antitumor activity was evaluated against HeLa, SK-OV-3, HL-7702, and MGC-803 cell lines. Most of these compounds demonstrated cytotoxicity, especially compound **30n** (IC_{50}: 6.58 ± 1.11 μM against HeLa) exhibited better inhibitory effect than 5-Fu (a commercial antitumor drug, IC_{50}: 36.58 ± 1.55 μM). Moreover, compound **30n** could block HeLa cells in the S phase and induce HeLa cell apoptosis via the mitochondrial pathway [66]. In 2020, Li et al. designed and synthesized compounds **31** and **32** (Scheme 2). Compared with cisplatin and oxaliplatin, compound **31** had better cytotoxic activity on A431 cells and a stronger ability to bind with DNA [67]. Moreover, compounds **33a**–**33m** (Scheme 2) were synthesized as antitumor and antimicrobial agents. Compound **33e**, a MIC value of 1.9 μg/mL, demonstrated the most potent inhibitory activity against *B. subtilis*. Meanwhile, compound **33m** showed anticancer activity comparable to the positive control etoposide [68].

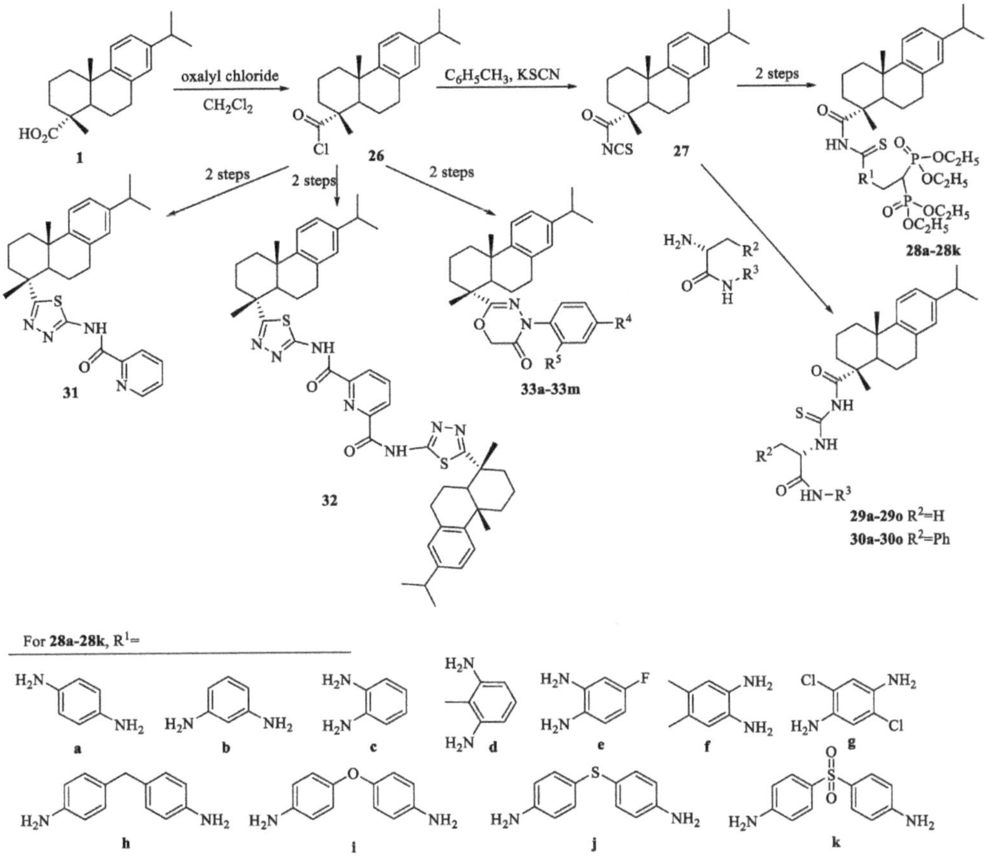

For **29a-29o** and **30a-30o**, R^3=

a. Ph **b.** o-Ph-CH_3 **c.** p-Ph-CH_3 **d.** o-Ph-Cl **e.** p-Ph-Cl **f.** o-Ph-OCH_3 **g.** p-Ph-OCH_3 **h.** o-Ph-F **i.** p-Ph-F **j.** o-Ph-Br **k.** p-Ph-Br **l.** 3,5-di-methyl-Ph **m.** 1-Naphthyl **n.** 2-pyridyl **o.** 3,4,5-tri-methoxyl-Ph

For **33a-33m**,

a. R^4 = H, R^5 = H; **b.** R^4 = CH_3, R^5 = H; **c.** R^4 = C_2H_5, R^5 = H; **d.** R^4 = F, R^5 = H; **e.** R^4 = Cl, R^5 = H; **f.** R^4 = Br, R^5 = H; **g.** R^4 = OCH_3, R^5 = H; **h.** R^4 = CF_3, R^5 = H; **i.** R^4 = NO_2, R^5 = H; **j.** R^4 = CN, R^5 = H; **k.** R^4 = H, R^5 = CH_3; **l.** R^4 = H, R^5 = Cl; **m.** R^4 = H, R^5 = OCH_3

Scheme 2. Synthesis of compounds **28a**–**28k**, **29a**–**29o**, **30a**–**30o**, **31**, **32** and **33a**–**33m**.

First, dehydroabietic acid (**1**) reacted with ethyl chloroacetate to obtain intermediate compound **34**. Aminolysis of **34** gave hydrazide **35**, which reacted with different aromatic aldehydes to obtain the target products **36a**–**36x** (Scheme 3). Compounds **36i**, **36k**, **36l**, **36w**, and **36x** displayed potent anticancer activity. Especially compound **36w** exhibited the promising inhibitory effects against HeLa and BEL-7402 cells with IC_{50} values of 2.21 and 14.46 μM, respectively [69].

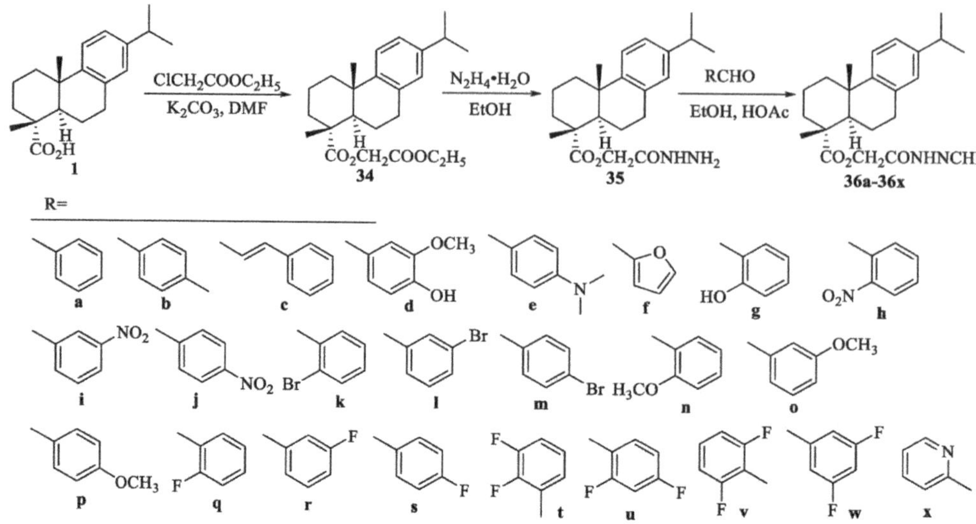

Scheme 3. Synthesis of compounds **36a**–**36x**.

As described in Scheme 4, Wang et al. designed and synthesized a novel class of dehydroabietic acid derivatives containing oxazolidinone moiety (**39a**–**39o**). Compound **39j** not only induced apoptosis of MGC-803 cells (IC_{50} = 3.82 ± 0.18 μM), but also arrested the cell cycle in the G1 phase [70].

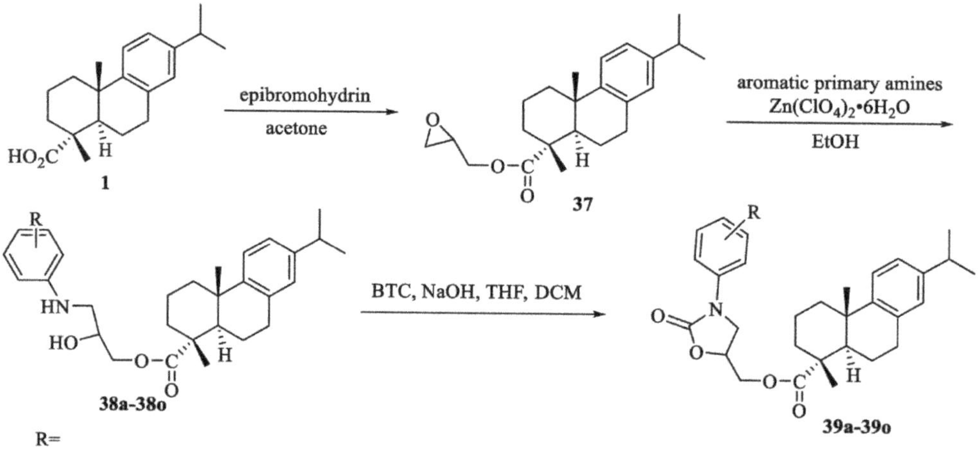

a. H **b.** 2-CH_3 **c.** 3-CH_3 **d.** 4-CH_3 **e.** 2-OCH_3 **f.** 3-OCH_3 **g.** 4-OCH_3 **h.** 2-F **i.** 3-F **j.** 4-F **k.** 3-Cl **l.** 4-Cl **m.** 3-Br **n.** 4-Br **o.** 3-ethynyl

Scheme 4. Synthesis of compounds **39a**–**39o**.

Additionally, dehydroabietic acid containing 1,2,3-triazole derivatives (**41a**–**41p**, Scheme 5) were prepared. The most potential compounds 41c (5.90 ± 0.41 μM) and 41k (6.25 ± 0.37 μM) exhibited better antiproliferation activity against HepG2 cells than that of cisplatin [71].

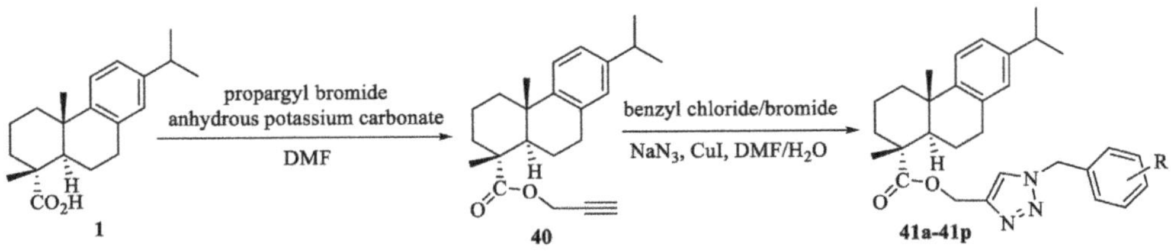

R=

a. H **b.** *o*-F **c.** *m*-F **d.** *p*-F **e.** *o*-Cl **f.** *m*-Cl **g.** *p*-Cl **h.** *o*-Br **i.** *m*-Br **j.** *p*-Br **k.** *o*-NO_2 **l.** *m*-NO_2 **m.** *p*-NO_2 **n.** *m*-CH_3 **o.** *p*-CH_3 **p.** *m*-OCH_3

Scheme 5. Synthesis of compounds **41a**–**41p**.

As depicted in Scheme 6, in the presence of anhydrous potassium carbonate, dehydroabietic acid (**1**) reacted with 1,2-dibromoethane to give intermediate **42**. Target compounds **43a**–**43o** were obtained by reacting compound **42** with substituted pyrimidine-2-thiol. Compound **43b** showed good inhibitory activity on the tested cells (IC_{50}: 7.00–11.93 μM), but it had a little toxic effect on normal cells. The mechanism of action of compound **43b** was to block MCF-7 cells in the S phase and induced apoptosis [72].

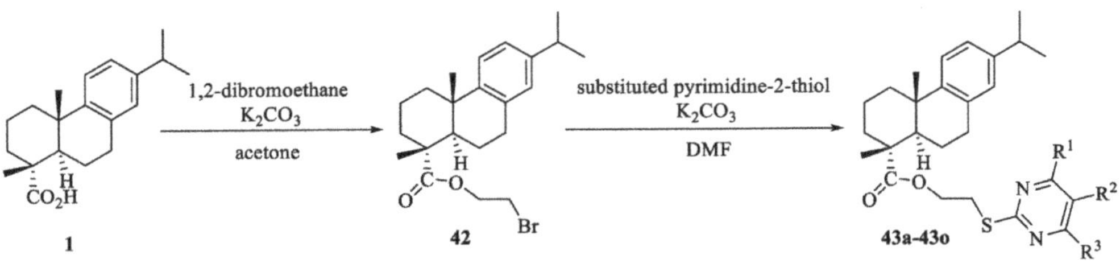

For **43a-43o**,

a. R^1 = H, R^2 = H, R^3 = OH **b.** R^1 = propyl, R^2 = H, R^3 = OH **c.** R^1 = methyl, R^2 = H, R^3 = methyl **d.** R^1 = methyl, R^2 = H, R^3 = OH **e.** R^1 = *o*-Fph, R^2 = CN, R^3 = OH **f.** R^1 = *p*-Fph, R^2 = CN, R^3 = OH **g.** R^1 = *p*-Brph, R^2 = CN, R^3 = OH **h.** R^1 = *o*-OCH_3ph, R^2 = CN, R^3 = OH **i.** R^1 = *m*-OCH_3ph, R^2 = CN, R^3 = OH **j.** R^1 = *p*-OCH_3ph, R^2 = CN, R^3 = OH **k.** R^1 = *o*-OHph, R^2 = CN, R^3 = OH **l.** R^1 = *m*-OHph, R^2 = CN, R^3 = OH **m.** R^1 = *p*-OHph, R^2 = CN, R^3 = OH **n.** R^1 = *m*-NO_2ph, R^2 = CN, R^3 = OH **o.** R^1 = *p*-CH_3ph, R^2 = CN, R^3 = OH

Scheme 6. Synthesis of compounds **43a**–**43o**.

As described in Scheme 7, Li et al. synthesized a class of dehydroabietic acid-nitrate conjugates (**47a**–**47r**) as antitumor agents. Compound **47n** showed the most promising activity against the BEL-7402 cell line (IC_{50}: 11.23 ± 0.21 μM). Compound **47j** exhibited the most potent cytotoxicity against the CNE-2 cell line (IC_{50}: 8.36 ± 0.14 μM) and evaluation of NO release indicated that the cytotoxic activity improved with the increase of the amount of NO produced in the CNE-2 cell line [73].

L-/D-amino acids and unusual amino acids were used as side chains to prepare new dehydroabietic acid derivatives (**49a**–**49f**, Scheme 8). Compounds **49b** and **49f** were the most effective anti-biofilm agents, which can quickly and effectively destroy membrane integrity [74].

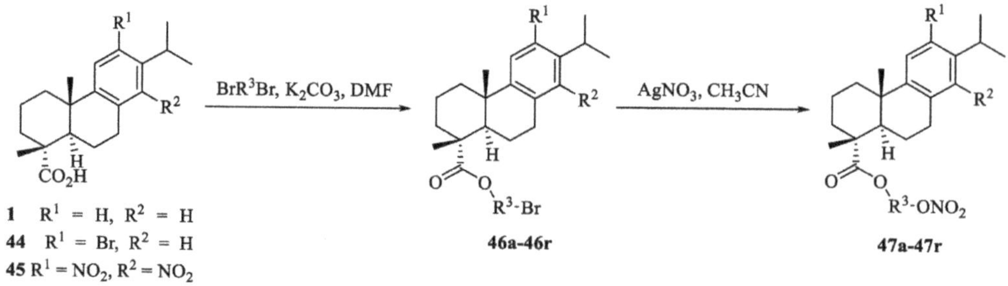

For **46(a-r)** and **47(a-r)**,

a. R^1, R^2 = H, R^3 = $(CH_2)_2$; **b.** R^1, R^2 = H, R^3 = $(CH_2)_3$; **c.** R^1, R^2 = H, R^3 = $(CH_2)_4$; **d.** R^1, R^2 = H, R^3 = $(CH_2)_5$; **e.** R^1, R^2 = H, R^3 = $(CH_2)_6$; **f.** R^1, R^2 = H, R^3 = $(CH_2)_8$; **g.** R^1, R^2 = H, R^3 = $(CH_2)_{12}$; **h.** R^1, R^2 = H, R^3 = $CH_2CH{=}CHCH_2$; **i.** R^1 = Br, R^2 = H, R^3 = $(CH_2)_2$; **j.** R^1 = Br, R^2 = H, R^3 = $(CH_2)_3$; **k.** R^1 = Br, R^2 = H, R^3 = $(CH_2)_4$; **l.** R^1 = Br, R^2 = H, R^3 = $(CH_2)_5$; **m.** R^1, R^2 = NO_2, R^3 = $(CH_2)_2$; **n.** R^1, R^2 = NO_2, R^3 = $(CH_2)_3$; **o.** R^1, R^2 = NO_2, R^3 = $(CH_2)_4$; **p.** R^1, R^2 = NO_2, R^3 = $(CH_2)_5$; **q.** R^1, R^2 = NO_2, R^3 = $(CH_2)_6$; **r.** R^1, R^2 = NO_2, R^3 = $CH_2CH{=}CHCH_2$

Scheme 7. Synthesis of compounds **47a**–**47r**.

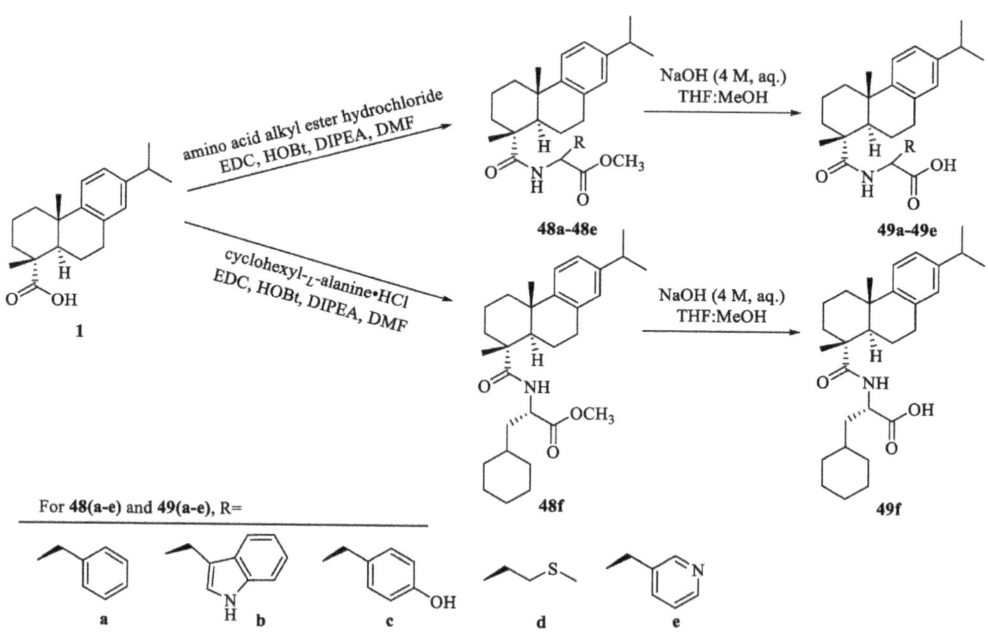

Scheme 8. Synthesis of compounds **49a**–**49f**.

As illustrated in Scheme 9, compounds **51a**–**51q** were smoothly synthesized by cyclization and Mannich-type reactions. At 50 µg/mL, compounds **51e**, **51f**, **51h**, and **51i** had better antifungal activity than azoxystrobin (a commercial antifungal drug) [75]. A series of dehydroabietic acid derivatives containing 1,3,4-thiadiazole-thiazolidinone (**52a**–**52p**) were synthesized as antifungal agents. They showed excellent antifungal activities against *Gibberella zeae* at a concentration of 50 µg/mL. Additionally, the antifungal activity of compounds **52c**, **52f**, and **52n** (inhibition rate: 91.3%) was comparable to that of azoxystrobin (positive control) [76]. Mo et al. designed and synthesized a series of novel dehydroabietic acid derivatives containing 1,3,4-thiadiazole thiourea (**53a**–**53k**) to determine their insecticidal activity against *Helicoverpa armigera, P. xylostella,* and *Ostrinia mubilalis.* Compounds **53a** and **53c** showed excellent insecticidal activity against *H. armigera* at 200 mg/L with mortality rates of 93.3% and 83.3%, respectively. The control effects of compounds **53b**, **53c**, **53d**, and **53i** against corn borer were 66.7%, 66.7%, 73.3%, and 60%, respectively [77].

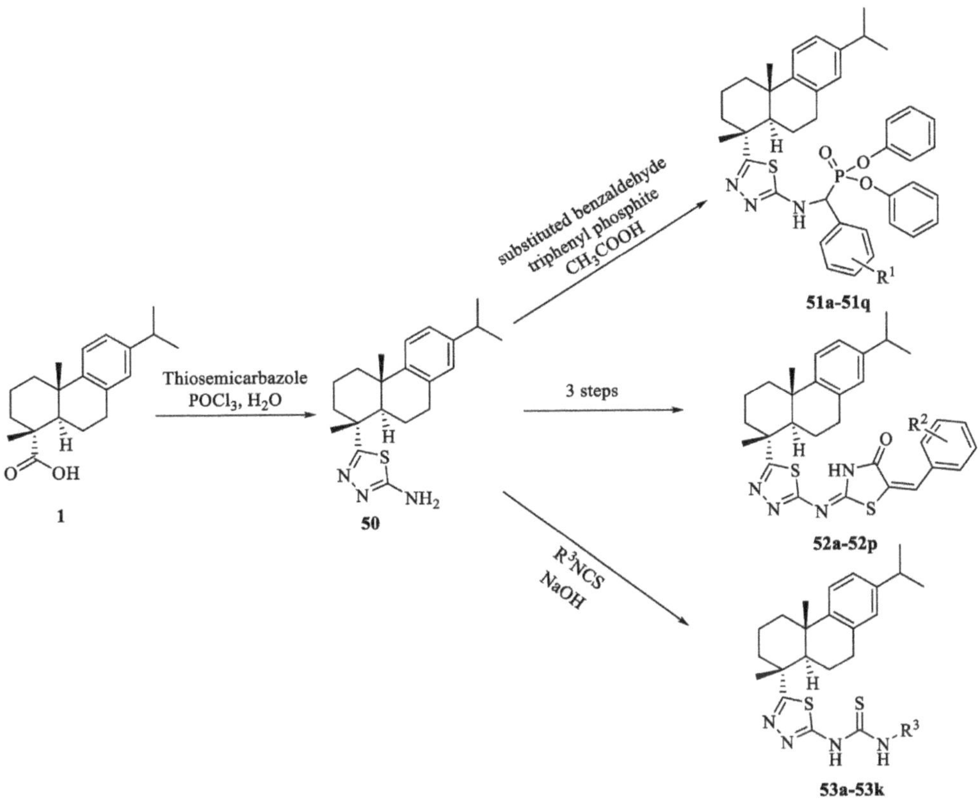

For **51a-51q**, R^1=

a. *o*-Br **b.** *p*-Br **c.** *o*-Cl **d.** *p*-Cl **e.** *o*-F **f.** *p*-F **g.** *o*-MeO **h.** *m*-MeO **i.** *p*-MeO **j.** H **k.** *o*-NO_2 **l.** *p*-NO_2 **m.** *o*-Me **n.** *m*-Me **o.** *p*-Me **p.** *m*-OH **q.** *p*-OH

For **52a-52p**, R^2=

a. H **b.** *o*-F **c.** *p*-F **d.** *o*-Cl **e.** *p*-Cl **f.** *o*-Br **g.** *p*-Br **h.** *o*-NO_2 **i.** *m*-NO_2 **j.** *p*-NO_2 **k.** *o*-Me **l.** *m*-Me **m.** *p*-Me **n.** *m*-OMe **o.** *o*-OMe **p.** *p*-OMe

For **53a-53k**, R^3=

a. C_6H_5 **b.** 2-$CH_3C_6H_4$ **c.** 3-$CH_3C_6H_4$ **d.** 4-$CH_3C_6H_4$ **e.** 2-$CH_3OC_6H_4$ **f.** 4-$CH_3OC_6H_4$ **g.** 4-FC_6H_4 **h.** 3-ClC_6H_4 **i.** 4-ClC_6H_4 **j.** 4-BrC_6H_4 **k.** 4-$CF_3C_6H_4$

Scheme 9. Synthesis of compounds **51a**–**51q**, **52a**–**52p** and **53a**–**53k**.

3.2. Structural Modification at B Ring of Dehydroabietic Acid

A series of 7-*N*-acylaminopropyloxime derivatives of dehydroabietic acid (**57a**–**57s**, Scheme 10) were synthesized, and their antibacterial activities against *S. aureus* Newman strain, NRS-1, NRS-70, NRS-100, NRS-108, and NRS-271 were studied. Compound **57j** showed high antibacterial activities against five kinds of multi-drug resistant *S. aureus* with MIC values of 1.56–3.13 μg/mL [78].

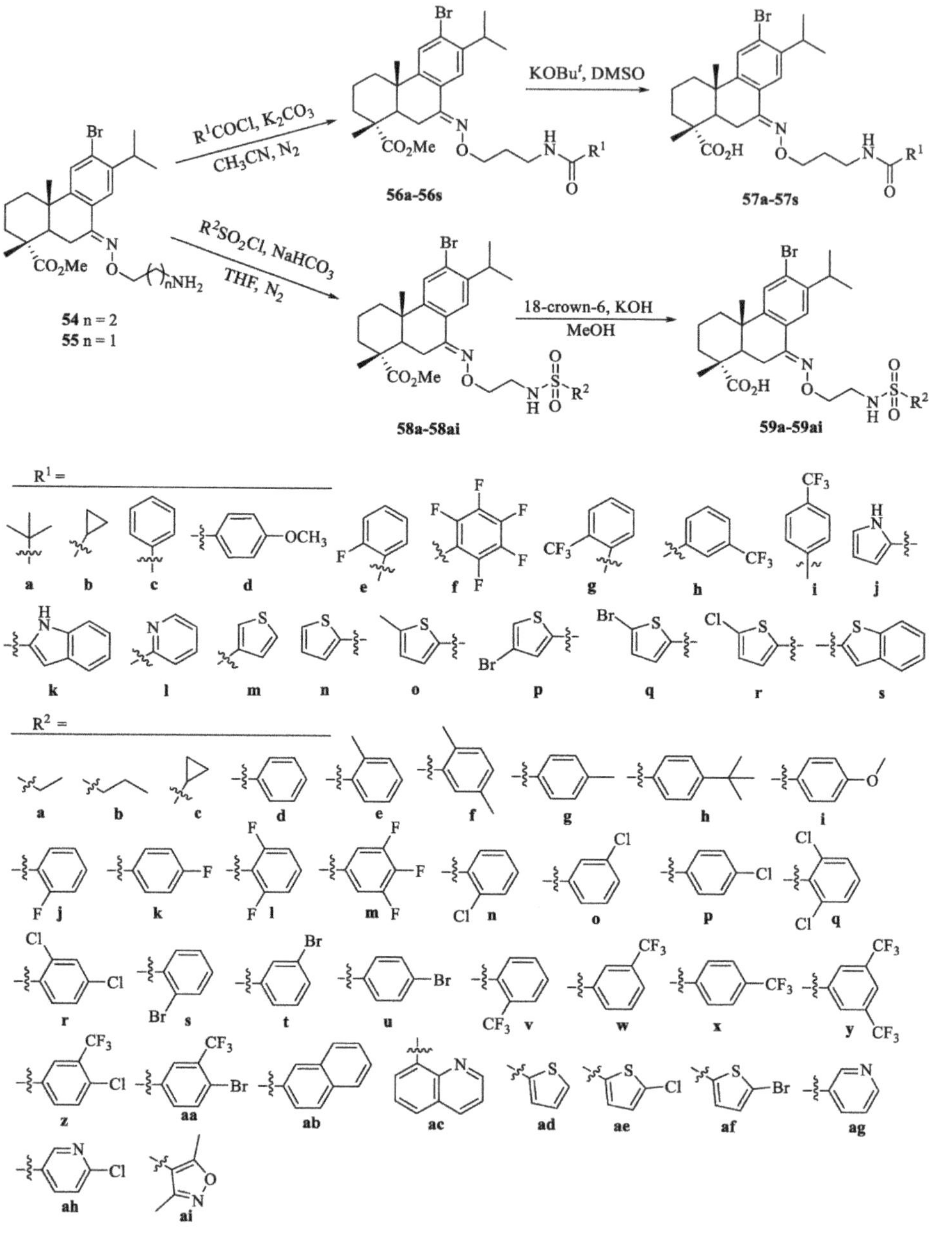

Scheme 10. Synthesis of compounds **57a**–**57s** and **59a**–**59ai**.

In 2018, Zhang et al. synthesized a series of *N*-sulfonaminoethyloxime derivatives of dehydroabietic acid from compound **55** (Scheme 10). The MIC value of analog **59w** against *S. aureus* Newman was 0.39–0.78 μg/mL, exhibiting the highest activity. In addition, the remaining analogs also showed better antibacterial activities against five multi-drug resistant *S. aureus*, with MIC values ranging from 0.78 to 1.56 μg/mL [79].

A series of hydrazone derivatives of dehydroabietic acid (**61a**–**61g**, Scheme 11) were synthesized. All compounds showed antibacterial activities against *Escherichia coli*, *S. aureus*, and *B. subtilis*. Especially compound **61d** was the potent antimicrobial agent against *B. subtilis* and *S. aureus* [80].

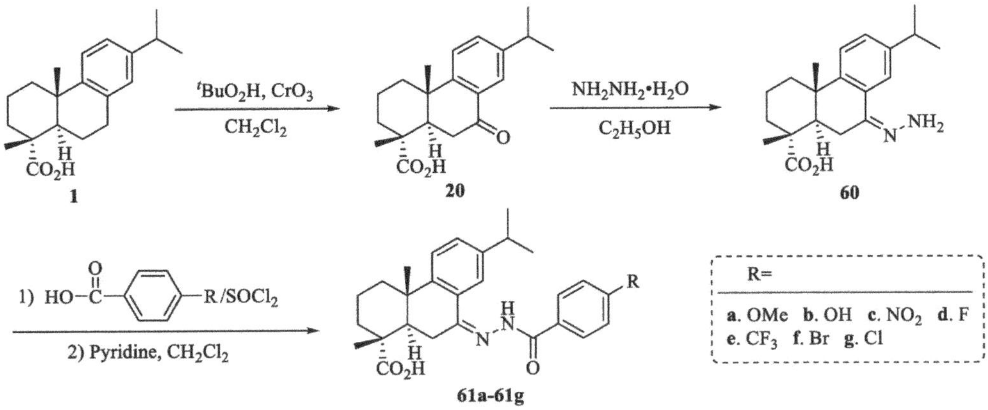

Scheme 11. Synthesis of compounds **61a**–**61g**.

Chen et al. synthesized a series of dehydroabietic acid derivatives (**63a**–**63t**, Scheme 12) to find novel and effective antitumor agents, and tested *in vitro* cytotoxic activity against HepG2, SCC9, and 293T by CCK-8 method. Compounds **63r** and **63s** had a certain cytotoxic activity to cancer cells, but were weak to normal cells. Two compounds could inhibit the PI3K/AKT/mTOR signaling pathway to exert anti-cancer effects. The results of molecular docking studies indicated that the compounds may inhibit the pathway through ATP competition [81].

As depicted in Scheme 12, compounds **64(a-h)**–**66(a-h)** and **67a**–**67j** were synthesized smoothly from dehydroabietic acid. Among these compounds, compound **67g** had excellent anti-proliferative activity against three liver cancer cell lines (SMMC-7721, HepG2, and Hep3B), with IC_{50} values of 0.51–1.39 μM. In addition, compound **67g** may inhibit MEK1 kinase activity, increase intracellular ROS levels, destroy cell membranes, and thus cause HepG2 cell apoptosis [82].

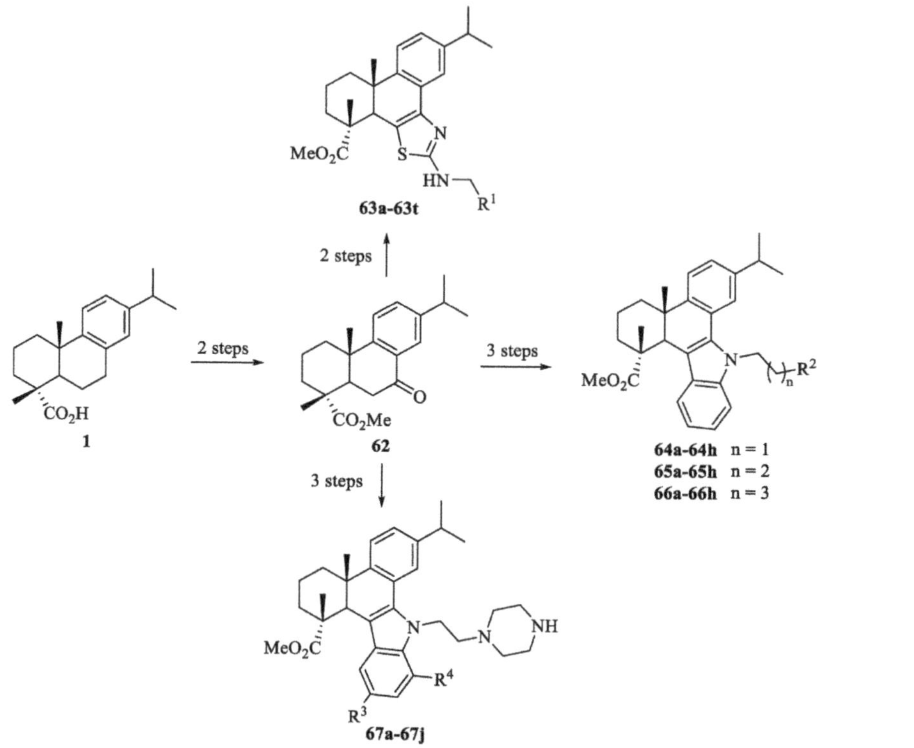

R^1=

a. phenyl; **b**. 2-CH$_3$-phenyl; **c**. 3-CH$_3$-phenyl; **d**. 4-CH$_3$-phenyl; **e**. 2-CH$_3$O-phenyl; **f**. 3-CH$_3$O-phenyl; **g**. 4-CH$_3$O-phenyl; **h**. 2-F-phenyl; **i**. 3-F-phenyl; **j**. 4-F-phenyl; **k**. 2-Cl-phenyl; **l**. 3-Cl-phenyl; **m**. 4-Cl-phenyl; **n**. 2-Br-phenyl; **o**. 3-Br-phenyl; **p**. 4-Br-phenyl; **q**. furan-2-yl; **r**. 1H-pyrrol-2-yl; **s**. pyridin-3-yl; **t**. 6-Cl-pyridin-3-yl

R^2=

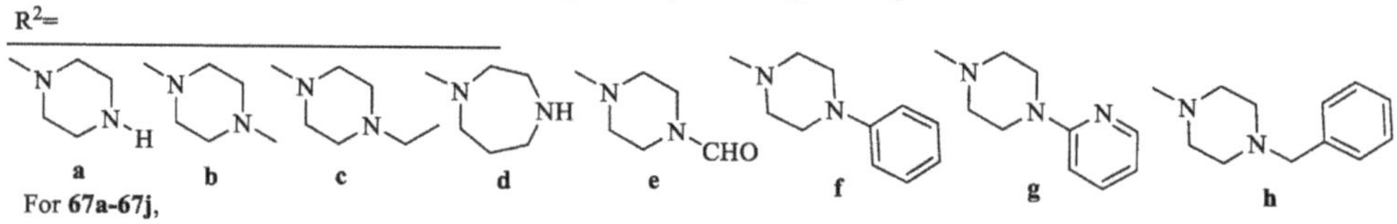

For **67a-67j**,

a. R^3 = Me, R^4 = H; **b**. R^3 = H, R^4 = Me; **c**. R^3 = Et, R^4 = H; **d**. R^3 = *i*-Pr, R^4 = H; **e**. R^3 = OMe, R^4 = H; **f**. R^3 = H, R^4 = OMe; **g**. R^3 = OEt, R^4 = H; **h**. R^3 = H, R^4 = OEt; **i**. R^3 = F, R^4 = H; **j**. R^3 = Cl, R^4 = H

Scheme 12. Synthesis of compounds **63a–63t, 64(a-h)–66(a-h)** and **67a–67j**.

3.3. Structural Modification at C Ring of Dehydroabietic Acid

In order to find effective new antimicrobial agents, a class of new derivatives of dehydroabietic acid (**69a–69o**, Scheme 13) was obtained. Compound **69o** showed excellent antibacterial activity against both Gram-negative and positive bacteria (MIC: 1.6–3.1 μg/mL) but had no obvious toxicity to mammalian cells. In addition, compound **69o** (containing 1,2,3-triazole moiety at the C-14 position) had good drug-like properties [83]. Additionally, compound **69p** (IC_{50} values from 0.7 to 1.2 μM) not only induced apoptosis of MDA-MB-231 cells but also had weak toxicity to normal cells. And its anti-proliferative activity was better than 5-Fu (average IC_{50}: 16.1 μM) [84].

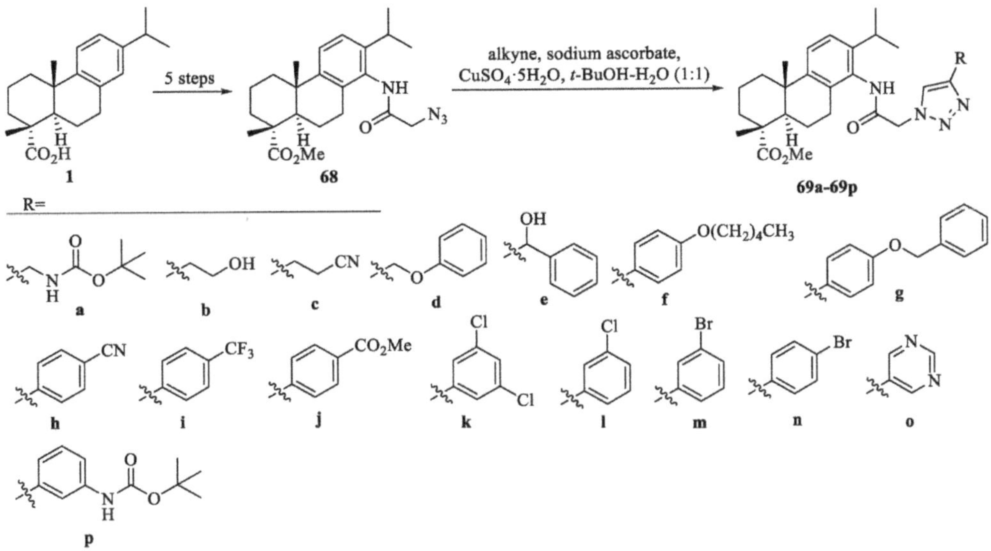

Scheme 13. Synthesis of compounds **69a**–**69p**.

Several derivatives of dehydroabietic acid containing 12-thiazole moiety (**71a**–**71e**) were designed and synthesized as shown in Scheme 14. In order to understand the importance of hABHD16A inhibition *in vivo*, the feasibility of these derivatives as a starting point for the design of selective ABHD16A (a new target for inflammation-mediated pain) inhibitors was investigated. Compound **71d** had an IC_{50} value of 3.4 ± 0.2 μM with good selectivity [85].

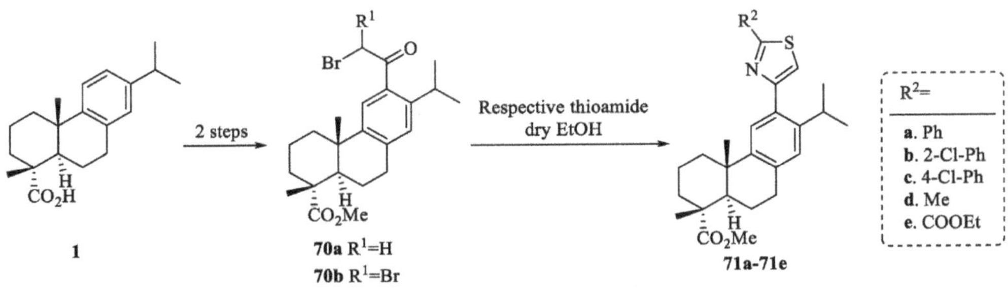

Scheme 14. Synthesis of compounds **71a**–**71e**.

Gu et al. studied the cytotoxicity of a series of newly synthesized dehydroabietic acid derivatives (**74(a-k)**–**75(a-k)**, Scheme 15) on liver cancer cells and found that most of the compounds had obvious cytotoxic activity on SMMC-7721 and HepG2 liver cancer cells with low toxicity on normal human liver cells. Among them, compounds **74b** and **74e** showed the best cytotoxicity against SMMC-7721 and HepG2 cells (IC_{50} values: 0.36 ± 0.13 and 0.12 ± 0.03 μM, respectively). In addition, compound **74b** not only caused SMMC-7721 cells' cell cycle arrest in the G2/M phase, but also induced apoptosis [86].

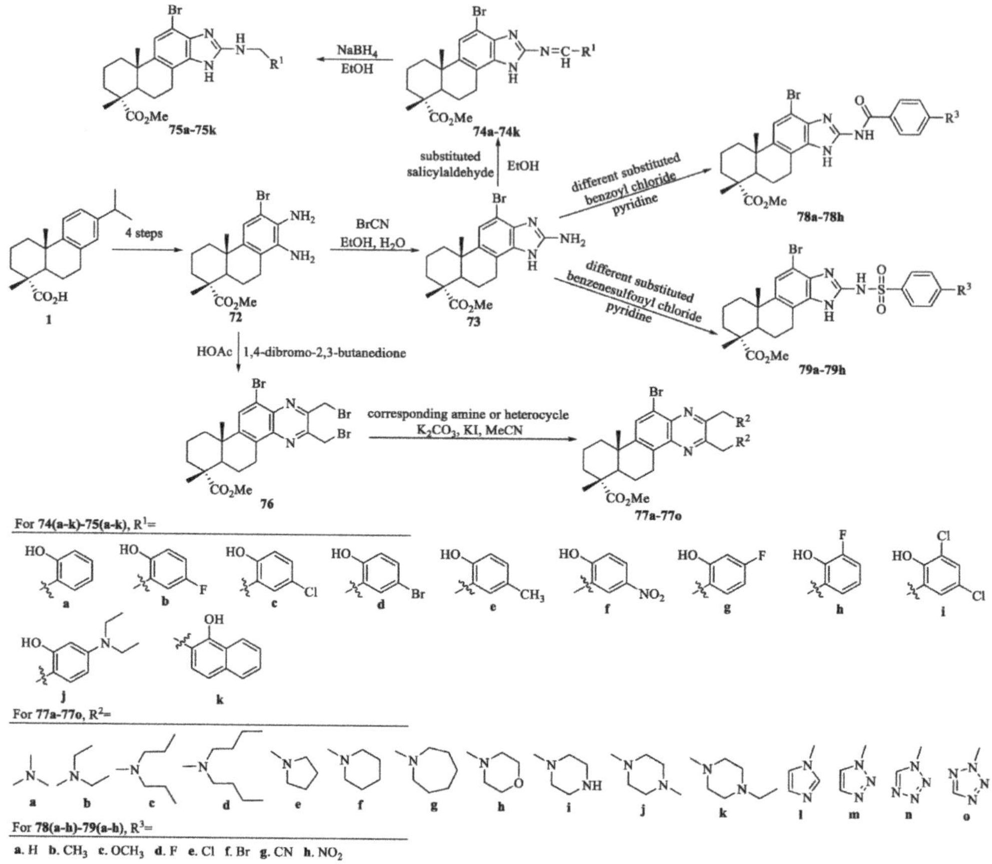

Scheme 15. Synthesis of compounds **74(a-k)–75(a-k), 77a–77o and 78(a-h)–79(a-h)**.

Subsequently, novel quinoxaline derivatives of dehydroabietic acid (**77a–77o**, Scheme 15) were synthesized and tested against MCF-7, SMMC-7721, and HeLa cancer cell lines. Compound **77b** caused SMMC-7721 cells to arrest in the G0/G1 phase of the cell cycle and induced their apoptosis in a dose-dependent manner. Its IC_{50} values against three different cancer cells were 0.72–1.78 μM, which had the best anti-cancer effect, while its toxicity to LO2 cells was reduced [87]. In 2020, a series of *N*-(1*H*-benzo[*d*]imidazol-2-yl)benzamide/benzenesulfonamide derivatives of dehydroabietic acid (**78(a-h)–79(a-h)**, Scheme 15) were prepared. Compounds **78a**, **78g**, **78h**, **79g**, and **79h** showed good anti-cancer effects on at least one cancer cell line (IC_{50} <10 μM), while compounds **1**, **78b**, **78c**, **78e**, and **78f** were inactive (IC_{50} > 50 μM). In addition, benzenesulfonamide derivatives (**79a–79h**) had stronger inhibitory activity than benzamide ones (**78a–78h**). The most cytotoxic compound **79h** (IC_{50}: 0.87–9.39 μM) can arrest MCF-7 cells in the S phase through ROS-mediated mitochondrial pathway, and finally induce MCF-7 cell apoptosis [88].

Miao et al. synthesized a series of novel 2-aryl-benzimidazole derivatives of dehydroabietic acid (**80(a-k)–81(a-k)**, Scheme 16). Compound **80j** exhibited the strongest cytotoxic activity with the IC_{50} value of 0.08–0.42 μM. In addition, further in-depth exploration elucidated that compound **80j** significantly inhibited the migration ability of SMMC-7721 cells, and induced cell cycle arrest and apoptosis of the cells in the G2/M phase. Through tubulin polymerization and immunofluorescence assays, it was found that compound **80j** not only inhibited tubulin polymerization but also destroyed the intracellular microtubule network [89].

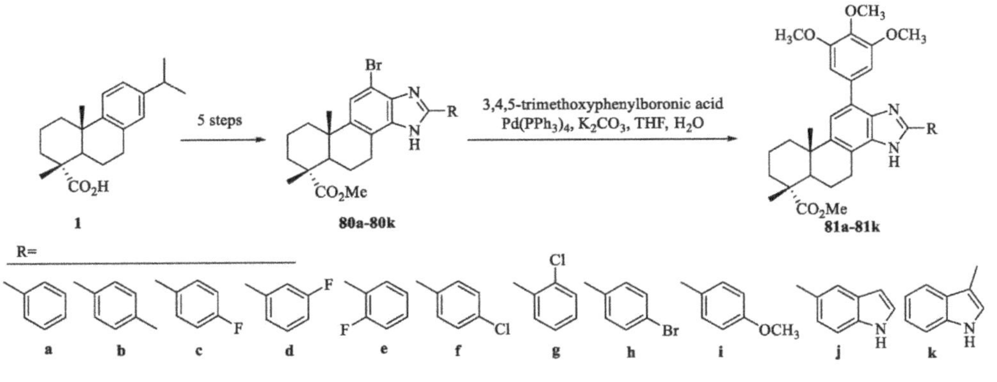

Scheme 16. Synthesis of compounds **80(a-k)–81(a-k)**.

As shown in Scheme 17, a series of sulfonylurea derivatives of dehydroabietic acid (**84a–84k**) were prepared. Compounds **84a, 84c, 84e**, and **84i** had higher anti-HCT-116 cell activity than 5-Fu. In particular, compound **84a**, an IC_{50} value of which was 1.18 ± 0.52 μM, showed the best anti-tumor proliferation activity [90].

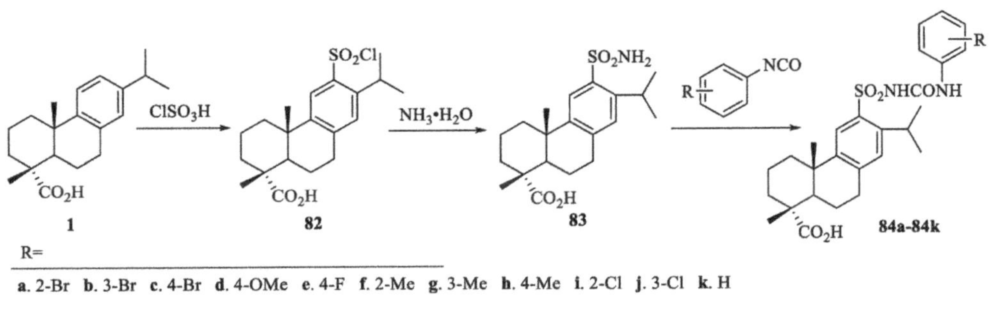

Scheme 17. Synthesis of compounds **84a–84k**.

Some IC_{50} values and MIC values of the most active derivatives were listed in Tables 2 and 3.

Table 2. Some IC_{50} values of the most active derivatives against different cancer cell lines.

Compound	Cancer Cell Line	IC_{50} Value
22f	HeLa	7.76 ± 0.98 μM [64]
28e	SK-OV-3	1.79 ± 0.43 μM [65]
30n	HeLa	6.58 ± 1.11 μM [66]
36w	HeLa	2.21 ± 0.04 μM [69]
36w	BEL-7402	14.46 ± 0.22 μM [69]
39j	MGC-803	3.82 ± 0.18 μM [70]
41c	HepG2	5.90 ± 0.41 μM [71]
41k	HepG2	6.25 ± 0.37μM [71]

Table 2. *Cont.*

Compound	Cancer Cell Line	IC_{50} Value
43b	MCF-7	7.00 ± 0.96 μM [72]
47n	BEL-7402	11.23 ± 0.21 μM [73]
47j	CNE-2	8.36 ± 0.14 μM [73]
63r	HepG2	24.41 ± 0.26 μM [81]
63s	HepG2	22.92 ± 0.24 μM [81]
67g	SMMC-7721	1.39 ± 0.13 μM [82]
67g	HepG2	0.51 ± 0.09 μM [82]
67g	Hep3B	0.73 ± 0.08 μM [82]
69p	MDA-MB-231	0.7 ± 0.1 μM [84]
74b	SMMC-7721	0.36 ± 0.13 μM [86]
74e	HepG2	0.12 ± 0.03 μM [86]
77b	SMMC-7721	0.72 ± 0.09 μM [87]
79h	MCF-7	0.87 ± 0.18 μM [88]
80j	SMMC-7721	0.08 ± 0.01 μM [89]
84a	HCT-116	1.18 ± 0.52 μM [90]

Table 3. Some MIC values of the most active compounds against different bacteria.

Compound	Bacteria	MIC Value
33e	*B. subtilis*	1.9 μg/mL [68]
57j	*S. aureus* NRS-70, NRS-100, NRS-108 and NRS-271	1.56–2.5 μg/mL [78]
69o	*S. aureus*, *E. coli* and *P. fluorescens*	1.6 μg/mL [83]

3.4. Biotransformation of Dehydroabietic Acid

Possible pathways for the biotransformation of dehydroabietic acid by Mucor circinelloides IT25, Mortierella isabellina HR32, Moraxella sp. HR6, Trametes versicolor, Phlebiopsis gigantea, Flavobacterium resinovorum, Fusarium oxyosporum/F. moniliforme and Fomes annosum were shown in Scheme 18. These microorganisms were some of the previously reported microorganisms with hydroxylated metabolites. Two molds (M. circinelloides IT 25 and M. isabellina HR32) regioselectively and stereoselectively hydroxylated dehydroabietic acid to 2α-hydroxydehydroabietic acid (**85**). Dehydroabietic acid was oxidized at the C-3 and C-7 positions, decarboxylated at the C-4 position by the action of bacteria Moraxella sp. HR6, and finally converted to 3,7-dioxodehydroabietin (**86**) [91]. The first step in the degradation of dehydroabietic acid was caused by two fungi, T. versicolor and P. gigantea, and a stereoselective hydroxylation was at the C-1 position, followed by further hydroxylation at the C-7 or C-16 position to form dihydroxylated compounds **89** and **91**. Compound **87** or **91** was further hydroxylated at the C-1 or C-7 position, resulting in trihydroxylated compound **92**. The hydroxyl group at the C-7 position of compounds **91** and **92** can be further oxidized to a carbonyl functional group to afford compounds **93** and **94** [92,93].

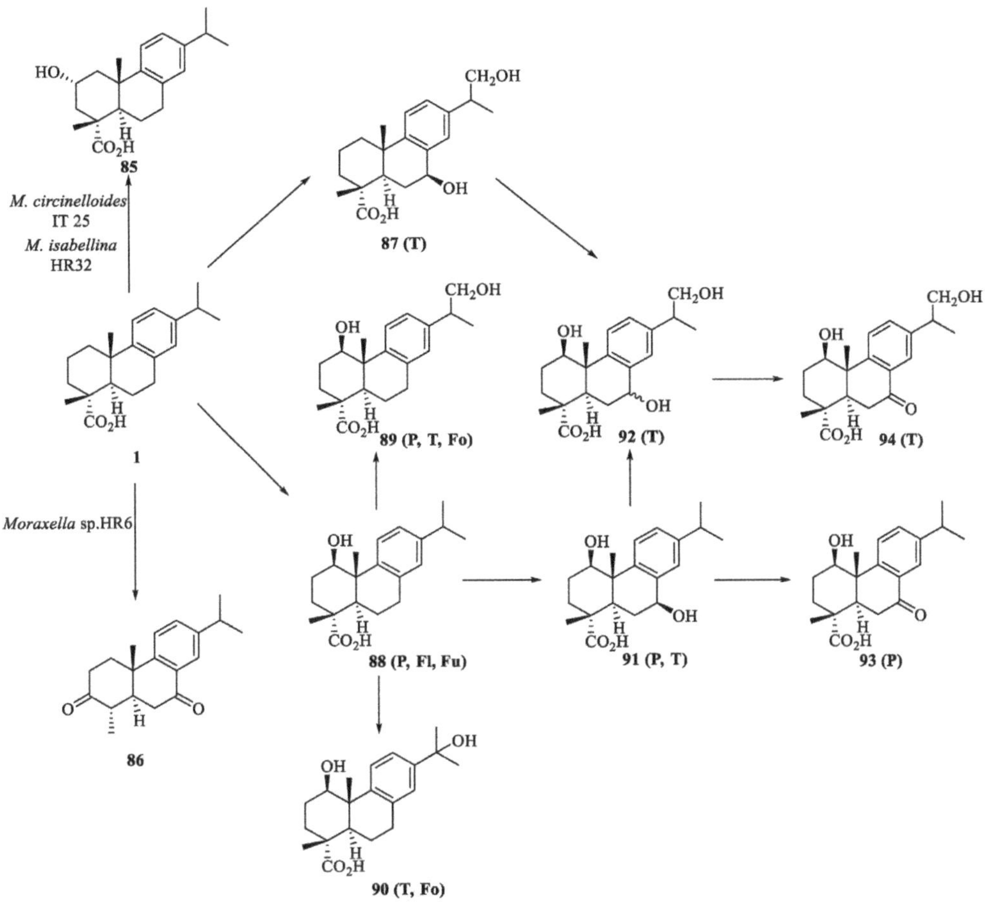

Scheme 18. Possible pathways for the biotransformation of dehydroabietic acid by Mucor circinelloides IT25, Mortierella isabellina HR32, *Moraxella* sp. HR6, Trametes versicolor (**T**), Phlebiopsis gigantea (**P**), Flavobacterium resinovorum (**Fl**), Fusarium oxysporum/F. moniliforme (**Fu**) and Fomes annosum (**Fo**).

4. Total Synthesis of Dehydroabietic Acid

Dehydroabietic acid has attracted the attention of researchers because of its diverse biological activities. However, separation and purification of dehydroabietic acid usually required expensive catalysts, and the organic amine salt method was more toxic and polluted the environment [10]. Therefore, it was particularly important to find other methods for synthesis of dehydroabietic acid. Some total synthesis processes of dehydroabietic acid were as follows.

In 1956, the first synthesis of dehydroabietic acid was begun with 2-isopropylnaphthalene (**95**) as the starting material (Scheme 19). The most important step was to define the C-4 center, that is, to alkylate phenanthrone (**96**) with ethyl bromoacetate. The enone (**96**) was then alkylated to obtain the keto ester (**97**), which reacted with $HSCH_2CH_2SH$ to form the thioketal (**98**). The thioketal (**98**) was converted into its methyl ester (**99**) by ester hydrolysis and diazotization reactions. After Raney nickel desulfurization, ester hydrolysis, and Pd/C hydrogenation, intermediate **100** was obtained, which reacted with CH_2N_2 to produce methyl ester compound **101**. Compound **103** was obtained by Barbiere Wieland degradation of **101**. Finally, dehydroabietic acid (**1**) was obtained by hydrogenation of **103** on Pd/C [94].

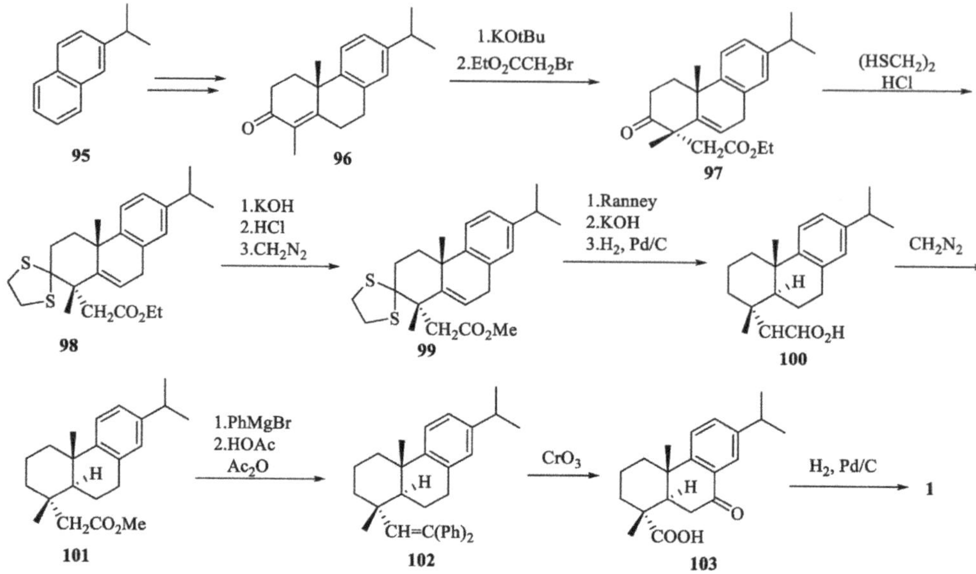

Scheme 19. Total synthesis of dehydroabietic acid from 2-isopropylnaphthalene.

The second method started with 2-methyl-2-(p-isopropylphenyl)cyclohexanone (**104**) was described in Scheme 20. The key steps were stereoselective alkylation of **105** with KO*t*-Bu and MeI, and reduction with Li in liquid ammonia to give decalone **107**, which reacted with HCO_2Et, O_3, and polyphosphoric acid (PPA) to give tricyclic keto acid **109**. After the hydrogenolysis reaction, the amine oxide pyrolyzed to afford the vinylic compound **110**, which reacted with OsO_4 and HIO_4 to give the corresponding aldehyde (**111**). Finally, dehydroabietic acid (**1**) was obtained by reaction of 111 with NH_2OH and KOH [94].

The third method was a short enantioselective synthesis starting from the geranyl acetate derivative (**112**) as shown in Scheme 21. Compound **112** was converted to alkene **113** by the Wittig reaction, which was further catalyzed by Li_2CuCl_4 to give compound **114**. The enantioselective cyclization of **114** and **115**/$SbCl_5$ complex afforded compound **116,** which was a key step in the reaction. Finally, compound **116** reacted with $KMnO_4$ and $NaIO_4$ to obtain dehydroabietic acid (**1**) [94].

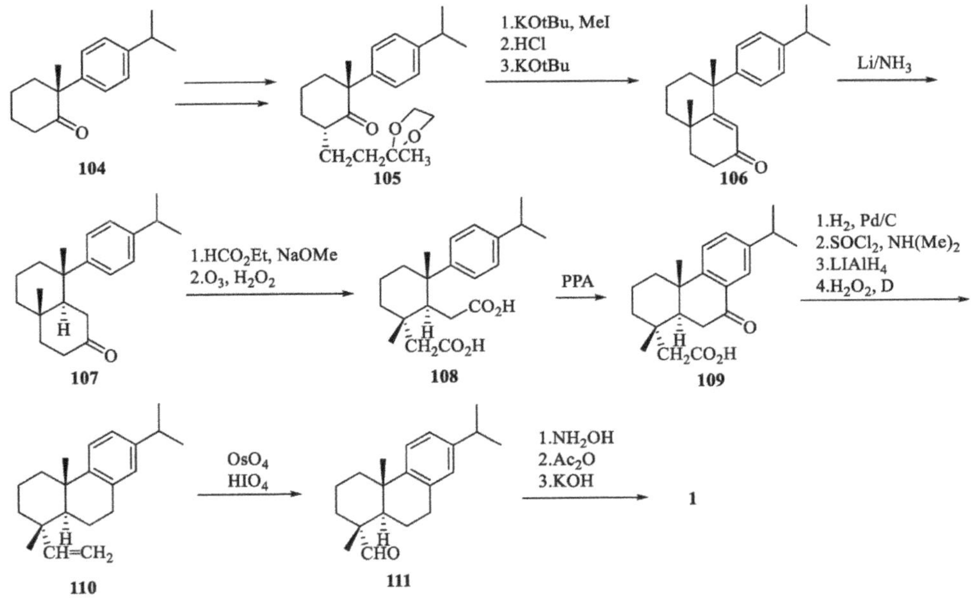

Scheme 20. Total synthesis of dehydroabietic acid from 2-methyl-2-(p-isopropylphenyl)cyclohexanone.

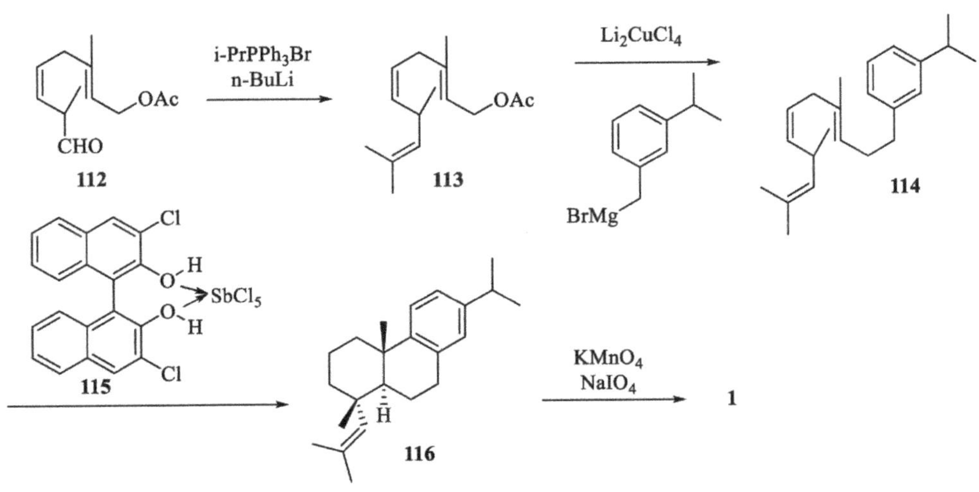

Scheme 21. Total synthesis of dehydroabietic acid from the geranyl acetate derivative.

5. Structure-Activity Relationships of Dehydroabietic Acid and Its Derivatives

The SARs analysis of dehydroabietic acid and its derivatives about the antibacterial and anticancer activities (Figure 6) were as follows:

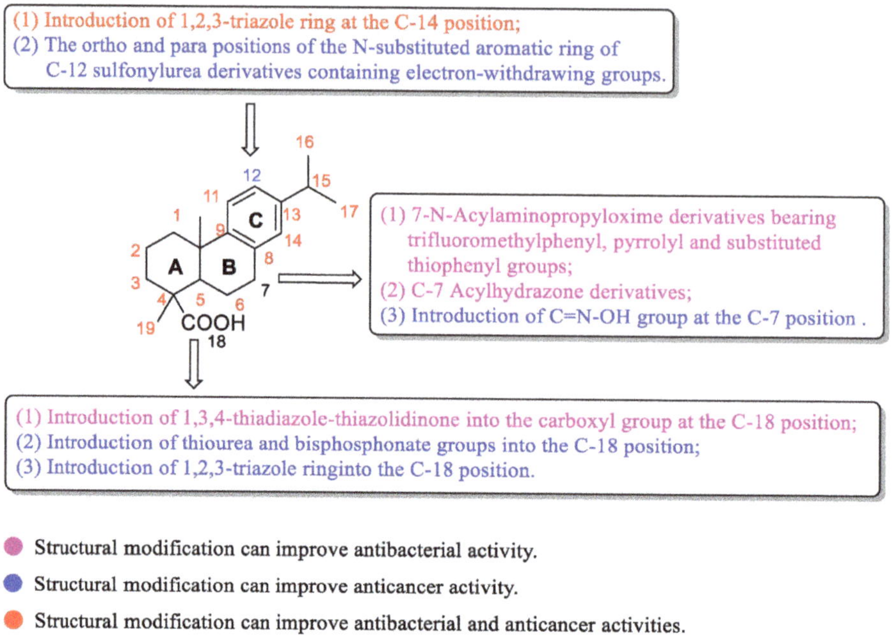

Figure 6. The overall SARs of dehydroabietic acid and its derivatives for antibacterial and anticancer activities.

(1) For antibacterial activity: (a) Introduction of trifluoromethyl phenyl, pyrrolyl, or substituted thiophenyl groups (**57h**–**57j** and **57p**–**57s**) into the oxime derivatives at the C-7 position of dehydroabietic acid was beneficial to the antibacterial activity [78]. C-7 Acylhydrazone derivatives of dehydroabietic acid (especially *p*-fluorobenzoyl hydrazone) can increase the antibacterial activity [80]. (b) Introduction of 1,3,4-thiadiazole-thiazolidinone into the C-18 carboxyl group of dehydroabietic acid can effectively improve the antibacterial activity [76]. (c) Introduction of 1,2,3-triazole ring at the C-14 position of dehydroabietic acid can improve the antibacterial activity [83].

(2) For anticancer activity: (a) Introduction of the C=N-OH group at the C-7 position can improve anti-tumor activity [64]. (b) Introduction of thiourea and bisphosphonate groups into the C-18 position of dehydroabietic acid was beneficial to the anticancer activity [65]. (c) 1,2,3-Triazole ring at the C-18 position played a key role for dehydroabietic acid derivatives showing the anti-cancer effect [71,84]. (d) C-12 sulfonylurea derivatives of dehydroabietic acid bearing electron-withdrawing groups at the ortho or para position on the N-substituted aromatic ring, can improve the antiproliferative activity [90].

Researchers expected to synthesize high bioactive compounds with low toxicity. Herein, we summarized some IC_{50} values of the most active derivatives against human normal cell lines (Table 4). Most of these derivatives exhibited more potent activity than dehydroabietic acid against cancer cell lines, and they showed low cytotoxicity against human normal cells. The IC_{50} value of compound **36w** against HeLa cells (2.21 ± 0.04 μM) was about 17-folds higher than that of the precursor (37.40 ± 0.64 μM), while the cytotoxicity of compound **36w** against normal HL-7702 human liver cell was lower (66.08 ± 1.84 μM) [69]. Compound **39j** exhibited good cytotoxicity against the tested cancer cell lines (IC_{50} values:.82–17.76 μM) compared to the lead compound, with no increase in cytotoxicity against normal cells (IC_{50} > 100 μM) [70]. In addition, the anticancer activity of compounds **41c**, **41k**, **47j**, and **47n** was significantly improved, but their toxicity against human normal cells was not increased compared to that of dehydroabietic acid;

that is, they exhibited excellent selectivity between normal and cancer cells [71,73]. The cytotoxicity of compound **77b** against normal human liver cell LO2 (IC_{50}: 11.09 ± 0.57 μM) was much lower than that of compound **77b** against MCF-7, SMMC-7721, and HeLa cancer cells (IC_{50} values: 0.72–1.78 μM). However, compared with that of dehydroabietic acid (IC_{50} > 50 μM), it was more toxic to normal cells [87]. Compound **79h** was also more cytotoxic against normal human LO2 cell (IC_{50}: 42.83 ± 3.18 μM) than dehydroabietic acid (IC_{50} > 50 μM) [88]. Based on the above results, when compared with the precursor, the bioactivity of some compounds was significantly improved after structural modifications, but their toxicity against normal human cells was also increased. Fortunately, compounds **39j**, **41c**, **41k**, **47j**, and **47n** not only showed promising biological activity, but also had good selectivity between normal and malignant cells; so they can be further studied and developed as potential anticancer drugs.

Table 4. Some IC_{50} values of the most active derivatives against human normal cell lines.

Compound	Human Normal Cell Line	IC_{50} Value
22f	HL-7702	53.78 ± 1.4 μM [64]
36w	HL-7702	66.08 ± 1.84 μM [69]
39j	LO2	>100 μM [70]
41c, 41k	HL-7702	>100 μM [71]
47j	HL-7702, NP69	>100 μM [73]
47n	NP69	>100 μM [73]
47n	HL-7702	88.18± 0.23 μM [73]
67g	QSG-7701	12.52 ± 0.58 μM [80]
74b	LO2	3.89 ± 0.29 μM [86]
74e	LO2	3.02 ± 0.21 μM [86]
77b	LO2	11.09 ± 0.57 μM [87]
79h	LO2	42.83 ± 3.18 μM [88]
80j	QSG-7701	5.82 ± 0.38 μM [89]

6. Conclusions

Dehydroabietic acid, a tricyclic diterpenoid resin acid, showed a wide range of biological activities. The structural characteristics suggested that dehydroabietic acid had multiple structural modification sites, and using it as a lead compound can improve its application value. This review gave an overview of the biological activities, structural modifications, biotransformation, total synthesis, and structure-activity relationships of dehydroabietic acid and its derivatives. Although several structural modifications of dehydroabietic acid and its derivatives were summarized, their properties were mainly focused on anticancer and antibacterial activities. Obviously, their agricultural and other activities needed to be expanded in the future. Therefore, we hope that this review can provide references for further research and development of dehydroabietic acid and its derivatives, as well as their applications as drugs and pesticides in the medical and agricultural fields.

Author Contributions: All persons who have made substantial contributions to present work are named in the manuscript. M.H., H.X. and M.L. designed the work, analyzed the data, and wrote and revised the paper. All authors have read and agreed to the published version of the manuscript.

Funding: This research was funded by the National Natural Science Foundation of China (Project No. 31872013).

Acknowledgments: M.L. want to thank the National Natural Science Foundation of China (No. 31872013), and State Key Laboratory of Elemento-Organic Chemistry, Nankai University (No. 202109).

Conflicts of Interest: The authors declare no conflict of interest.

Abbreviations

AP-1	Activating protein-1
5-Fu	5-Fluorouracil
G6Pase	Glucose-6-Phosphatase
GS	Glycogen Synthase
IC_{50}	Half-inhibitory concentration
MCP-1	Monocyte chemoattractant protein-1
MIC	Minimum inhibitory concentration
NF-κB	Nuclear factor-κB
Nrf2	Nuclear factor erythroid 2-related factor 2
NO	Nitric oxide
PARP	Poly (ADP-ribose) polymerase
PPA	Polyphosphoric acid
PPAR	Peroxisome proliferator-activated receptors
ROS	Reactive oxygen species
TAK1	Transforming growth factor β-activated kinase 1
TNF-α	Tumor necrosis factor-α

References

1. Zhang, Y.Y.; Xu, H. Recent progress in the chemistry and biology of limonoids. *RSC Adv.* **2017**, *7*, 35191–35220. [CrossRef]
2. Li, Q.; Huang, X.; Li, S.; Ma, J.; Lv, M.; Xu, H. Semisynthesis of esters of fraxinellone C4/10-oxime and their pesticidal activities. *J. Agric. Food Chem.* **2016**, *64*, 5472–5478. [CrossRef]
3. Nagini, S.; Nivetha, R.; Palrasu, M.; Mishra, R. Nimbolide, a neem limonoid, is a promising candidate for the anticancer drug arsenal. *J. Med. Chem.* **2021**, *64*, 3560–3577. [CrossRef] [PubMed]
4. Huang, X.B.; Li, T.; Shan, X.; Lu, R.; Hao, M.; Lv, M.; Sun, Z.; Xu, H. High value-added use of citrus industrial wastes in agriculture: Semisynthesis and anti-tobacco mosaic virus/insecticidal activities of ester derivatives of limonin modified in the B ring. *J. Agric. Food Chem.* **2020**, *68*, 12241–12251. [CrossRef] [PubMed]
5. Huang, X.B.; Lv, M.; Ma, Q.; Zhang, Y.; Xu, H. High value-added application of natural products in crop protection: Semisynthesis and acaricidal activity of limonoid-type derivatives and investigation of their biocompatible O/W nanoemulsions as agronanopesticide candidates. *J. Agric. Food Chem.* **2021**, *69*, 14488–14500. [CrossRef]
6. Sun, Y.; Yin, Y.; Sun, Y.; Li, Q.; Cui, L.; Xu, W.; Kong, L.; Luo, J. Aglatestine A, a rearranged limonoid with a 3/6/6 tricarbocyclic framework from the fruits of *Aglaia edulis*. *J. Org. Chem.* **2021**, *86*, 11263–11268. [CrossRef] [PubMed]
7. Huang, X.C.; Wang, M.; Pan, Y.M.; Yao, G.Y.; Wang, H.S.; Tian, X.Y.; Qin, J.K.; Zhang, Y. Synthesis and antitumor activities of novel thiourea α-aminophosphonates from dehydroabietic acid. *Eur. J. Med. Chem.* **2013**, *69*, 508–520. [CrossRef]
8. Xu, H.; Liu, L.; Fan, X.; Zhang, G.; Li, Y.; Jiang, B. Identification of a diverse synthetic abietane diterpenoid library for anticancer activity. *Bioorganic Med. Chem. Lett.* **2017**, *27*, 505–510. [CrossRef] [PubMed]
9. Gonzalez, M.A.; Perez-Guaita, D.; Correa-Royero, J.; Zapata, B.; Agudelo, L.; Mesa-Arango, A.; Betancur-Galvis, L. Synthesis and biological evaluation of dehydroabietic acid derivatives. *Eur. J. Med. Chem.* **2010**, *45*, 811–816. [CrossRef] [PubMed]
10. Han, C.R.; Song, Z.Q.; Shang, S.B. Research progress on abietic acid, dehydroabietic acid and their bioactive derivatives. *Chem. Ind. Eng. Prog.* **2007**, *26*, 490–495. [CrossRef]
11. Costa, M.S.; Rego, A.; Ramos, V.; Afonso, T.B.; Freitas, S.; Preto, M.; Lopes, V.; Vasconcelos, V.; Magalhaes, C.; Leao, P.N. The conifer biomarkers dehydroabietic and abietic acids are widespread in Cyanobacteria. *Sci. Rep.* **2016**, *6*, 23436. [CrossRef]
12. Tagat, J.R.; Nazareno, D.V.; Puar, M.S.; McCombie, S.W.; Ganguly, A.K. Synthesis and anti-herpes activity of some A-ring functionalized dehydroabietane derivatives. *Bioorganic Med. Chem. Lett.* **1994**, *4*, 1101–1104. [CrossRef]
13. Kim, W.-J.; Kang, H.-G.; Kim, S.-J. Dehydroabietic acid inhibits the gastric cancer cell growth via induced apoptosis and cell cycle arrest. *Mol. Cell. Toxicol.* **2021**, *17*, 133–139. [CrossRef]
14. Jokinen, J.J.; Sipponen, A. Refined spruce resin to treat chronic wounds: Rebirth of an old folkloristic therapy. *Adv. Wound Care* **2016**, *5*, 198–207. [CrossRef]
15. Gonzalez, M.A. Aromatic abietane diterpenoids: Their biological activity and synthesis. *Nat. Prod. Rep.* **2015**, *32*, 684–704. [CrossRef]
16. Sepulveda, B.; Astudillo, L.; Rodriguez, J.A.; Yanez, T.; Theoduloz, C.; Schmeda-Hirschmann, G. Gastroprotective and cytotoxic effect of dehydroabietic acid derivatives. *Pharmacol. Res.* **2005**, *52*, 429–437. [CrossRef] [PubMed]
17. Tolmacheva, I.A.; Taranfin, A.V.; Boteva, A.A.; Anikina, L.V.; Vikharev, Y.B.; Grishko, V.V.; Tolstikov, A.G. Synthesis and biological activity of nitrogen-containing derivatives of methyl dehydroabietate. *Pharm. Chem. J.* **2006**, *40*, 489–493. [CrossRef]

18. Duan, W.G.; Li, X.R.; Mo, Q.J.; Huang, J.X.; Cen, B.; Xu, X.T.; Lei, F.H. Synthesis and herbicidal activity of 5-dehydroabietyl-1,3,4-oxadiazole derivatives. *Holzforschung* **2011**, *65*, 191–197. [CrossRef]
19. Helfenstein, A.; Vahermo, M.; Nawrot, D.A.; Demirci, F.; Iscan, G.; Krogerus, S.; Yli-Kauhaluoma, J.; Moreira, V.M.; Tammela, P. Antibacterial profiling of abietane-type diterpenoids. *Bioorganic Med. Chem.* **2017**, *25*, 132–137. [CrossRef] [PubMed]
20. Xie, Y.S.; Isman, M.B.; Yi, F.; Wong, A. Diterpene resin acids: Major active principles in tall oil against Variegated cutworm, *Peridroma saucia* (Lepidoptera, Noctuidae). *J. Chem. Ecol.* **1993**, *19*, 1075–1084. [CrossRef] [PubMed]
21. Robert, J.A.; Madilao, L.L.; White, R.; Yanchuk, A.; King, J.; Bohlmann, J. Terpenoid metabolite profiling in Sitka spruce identifies association of dehydroabietic acid, (+)-3-carene, and terpinolene with resistance against white pine weevil. *Botany* **2010**, *88*, 810–820. [CrossRef]
22. Oh, H.W.; Yun, C.S.; Jeon, J.H.; Kim, J.A.; Park, D.S.; Ryu, H.W.; Oh, S.R.; Song, H.H.; Shin, Y.; Jung, C.S.; et al. Conifer diterpene resin acids disrupt juvenile hormone-mediated endocrine regulation in the Indian meal moth *Plodia interpunctella*. *J. Chem. Ecol.* **2017**, *43*, 703–711. [CrossRef]
23. Rao, X.; Song, Z.; Han, Z.; Jiang, Z. Synthesis and insect attractant activity of fluorine-containing *Pinus* diterpenic amides and imines. *Nat. Prod. Res.* **2009**, *23*, 851–860. [CrossRef] [PubMed]
24. Xin, C.; Zhang, Y.; Bao, M.; Yu, C.; Hou, K.; Wang, Z. Novel carrier-free, charge-reversal and DNA-affinity nanodrugs for synergistic cascade cancer chemo-chemodynamic therapy. *J. Colloid Interface Sci.* **2022**, *606*, 1488–1508. [CrossRef] [PubMed]
25. Huang, X.; Liu, J.; Zhao, B.; Bai, Y.; Peng, Z.; Zhou, J.; Wang, C.; Zhao, X.; Han, S.; Zhang, C. One-step synthesis of biomass-based carbon dots for detection of metal ions and cell imaging. *Front. Energy Res.* **2022**, *10*, 871617. [CrossRef]
26. Singh, A.K.; Chandra, R. Pollutants released from the pulp paper industry: Aquatic toxicity and their health hazards. *Aquat. Toxicol.* **2019**, *211*, 202–216. [CrossRef] [PubMed]
27. Pandelides, Z.; Guchardi, J.; Holdway, D. Dehydroabietic acid (DHAA) alters metabolic enzyme activity and the effects of 17β-estradiol in rainbow trout (*Oncorhynchus mykiss*). *Ecotoxicol. Environ. Saf.* **2014**, *101*, 168–176. [CrossRef]
28. Lee, S.; Lee, S.; Roh, H.S.; Song, S.S.; Ryoo, R.; Pang, C.; Baek, K.H.; Kim, K.H. Cytotoxic constituents from the sclerotia of *Poria cocos* against human lung adenocarcinoma cells by inducing mitochondrial apoptosis. *Cells* **2018**, *7*, 116. [CrossRef]
29. Luo, D.; Ni, Q.; Ji, A.; Gu, W.; Wu, J.; Jiang, C. Dehydroabietic acid derivative QC4 induces gastric cancer cell death via oncosis and apoptosis. *Biomed. Res. Int.* **2016**, *2016*, 2581061. [CrossRef]
30. Huang, R.Z.; Liang, G.B.; Huang, X.C.; Zhang, B.; Zhou, M.M.; Liao, Z.X.; Wang, H.S. Discovery of dehydroabietic acid sulfonamide based derivatives as selective matrix metalloproteinases inactivators that inhibit cell migration and proliferation. *Eur. J. Med. Chem.* **2017**, *138*, 979–992. [CrossRef]
31. Kolsi, L.E.; Leal, A.S.; Yli-Kauhaluoma, J.; Liby, K.T.; Moreira, V.M. Dehydroabietic oximes halt pancreatic cancer cell growth in the G1 phase through induction of p27 and downregulation of cyclin D1. *Sci. Rep.* **2018**, *8*, 15923. [CrossRef] [PubMed]
32. Zaki, H.; Belhassan, A.; Benlyas, M.; Lakhlifi, T.; Bouachrine, M. New dehydroabietic acid (DHA) derivatives with anticancer activity against HepG2 cancer cell lines as a potential drug targeting EGFR kinase domain. CoMFA study and virtual ligand-based screening. *J. Biomol. Struct. Dyn.* **2021**, *39*, 2993–3003. [CrossRef] [PubMed]
33. Kim, E.; Kang, Y.G.; Kim, Y.J.; Lee, T.R.; Yoo, B.C.; Jo, M.; Kim, J.H.; Kim, J.H.; Kim, D.; Cho, J.Y. Dehydroabietic acid suppresses inflammatory response via suppression of Src-, Syk-, and TAK1-mediated pathways. *Int. J. Mol. Sci.* **2019**, *20*, 1593. [CrossRef] [PubMed]
34. Kang, M.S.; Hirai, S.; Goto, T.; Kuroyanagi, K.; Lee, J.Y.; Uemura, T.; Ezaki, Y.; Takahashi, N.; Kawada, T. Dehydroabietic acid, a phytochemical, acts as ligand for PPARs in macrophages and adipocytes to regulate inflammation. *Biochem. Biophys. Res. Commun.* **2008**, *369*, 333–338. [CrossRef] [PubMed]
35. Tretyakova, E.V.; Salimova, E.V.; Parfenova, L.V. Synthesis and antimicrobial and antifungal activity of resin acid acetylene derivatives. *Russ. J. Bioorganic Chem.* **2020**, *45*, 545–551. [CrossRef]
36. Chen, H.; Geng, Y.; Wang, S.F.; Gu, W. Syntheses, crystal structures and antibacterial activities of two new methyl 12-alkylamino-13,14-dinitrodeisopropyl-dehydroabietates. *Chin. J. Struct. Chem.* **2019**, *38*, 257–262. [CrossRef]
37. Sakunpak, A.; Sueree, L. Thin-layer chromatography–contact bioautography as a tool for bioassay-guided isolation of anti-*Streptococcus mutans* compounds from *Pinus merkusii* heartwood. *JPC-J. Planar Chromat.* **2018**, *31*, 355–359. [CrossRef]
38. Hassan, G.; Forsman, N.; Wan, X.; Keurulainen, L.; Bimbo, L.M.; Stehl, S.; van Charante, F.; Chrubasik, M.; Prakash, A.S.; Johansson, L.-S.; et al. Non-leaching, highly biocompatible nanocellulose surfaces that efficiently resist fouling by bacteria in an artificial dermis model. *ACS Appl. Bio Mater.* **2020**, *3*, 4095–4108. [CrossRef]
39. Berger, M.; Roller, A.; Maulide, N. Synthesis and antimicrobial evaluation of novel analogues of dehydroabietic acid prepared by C-H-Activation. *Eur. J. Med. Chem.* **2017**, *126*, 937–943. [CrossRef] [PubMed]
40. Chaban, M.F.; Antoniou, A.I.; Karagianni, C.; Toumpa, D.; Joray, M.B.; Bocco, J.L.; Sola, C.; Athanassopoulos, C.M.; Carpinella, M.C. Synthesis and structure-activity relationships of novel abietane diterpenoids with activity against *Staphylococcus aureus*. *Future Med. Chem.* **2019**, *11*, 3109–3124. [CrossRef] [PubMed]
41. Liu, M.L.; Pan, X.Y.; Yang, T.; Zhang, W.M.; Wang, T.Q.; Wang, H.Y.; Lin, H.X.; Yang, C.G.; Cui, Y.M. The synthesis and antistaphylococcal activity of dehydroabietic acid derivatives: Modifications at C-12. *Bioorganic Med. Chem. Lett.* **2016**, *26*, 5492–5496. [CrossRef]
42. Touré, S.; Dusfour, I.; Stien, D.; Eparvier, V. Two new tetrahydrofuran derivatives from the fungus *Mucor* spp. SNB-VECD11D isolated from the Chrysomelidae *Acalymma bivittula*. *Tetrahedron Lett.* **2017**, *58*, 3727–3729. [CrossRef]

43. Gao, Y.Q.; Li, J.; Song, Z.Q.; Song, J.; Shang, S.B.; Xiao, G.M.; Wang, Z.D.; Rao, X.P. Turning renewable resources into value-added products: Development of rosin-based insecticide candidates. *Ind. Crops Prod.* **2015**, *76*, 660–671. [CrossRef]
44. Liu, L.; Yan, X.Y.; Gao, Y.Q.; Rao, X.P. Synthesis and antifeedant activities of rosin-based esters against *armyworm*. *Comb. Chem. High Throughput Screen.* **2016**, *19*, 193–199. [CrossRef] [PubMed]
45. Chiu, C.C.; Keeling, C.I.; Henderson, H.M.; Bohlmann, J. Functions of mountain pine beetle cytochromes P450 CYP6DJ1, CYP6BW1 and CYP6BW3 in the oxidation of pine monoterpenes and diterpene resin acids. *PLoS ONE* **2019**, *14*, e0216753. [CrossRef]
46. Goncalves, M.D.; Bortoleti, B.T.S.; Tomiotto-Pellissier, F.; Miranda-Sapla, M.M.; Assolini, J.P.; Carloto, A.C.M.; Carvalho, P.G.C.; Tudisco, E.T.; Urbano, A.; Ambrosio, S.R.; et al. Dehydroabietic acid isolated from *Pinus elliottii* exerts *in vitro* antileishmanial action by pro-oxidant effect, inducing ROS production in promastigote and downregulating Nrf2/ferritin expression in amastigote forms of *Leishmania amazonensis*. *Fitoterapia* **2018**, *128*, 224–232. [CrossRef]
47. Pertino, M.W.; Vega, C.; Rolon, M.; Coronel, C.; Rojas de Arias, A.; Schmeda-Hirschmann, G. Antiprotozoal activity of triazole derivatives of dehydroabietic acid and oleanolic acid. *Molecules* **2017**, *22*, 369. [CrossRef] [PubMed]
48. Vahermo, M.; Krogerus, S.; Nasereddin, A.; Kaiser, M.; Brun, R.; Jaffe, C.L.; Yli-Kauhaluoma, J.; Moreira, V.M. Antiprotozoal activity of dehydroabietic acid derivatives against *Leishmania donovani* and *Trypanosoma cruzi*. *MedChemComm* **2016**, *7*, 457–463. [CrossRef]
49. Xie, Z.; Gao, G.; Wang, H.; Li, E.; Yuan, Y.; Xu, J.; Zhang, Z.; Wang, P.; Fu, Y.; Zeng, H.; et al. Dehydroabietic acid alleviates high fat diet-induced insulin resistance and hepatic steatosis through dual activation of PPAR-γ and PPAR-α. *Biomed. Pharmacother.* **2020**, *127*, 110155. [CrossRef] [PubMed]
50. Kang, M.S.; Hirai, S.; Goto, T.; Kuroyanagi, K.; Kim, Y.I.; Ohyama, K.; Uemura, T.; Lee, J.Y.; Sakamoto, T.; Ezaki, Y.; et al. Dehydroabietic acid, a diterpene, improves diabetes and hyperlipidemia in obese diabetic KK-Ay mice. *Biofactors* **2009**, *35*, 442–448. [CrossRef]
51. Islam, M.T.; Ali, E.S.; Mubarak, M.S. Anti-obesity effect of plant diterpenes and their derivatives: A review. *Phytother. Res.* **2020**, *34*, 1216–1225. [CrossRef]
52. Gao, G.; Xie, Z.; Li, E.W.; Yuan, Y.; Fu, Y.; Wang, P.; Zhang, X.; Qiao, Y.; Xu, J.; Holscher, C.; et al. Dehydroabietic acid improves nonalcoholic fatty liver disease through activating the Keap1/Nrf2-ARE signaling pathway to reduce ferroptosis. *J. Nat. Med.* **2021**, *75*, 540–552. [CrossRef]
53. Park, J.; Kim, W.J.; Kim, W.; Park, C.; Choi, C.Y.; Cho, J.H.; Kim, S.J.; Cheong, H. Antihypertensive effects of dehydroabietic and 4-*epi-trans*-communic acid isolated from *Pinus densiflora*. *J. Med. Food* **2021**, *24*, 50–58. [CrossRef] [PubMed]
54. Kim, J.; Kang, Y.G.; Lee, J.Y.; Choi, D.H.; Cho, Y.U.; Shin, J.M.; Park, J.S.; Lee, J.H.; Kim, W.G.; Seo, D.B.; et al. The natural phytochemical dehydroabietic acid is an anti-aging reagent that mediates the direct activation of SIRT1. *Mol. Cell. Endocrinol.* **2015**, *412*, 216–225. [CrossRef]
55. Park, N.H.; Kang, Y.G.; Kim, S.H.; Bae, I.H.; Lee, S.H.; Kim, D.Y.; Hwang, J.S.; Kim, Y.J.; Lee, T.R.; Lee, E.S. Dehydroabietic acid induces regeneration of collagen fibers in ultraviolet B-irradiated human dermal fibroblasts and skin equivalents. *Ski. Pharmacol. Physiol.* **2019**, *32*, 109–116. [CrossRef]
56. Liu, X.; Hu, T.; Lin, G.; Wang, X.; Zhu, Y.; Liang, R.; Duan, W.; Wei, M. The synthesis of a DHAD/ZnAlTi-LDH composite with advanced UV blocking and antibacterial activity for skin protection. *RSC Adv.* **2020**, *10*, 9786–9790. [CrossRef]
57. Lee, S.; Choi, E.; Yang, S.M.; Ryoo, R.; Moon, E.; Kim, S.H.; Kim, K.H. Bioactive compounds from sclerotia extract of *Poria cocos* that control adipocyte and osteoblast differentiation. *Bioorganic Chem.* **2018**, *81*, 27–34. [CrossRef]
58. Nachar, A.; Saleem, A.; Arnason, J.T.; Haddad, P.S. Regulation of liver cell glucose homeostasis by dehydroabietic acid, abietic acid and squalene isolated from balsam fir (*Abies balsamea* (L.) Mill.) a plant of the Eastern James Bay Cree traditional pharmacopeia. *Phytochemistry* **2015**, *117*, 373–379. [CrossRef]
59. Sacripante, G.G.; Zhou, K.; Farooque, M. Sustainable polyester resins derived from rosins. *Macromolecules* **2015**, *48*, 6876–6881. [CrossRef]
60. Ohwada, T.; Nonomura, T.; Maki, K.; Sakamoto, K.; Ohya, S.; Muraki, K.; Imaizumi, Y. Dehydroabietic acid derivatives as a novel scaffold for large-conductance calcium-activated K^+ channel openers. *Bioorganic Med. Chem. Lett.* **2003**, *13*, 3971–3974. [CrossRef] [PubMed]
61. Cui, Y.M.; Liu, X.L.; Zhang, W.M.; Lin, H.X.; Ohwada, T.; Ido, K.; Sawada, K. The synthesis and BK channel-opening activity of N-acylaminoalkyloxime derivatives of dehydroabietic acid. *Bioorganic Med. Chem. Lett.* **2016**, *26*, 283–287. [CrossRef]
62. Li, A.L.; Wang, Z.L.; Wang, W.Y.; Liu, Q.S.; Sun, Y.; Wang, S.F.; Gu, W. A novel dehydroabietic acid-based fluorescent probe for detection of Fe^{3+} and Hg^{2+} ions and its application in live-cell imaging. *Microchem. J.* **2021**, *160*, 105682. [CrossRef]
63. Cai, X.M.; Mu, T.Q.; Lin, Y.T.; Zhang, X.D.; Tang, Z.G.; Huang, S.L. Syntheses and photophysical properties of natural dehydroabietic acid-based ligands and their zinc complexes. *J. Mol. Struct.* **2021**, *1229*, 129793. [CrossRef]
64. Huang, X.C.; Jin, L.; Wang, M.; Liang, D.; Chen, Z.F.; Zhang, Y.; Pan, Y.M.; Wang, H.S. Design, synthesis and in vitro evaluation of novel dehydroabietic acid derivatives containing a dipeptide moiety as potential anticancer agents. *Eur. J. Med. Chem.* **2015**, *89*, 370–385. [CrossRef]
65. Huang, X.; Huang, R.; Liao, Z.; Pan, Y.; Gou, S.; Wang, H. Synthesis and pharmacological evaluation of dehydroabietic acid thiourea derivatives containing bisphosphonate moiety as an inducer of apoptosis. *Eur. J. Med. Chem.* **2016**, *108*, 381–391. [CrossRef] [PubMed]

66. Jin, L.; Qu, H.E.; Huang, X.C.; Pan, Y.M.; Liang, D.; Chen, Z.F.; Wang, H.S.; Zhang, Y. Synthesis and biological evaluation of novel dehydroabietic acid derivatives conjugated with acyl-thiourea peptide moiety as antitumor agents. *Int. J. Mol. Sci.* **2015**, *16*, 14571–14593. [CrossRef]
67. Li, L.Y.; Fei, B.L.; Wang, P.; Kong, L.Y.; Long, J.Y. Discovery of novel dehydroabietic acid derivatives as DNA/BSA binding and anticancer agents. *Spectrochim. Acta A Mol. Biomol. Spectrosc.* **2021**, *246*, 118944. [CrossRef] [PubMed]
68. Jin, X.Y.; Zhang, K.P.; Chen, H.; Miao, T.T.; Wang, S.F.; Gu, W. Synthesis, in vitro antimicrobial, and cytotoxic activities of new 1,3,4-oxadiazin-5(6*H*)-one derivatives from dehydroabietic acid. *J. Chin. Chem. Soc.* **2018**, *65*, 538–547. [CrossRef]
69. Li, F.Y.; Wang, X.; Duan, W.G.; Lin, G.S. Synthesis and in vitro anticancer activity of novel dehydroabietic acid-based acylhydrazones. *Molecules* **2017**, *22*, 1087. [CrossRef]
70. Wang, X.; Pang, F.H.; Huang, L.; Yang, X.P.; Ma, X.L.; Jiang, C.N.; Li, F.Y.; Lei, F.H. Synthesis and biological evaluation of novel dehydroabietic acid-oxazolidinone hybrids for antitumor properties. *Int. J. Mol. Sci.* **2018**, *19*, 3116. [CrossRef]
71. Li, F.Y.; Huang, L.; Li, Q.; Wang, X.; Ma, X.L.; Jiang, C.N.; Zhou, X.Q.; Duan, W.G.; Lei, F.H. Synthesis and antiproliferative evaluation of novel hybrids of dehydroabietic acid bearing 1,2,3-triazole moiety. *Molecules* **2019**, *24*, 4191. [CrossRef]
72. Huang, L.; Huang, R.; Pang, F.; Li, A.; Huang, G.; Zhou, X.; Li, Q.; Li, F.; Ma, X. Synthesis and biological evaluation of dehydroabietic acid-pyrimidine hybrids as antitumor agents. *RSC Adv.* **2020**, *10*, 18008–18015. [CrossRef]
73. Li, F.; Huang, L.; Zhou, X.; Li, Q.; Ma, X.; Duan, W.; Wang, X. Synthesis and cytotoxicity evaluation of dehydroabietic acid derivatives bearing nitrate moiety. *Chin. J. Org. Chem.* **2020**, *40*, 2845–2854. [CrossRef]
74. Manner, S.; Vahermo, M.; Skogman, M.E.; Krogerus, S.; Vuorela, P.M.; Yli-Kauhaluoma, J.; Fallarero, A.; Moreira, V.M. New derivatives of dehydroabietic acid target planktonic and biofilm bacteria in *Staphylococcus aureus* and effectively disrupt bacterial membrane integrity. *Eur. J. Med. Chem.* **2015**, *102*, 68–79. [CrossRef]
75. Chen, N.-Y.; Duan, W.-G.; Liu, L.-Z.; Li, F.-Y.; Lu, M.-P.; Liu, B.-M. Synthesis and antifungal activity of dehydroabietic acid-based thiadiazole-phosphonates. *Holzforschung* **2015**, *69*, 1069–1075. [CrossRef]
76. Chen, N.; Duan, W.; Lin, G.; Liu, L.; Zhang, R.; Li, D. Synthesis and antifungal activity of dehydroabietic acid-based 1,3,4-thiadiazole-thiazolidinone compounds. *Mol. Divers.* **2016**, *20*, 897–905. [CrossRef]
77. Mo, Q.J.; Liu, L.Z.; Duan, W.G.; Cen, B.; Lin, G.S.; Chen, N.Y.; Huang, Y.; Liu, B.M. Synthesis and insecticidal activities of N-(5-dehydroabietyl-1,3,4-thiadiazol-2-yl)-N′-substituted thioureas. *Chem. Ind. Forest Prod.* **2015**, *35*, 8–16. [CrossRef]
78. Zhang, W.M.; Yang, T.; Pan, X.Y.; Liu, X.L.; Lin, H.X.; Gao, Z.B.; Yang, C.G.; Cui, Y.M. The synthesis and antistaphylococcal activity of dehydroabietic acid derivatives: Modifications at C12 and C7. *Eur. J. Med. Chem.* **2017**, *127*, 917–927. [CrossRef]
79. Zhang, W.M.; Yao, Y.; Yang, T.; Wang, X.Y.; Zhu, Z.Y.; Xu, W.T.; Lin, H.X.; Gao, Z.B.; Zhou, H.; Yang, C.G.; et al. The synthesis and antistaphylococcal activity of N-sulfonaminoethyloxime derivatives of dehydroabietic acid. *Bioorganic Med. Chem. Lett.* **2018**, *28*, 1943–1948. [CrossRef]
80. Zhou, Z.; Zhou, T.T. Synthesis and antibacterial activity of C-7 acylhydrazone derivatives of dehydroabietic acid. *J. Chem. Res.* **2018**, *8*, 405–407. [CrossRef]
81. Chen, N.Y.; Xie, Y.L.; Lu, G.D.; Ye, F.; Li, X.Y.; Huang, Y.W.; Huang, M.L.; Chen, T.Y.; Li, C.P. Synthesis and antitumor evaluation of (aryl)methyl-amine derivatives of dehydroabietic acid-based B ring-fused-thiazole as potential PI3K/AKT/mTOR signaling pathway inhibitors. *Mol. Divers.* **2021**, *25*, 967–979. [CrossRef]
82. Chen, H.; Qiao, C.; Miao, T.T.; Li, A.L.; Wang, W.Y.; Gu, W. Synthesis and biological evaluation of novel *N*-(piperazin-1-yl)alkyl-1*H*-dibenzo[*a*,*c*]carbazole derivatives of dehydroabietic acid as potential MEK inhibitors. *J. Enzym. Inhib. Med. Chem.* **2019**, *34*, 1544–1561. [CrossRef] [PubMed]
83. Hou, W.; Zhang, G.; Luo, Z.; Li, D.; Ruan, H.; Ruan, B.H.; Su, L.; Xu, H. Identification of a diverse synthetic abietane diterpenoid library and insight into the structure-activity relationships for antibacterial activity. *Bioorg. Med. Chem. Lett.* **2017**, *27*, 5382–5386. [CrossRef] [PubMed]
84. Hou, W.; Luo, Z.; Zhang, G.; Cao, D.; Li, D.; Ruan, H.; Ruan, B.H.; Su, L.; Xu, H. Click chemistry-based synthesis and anticancer activity evaluation of novel C-14 1,2,3-triazole dehydroabietic acid hybrids. *Eur. J. Med. Chem.* **2017**, *138*, 1042–1052. [CrossRef]
85. Ahonen, T.J.; Savinainen, J.R.; Yli-Kauhaluoma, J.; Kalso, E.; Laitinen, J.T.; Moreira, V.M. Discovery of 12-thiazole abietanes as selective inhibitors of the human metabolic serine hydrolase hABHD16A. *ACS Med. Chem. Lett.* **2018**, *9*, 1269–1273. [CrossRef] [PubMed]
86. Gu, W.; Miao, T.T.; Hua, D.W.; Jin, X.Y.; Tao, X.B.; Huang, C.B.; Wang, S.F. Synthesis and *in vitro* cytotoxic evaluation of new 1*H*-benzo[*d*]imidazole derivatives of dehydroabietic acid. *Bioorganic Med. Chem. Lett.* **2017**, *27*, 1296–1300. [CrossRef]
87. Gu, W.; Wang, S.; Jin, X.; Zhang, Y.; Hua, D.; Miao, T.; Tao, X.; Wang, S. Synthesis and evaluation of new quinoxaline derivatives of dehydroabietic acid as potential antitumor agents. *Molecules* **2017**, *22*, 1154. [CrossRef]
88. Li, A.L.; Yang, Y.Q.; Wang, W.Y.; Liu, Q.S.; Sun, Y.; Gu, W. Synthesis, cytotoxicity and apoptosis-inducing activity of novel 1*H*-benzo[*d*]imidazole derivatives of dehydroabietic acid. *J. Chin. Chem. Soc.* **2020**, *67*, 1668–1678. [CrossRef]
89. Miao, T.-T.; Tao, X.-B.; Li, D.-D.; Chen, H.; Jin, X.-Y.; Geng, Y.; Wang, S.-F.; Gu, W. Synthesis and biological evaluation of 2-arylbenzimidazole derivatives of dehydroabietic acid as novel tubulin polymerization inhibitors. *RSC Adv.* **2018**, *8*, 17511–17526. [CrossRef] [PubMed]
90. Jun, L.; Ming, L.Y.; Qun, S.L.; Qun, W.A. Synthesis and antitumor activity evaluation of dehydroabietic acid sulfonylureas derivatives. *Chin. J. New Drugs* **2019**, *28*, 1778–1783.

91. Mitsukura, K.; Imoto, T.; Nagaoka, H.; Yoshida, T.; Nagasawa, T. Regio- and stereo-selective hydroxylation of abietic acid derivatives by *Mucor circinelloides* and *Mortierella isabellina*. *Biotechnol. Lett.* **2005**, *27*, 1305–1310. [CrossRef] [PubMed]
92. Rico-Martínez, M.; Medina, F.G.; Marrero, J.G.; Osegueda-Robles, S. Biotransformation of diterpenes. *RSC Adv.* **2014**, *4*, 10627–10647. [CrossRef]
93. van Beek, T.A.; Claassen, F.W.; Dorado, J.; Godejohann, M.; Sierra-Alvarez, R.; Wijnberg, J.B.P.A. Fungal biotransformation products of dehydroabietic acid. *J. Nat. Prod.* **2007**, *70*, 154–159. [CrossRef] [PubMed]
94. González, M.A. Aromatic abietane diterpenoids: Total syntheses and synthetic studies. *Tetrahedron* **2015**, *71*, 1883–1908. [CrossRef]

insects

MDPI

Article

Toxicity, Behavioral Effects, and Chitin Structural Chemistry of *Reticulitermes flaviceps* Exposed to *Cymbopogon citratus* EO and Its Major Constituent Citral

Chunzhe Jin [1], Hui Han [1], Yongjian Xie [1,*], Baoling Li [1], Zhilin Zhang [2] and Dayu Zhang [1]

1 College of Advanced Agricultural Sciences, Zhejiang A&F University, Hangzhou 311300, China
2 Hubei Key Laboratory of Quality Control of Characteristic Fruits and Vegetables, Hubei Engineering University, Xiaogan 432000, China
* Correspondence: yjxie@zafu.edu.cn

Simple Summary: *Reticulitermes flaviceps*, as a main wood-boring pest, causes economically significant damage to wood materials. In this study, gas chromatography–mass spectrometry was used to detect and characterize the chemical constituents of lemongrass essential oil (*Cymbopogon citratus* (DC.) Stapf.), and we evaluated the associated vapor insecticidal effect on *R. flaviceps* worker adults. Lemongrass EO and its major constituent, citral, presented significant vapor toxicity against *R. flaviceps*, where the walking and gripping abilities of treated insects were reduced. Chitin content, thermal stability, and crystallinity were also reduced in the treated worker adults. The results of this study suggest that lemongrass EO can potentially be used to develop eco-friendly natural remedies for the management of *R. flaviceps*.

Abstract: Botanical pesticides are considered the most promising alternative to synthetic pesticides, considering their less negative impacts on the environment and human health. Here, we analyzed the components of lemongrass *Cymbopogon citratus* essential oil (EO) and evaluated its vapor activity against *Reticulitermes flaviceps*, in terms of the walking and gripping abilities of workers. In addition, the effects of lemongrass EO and its major component on the cuticular content and structure of chitin in termites were also observed. Our results indicate that cis-citral (36.51%) was the main constituent of lemongrass. In the vapor toxicity assay, the LC_{50} values of lemongrass EO and citral were 0.328 and 0.177 μL/L, respectively. When worker antennae were treated with lemongrass EO and citral, their walking and gripping capabilities were significantly inhibited. In addition, the cuticular content, thermal stability, and crystallinity of chitin in the termites were decreased after treatment with citral. Collectively, this study provides a basis for developing and utilizing lemongrass and citral as a new environment-friendly insecticide resource to control *R. flaviceps*.

Keywords: *Reticulitermes flaviceps*; *Cymbopogon citratus*; essential oil; citral; walking and gripping behavior; chitin structure

1. Introduction

Termites are important agricultural and forestry pests in tropical and sub-tropical regions, which damage crops, forests, garden trees, houses, and ancient buildings, thus causing an economic loss of more than USD 40 billion annually worldwide [1]. *Reticulitermes flaviceps* is widely distributed in China, and causes losses for the Chinese economy [2]. Synthetic pesticides are currently the most commonly used method for termite control, which often have a negative impact on the natural environment [3]. Due to the low toxicity, biodegradability, and environmentally friendly nature of essential oils (EOs), there is growing interest in their use as novel alternative synthetic insecticides [4].

The genus *Cymbopogon* (family Poaceae) comprises about 144 species, widely spread throughout the tropical and sub-tropical regions [5]. Plants in the genus *Cymbopogon* are good sources of EOs and are rich in bioactive compounds, including alcohols, aldehydes, and phenolic compounds. The chemical compositions of *Cymbopogon* spp. EOs, such as those from *Cymbopogon citratus* [6–8], *Cymbopogon distans* [9], *Cymbopogon flexuosus* [10], *Cymbopogon martinii* [11,12], *Cymbopogon nardus* [13–15], *Cymbopogon nervatus* [5,16], *Cymbopogon schoenanthus* [15,17,18], and *Cymbopogon winterianus* [19,20] were assessed in previous studies.

Cymbopogon spp. EOs present significant inter/intra-species differences in their chemical compositions. An EO of *C. citratus* from China is reported to be rich in citronellal (38.16%) [7] and geraniol (25.19%) [8], whereas cis-citral (Italy, 59.19%) [21], geranial (Brazil, 50.18%) [6] and neral (Benin, 24.6%) [22] were the major components in *C. citratus* oil in other countries. Similarly, citronellal (22.15–41.7%) [14,15,22,23] and citral (38.75%) [13] are reported as major components in *C. nardus* oil. Piperitone (59.2–71.5%) [15,17], α-eudesmol (17.89%) [18], and cis-p-Menth-2-en-1-ol (28.5%) [24] were the major components in *C. schoenanthus* oil. The major components of *C. flexuosus* oil were neral (30.4%) [10] and geranial (38.44%) [25]. A previous study reported that citronellal (24.0–55.4%) [19,26] was the main constituent in *C. winterianus* oil. Trans-p-mentha-2, 8-dien-1-ol (20.70%) [11] and geraniol (76.9%) [12] are reported as major constituents in *C. martinii* oil. Trans-p-mentha-1 (7), 8-dien-2-ol (32.6%) [16] and trans-p-mentha-2, 8-dien-1-ol (13.6%) [5] are reported as major components in *C. nervatus* oil.

According to previous reports, *Cymbopogon* EOs present good insecticidal activity against vector pests, such as *Anopheles funestus* [27], *Aedes aegypti* [28,29], *Aedes albopictus* [30], *Cochliomyia hominivorax* [31], *Culex quinquefasciatus* [32], *Haemaphysalis longicornis* [33], *Musca domestica* [34], *Rhipicephalus microplus* [35], and *Sarcoptes scabiei* [36]; storage pests, including *Acanthoscelides obtectus* [37], *Callosobruchus maculatus* [25,38,39], *Dinoderus porcellus* [22], *Rhyzopertha dominica* [40], *Sitophilus granaries* [41], *Sitophilus oryzae* [42–44], *Sitophilus zeamais* [45,46], *Tenebrio molitor* [47], *Trogoderma granarium* [48], *Tribolium castaneum* [49–53], and *Ulomoides dermestoides* [23,54]; and agricultural insect pests, such as *Bemisia tabaci* [55], *Euprosterna elaeasa* [56], *Megalurothrips sjostedti* [57], *Phthorimaea operculella* [58], *Spodoptera exigua* [59], *Spodoptera frugiperda* [60,61], *Trichoplusia ni* [62,63], and *Tuta absoluta* [64].

However, there exist almost no reports on the vapor toxicity efficacy of *Cymbopogon* spp. EOs against *Reticulitermes flaviceps*. Thus, the objectives of the present study were to: (1) evaluate the constituents of lemongrass EO; (2) evaluate the vapor activity of lemongrass EO and its major constituent against *R. flaviceps* in terms of their behavioral effect on the walking and gripping capability of workers; and (3) investigate the effects of citral on the chemical structure of chitin in *R. flaviceps*.

2. Materials and Methods

2.1. Insects

Two colonies of subterranean termite *R. flaviceps* were collected at Linglong Mountain in Lin'an, Hangzhou, and three colonies of *R. flaviceps* were collected from ZAFU campus, and reared with water and Mason's pine (*Pinus massoniana* L.) in the laboratory. We selected healthy and active workers of uniform size for further experiments.

2.2. Lemongrass EO and the Constituents

Lemongrass EO was obtained from the Moellhausen Flagship Store (Shanghai, China) and its major constituent, citral (96%), was purchased from TCI Shanghai (Shanghai, China) and kept at 4 °C until further use.

2.3. GC-MS Analysis

The lemongrass EO was determined using a gas chromatograph (Agilent 6890A, Santa Clara, CA, USA), equipped with a mass spectrometer detector (Agilent 5975C, Santa Clara, CA, USA). The injector temperature was set at 250 °C, the oven temperature was programmed at 50–250 °C (10 °C/min), the He carrier gas flow was 1.0 mL/min (split ratio of 1:50), and a sample volume of 1.0 μL was injected. Compounds were identified using NIST11.LIB, through comparison of retention indices (RI) with respect to those reported in the Adams [65] library.

2.4. Vapor Toxicity

To conduct fumigations [3], filter paper strips were stuck to the lids of 1 L glass jars and 0.12–0.22 μL of lemongrass EO; its major component citral, or (acetone as a control) was added. Twenty healthy workers were put into a glass bottle, the bottle cap was quickly closed, and a moist filter paper was placed on the bottom of the bottle as food. The experiment was repeated three times with three colonies, and the glass jars were kept at 25 ± 1 °C and 80% RH. A portion of moistened filter paper was placed at the bottom of the bottle for water and food. After 24 h, the number of dead termites was observed and recorded.

2.5. Behavior Effect

2.5.1. Walking Behavior

R. flaviceps workers were anesthetized with carbon dioxide for 5 s, then 1 μL of the lemongrass EO or citral was applied to their antennae. The treated workers were transferred to petri dishes lined with moist filter paper and, after 2 h, we observed their walking ability. Workers who could walk continuously for more than 5 s were considered to have normal walking ability, and the number of workers with normal walking ability was recorded. Acetone treatment was used as control, and three replicates with three colonies of 20 workers were used for each dose.

2.5.2. Gripping Behavior

R. flaviceps workers were anesthetized with carbon dioxide for 5 s, then 1 μL of the lemongrass EO or citral was applied to their antennae. The treated workers were transferred to petri dishes lined with moist filter paper. After 2 h, the petri dish was covered with a piece of filter paper and gently inverted (with the mouth facing downward) for 5 s. Then, the petri dish was gently inverted back up again, and we immediately recorded the number of workers stuck to the paper. Acetone treatment was used as control, and three replicates with three colonies of 20 workers were used for each dose.

2.6. Effect of Chitin Structural Chemistry

2.6.1. Insect Treatment

Fifty healthy and active adult workers were placed in glass vials and treated with a sub-lethal concentration (LC_{20} = 0.16 μL/L) of citral for 6 h, as described in Section 2.4.

2.6.2. Chitin Extraction

Chitin was extracted according to the procedure of Shah et al. [66], including three steps of demineralization, deproteinization, and decolorization. In brief, samples were first demineralized by treatment with 1 M HCl for 20 min at 100 °C. Then, the washed samples were deproteinized with 1 M NaOH solution for 24 h at 80 °C. Finally, the obtained chitin samples were decolorized through incubation in a mixture of chloroform, methanol, and water (1:2:4, *v*/*v*/*v*) for 20 min. The percentage of chitin weight was calculated based on the formula:

$$W = \frac{W2}{W1} \times 100,$$

where W_1 represents the weight of the raw sample and W_2 represents the weight of the chitin.

2.6.3. Fourier Transform Infrared Spectroscopy (FTIR)

Fourier transform infrared spectroscopy (Prestige-21, SHIMADZU, Kyoto, Japan) was conducted to measure absorbance values between 250 and 4000 cm^{-1}. Before measurement, a 1 mg sample of chitin was added to 100 mg of purified potassium bromide (KBr) powder. The effects of citral treatments on the molecular structure and composition of chitin were evaluated by observing the changes in the infrared band. Each experiment was repeated three times.

2.6.4. Thermogravimetric Analysis (TGA)

Citral-treated chitin samples were analyzed using a NETZSCH TG 209 F1 Libra thermal gravimetric analyzer (Selb, Germany). A 10 mg chitin sample was heated from 30 °C to 500 °C in nitrogen at a rate of 10 °C/min. The effect of citral on the thermal stability of the extracted chitin was analyzed by thermogravimetric (TG) and differential thermogravimetric (DTG) analyses.

2.6.5. X-ray Diffraction (XRD)

X-ray diffraction spectra were obtained using an X'Pert-Pro MPD X-ray diffractometer (Almelo, Holland). Data were collected on a copper target at a scan rate of 1°/min with a scan angle of 5–40°. The crystallinity index was calculated using the following equation:

$$C_rI = \frac{I110 - Iam}{I110} \times 100,$$

where I_{110} is the maximum intensity at $2\theta \cong 20°$ and I_{am} is the intensity of amorphous diffraction at $2\theta \cong 16°$.

2.6.6. Differential Scanning Calorimetry (DSC)

DSC analysis of 10 mg control and citral-treated chitin samples was conducted using a NETZSCH DSC 204F1 apparatus (Selb, Germany).

2.7. Statistical Analysis

Toxicity data were subjected to probit analysis in order to estimate the LC_{50} values of lemongrass EO and the major constituent. The mortality rate, chitin content, and walking and gripping ability data were subjected to one-way ANOVA and Tukey's HSD test ($p < 0.05$).

3. Results

3.1. Chemical Composition of Lemongrass EO

The chemical compositions of the lemongrass s EO are shown in Table 1. Nine components comprising 96.84% of the total lemongrass EO composition were identified. cis-citral (36.51%), trans-citral (31.42%), and geraniol (8.78%) were identified as major components of the lemongrass oil.

Table 1. Chemical composition of lemongrass EO.

No.	Constituents	RI [a]	RI [b]	%
1	α-Pinene			1.83
2	β-Pinene	935	932	0.64
3	Limonene	979	977	2.46
4	1,8-Cineole	1029	1025	6.52
5	Linalool	1038	1036	-
6	trans-Citral	1097	1095	31.42
7	Geraniol	1240	1235	8.78
8	Citronellyl formate	1250	1249	-

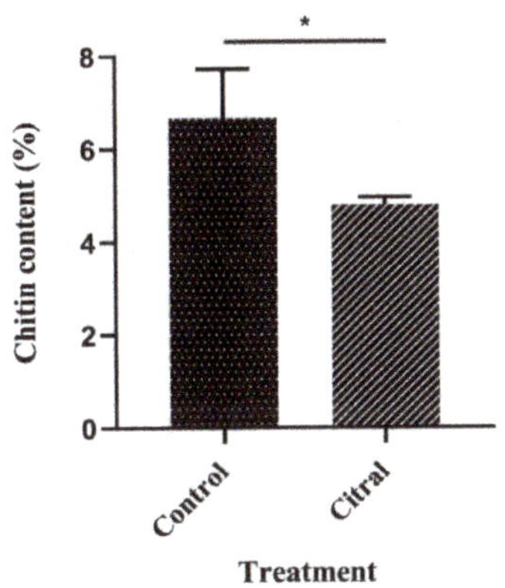

Figure 3. Citral decreases chitin content of *R. flaviceps* workers. Mean (±SD) values with * show significant difference ($p < 0.01$), according to Tukey's HSD test.

3.6. Fourier Transform Infrared Radiation

The FTIR spectra are presented in Figure 4 and Table 3. Significant changes in chitin chemical structure properties were observed in the citral-treated group. The chitin chemical structures of the control and citral-treated *R. flaviceps* workers presented three important amide bands at 1560, 1630, and 1656 cm^{-1}, corresponding to N–H bending and C–N stretching (amide II), C=O secondary amide stretching (amide I) and C=O secondary amide stretching (amide I), respectively.

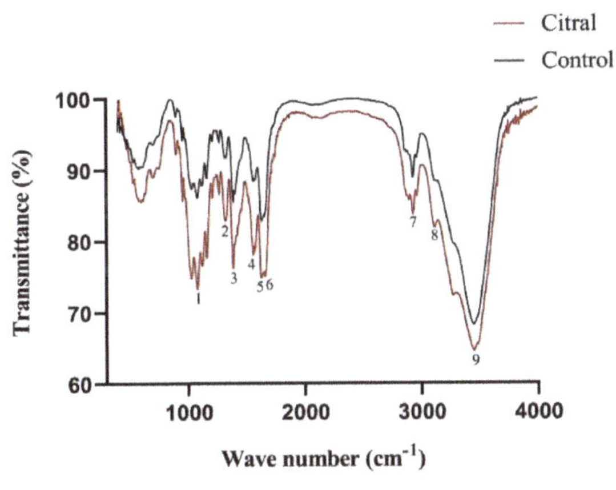

Figure 4. FTIR spectra of α-chitin from *R. flaviceps* in the control and citral-treated groups.

3.7. Thermogravimetric Analysis

The thermal stability and degradability of chitin were analyzed by thermogravimetric analysis (TGA; Figure 5). As shown in Figure 5A, three degradation processes occurred: the first one occurred at 30–110 °C, with a weight loss of 5.5%. The second degradation process occurred at 280–370 °C; the degradation rate in this process was accelerated, and the weight loss was 67.8%. Finally, the third degradation process occurred at 403–500 °C, and the degradation tended to be stable in this process. Figure 5B shows the maximum degradation rates of chitin content by citral. The DTG_{max} of control chitin was at 352.1 °C, while the DTG_{max} of citral-treated chitin was at 351.8 °C, indicating that the thermal stability of chitin was affected by citral treatment.

Table 3. Characteristics and variations in bands in the FTIR spectra of the chitin of *R. flaviceps* treated with citral and control insects.

No.	Wave Number (cm^{-1})	Functional Group and Vibration Modes	Band Assignment	Control	Citral
1	1050	C–O asym. stretch in phase ring	-	88.21	77.54
2	1315	CH_2 wagging	Amide III, components of proteins	91.65	82.77
3	1385	C–H bend, CH_3 sym. Deformation	-	85.77	76.88
4	1560	N–H bend, C–N stretch	Amide II	88.81	79.32
5	1630	C=O secondary amide stretch	Amide I	82.92	75.05
6	1656	C=O secondary amide stretch	Amide I	93.78	75.28
7	2932	CH_3 sym. stretch and CH_2 asym. stretch	Aliphatic compounds	89.53	84.08
8	3111	N–H secondary amine asym. stretch	Amide II	88.46	81.86
9	3440	O–H hydroxyl stretching	-	68.31	64.70

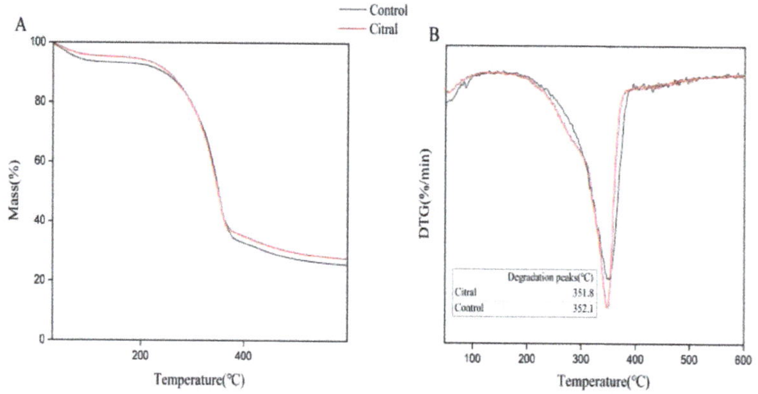

Figure 5. (**A**) TG curves of chitin in the control and citral-treated groups; (**B**) DTG profiles of the chitin of *R. flaviceps* in the control and citral-treated groups.

3.8. X-ray Diffraction

The α-crystalline structure of chitin samples was analyzed by X-ray diffraction. As shown in Figure 6, the crystal reflection peaks were located at 9 and 19°. The chitin crystallinity indices of the control and citral-treated samples were 60.8 and 41.2%, respectively. The results indicate that citral destroyed the chemical structure of chitin to varying degrees.

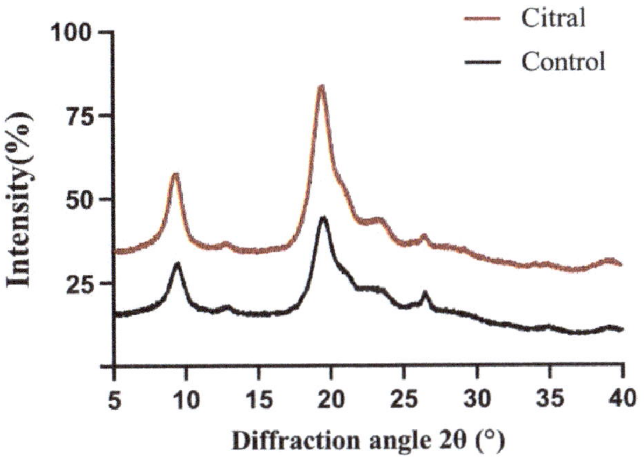

Figure 6. Comparison of X-ray powder diffractograms of the chitin of *R. flaviceps* in the control and citral-treated groups.

3.9. Differential Scanning Calorimetry

The thermal stability of the chitin samples was analyzed by differential scanning calorimetry. As shown in Figure 7, the heat value of the control group was 201.2 J/g, which was significantly higher than that of the treatment group. The maximum decomposition temperature was 109.3 °C in the control group and 108.6 °C in the citral-treated group. These results indicate that treatment with citral changed the thermal stability of chitin, which was consistent with the results of the previous two tests.

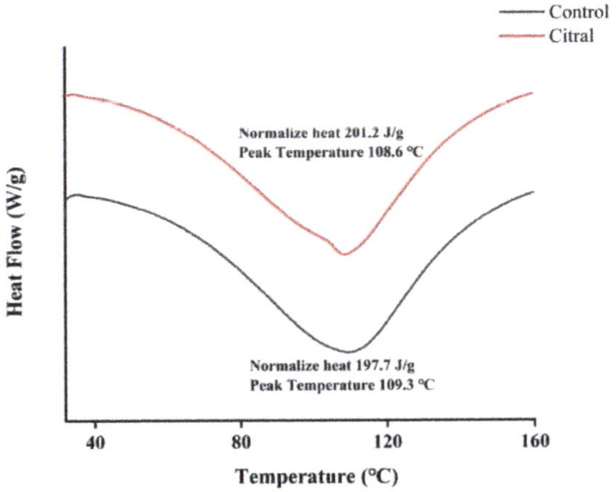

Figure 7. DSC thermogram of the chitin of *R. flaviceps* in the control and citral-treated groups.

4. Discussion

The findings presented here reveal that the main components of lemongrass EO are cis-citral (36.51%), trans-citral (31.42%), and geraniol (8.78%); see Table 1. In general, citral (i.e., a mixture of cis- and trans-citral) was the major component of *C. citratus* EO, in agreement with the results of Boukhatem et al. [67], Pinto et al. [68], Feriotto et al. [21], Brugger et al. [69], Manh et al. [28], Soonwera and Sittichok [30], Aungtikun et al. [34], and Loko et al. [22], with slight differences in the relative content. However, previous studies also show that citronellal [7] and geraniol [8] are the major components of *C. citratus* EO, which is not in accordance with our results (Table S1). Genotypic variations, cultivation techniques, extraction method, and agricultural and environmental conditions can influence the chemical composition of plant EOs of the same species [30,34].

In this study, the results clearly demonstrate that the lemongrass EO had insecticidal efficacy against *R. flaviceps* (Table 2). This is in agreement with Xie et al. [70], who demonstrated the antitermitic activity of *Syzgium aromaticum* EO against *R. chinensis* (LC_{50} = 12.5 μg/g) after 7 d. Studies also reported the effectiveness of *Eugenia caryophyllata* EO against *Coptotermes formosanus* [71], *R. speratus* [72], and *Odontotermes obesus* [73]. Similarly, Pandey et al. [74] also reported the antitermitic activity of *Sy. aromaticum* EO on *O. assamensis*. Yang et al. [3] recently demonstrated that the LC_{50} value of spearmint EO against *R. dabieshanensis* was 0.194 μL/L.

There are no previous studies on the insecticidal activities of lemongrass EO against *R. flaviceps*; however, there are previous reports on the insecticidal potential of *C. citratus* EO. In the previous reports, *C. citratus* EO exhibited insecticidal activities against a variety of pest insects belonging to the orders Hemiptera [55], Coleoptera [75], Lepidoptera [59,60], and Diptera [28,30]. These bioassay results demonstrate that *C. citratus* EO has a significant insecticidal effect, which is worthy of further development in the future.

Our results demonstrate that the lemongrass EO and its major component, citral, had strong vapor activity in *R. flaviceps*. Similar results were obtained for citral (LC_{50} of 0.01 μL/L), in terms of its good termiticidal activity against *R. chinensis* [76]. Additionally, citral is shown to possess strong vapor toxicity against *M. domestica* (LC_{50} = 0.74 μL/L) [77]. Similarly, Lee et al. [78], Palacios et al. [79], and Kumar et al. [80] found that citral presents an insecticidal effect against *M. domestica*. The results of the above studies suggest that citral presents similar trends in toxicity for various insect species.

Interestingly, not only did lemongrass EO and its major component show excellent fumigation activity against *R. flaviceps* workers, but they also significantly affected the walking and gripping ability of workers when applied to their antennae. Similarly, Zhang et al. [81] reported that *Citrus paradisi* EO and its main compounds significantly suppressed the walking and gripping abilities of *Solenopsis invicta* workers. Fu et al. [82] also reported that camphor EO affects attacking, feeding, and climbing behaviors in *S. invicta* workers. Any reduced walking, gripping, and climbing abilities of social insects results in their inability to adapt to the environment. The above conclusions indicate that the vapor activity of *C. citratus* EO and citral may be related to their effects on the social behaviors of the insects.

We observed a significant decrease in chitin content in the citral-treated group. Further, through the FTIR analysis results, it was found that the chitin of *R. flaviceps* exists in the α-form (α-chitin). This is consistent with the previous reports of Zhang et al. [83] and Shah et al. [66], who showed that most insect chitin exists in the α-form. Crystalline α-chitin usually exhibits FTIR bands in the range of 1550, 1620, and 1650 cm^{-1} [66,84]. Based on this, it was found that citral treatment induced changes in the chemical structure of *R. flaviceps* chitin.

In the TGA analysis plot of the extracted chitin structures, mass loss was seen in two steps (Figure 5). In the first step, 8% to 10% weight loss was recorded due to water vaporization [66,85,86]; meanwhile, in the second step, the decomposition could be attributed to the decomposition of chitin saccharide structures [87]. In a previous study, the DTG_{max} decomposition temperature of α-chitin extracted from *Sitotroga cerealella* was 388 °C [66]. In addition, Wang et al. [88] found that the DTG_{max} decomposition temperature of α-chitin extracted from organisms such as crab, shrimp, krill, and crayfish ranged between 350 and 381 °C. Therefore, the DTG_{max} disintegration temperatures of α-chitin structures extracted from different organisms are different. According to Aranaz et al. [89], the decomposition temperature of chitin affects its usefulness. In the present study, the disintegration temperature of chitin extracted from the control and citral-treated groups was similar to that reported in other studies, varying between 385 and 389 °C [66]. TGA showed that treatment with citral slightly reduced the thermal stability of chitin, which may also cause changes in its chemical structure and reduce its crystallinity. These differences may be due to the N-acetylated polymer units of chitin being more stable in the control group than in the treated group. This is consistent with the results of the previous FTIR analysis. Zia et al. [90] reported that chitin has a highly ordered crystal structure. The XRD analysis results in this study were consistent with previous findings [66,91]. As shown in Figure 6, a broad signal centered at $2\theta = 9^{\circ}$ was presented, which was attributed to the GlcNAc sequence [92]. Furthermore, due to the GlcN sequence, the intensity of the broad signal was centered at $2\theta = 19^{\circ}$ [66].

In addition, it is known in the literature that most of the EOs and their major components can exert their toxic efficacy on insects notably through inhibition of P450 cytochromes (CYPs), GABA receptors, octopamine synapses, and the inhibition of acetylcholinesterase (AchE) [3]. Alves et al. [38] found that lemongrass EO and citral showed AChE inhibitory efficacy in *Callosobruchus maculatu*. Decreased AChE activity causes a direct change in insect behavior, such as flight, copulation, and oviposition, as well as changes in many other biological processes of insects [38]. Therefore, the exact toxicity mode of action and target of lemongrass EO and citral against the tested *R. flaviceps* in this study need to be revealed and confirmed by further experiments.

5. Conclusions

In this study, we clearly demonstrated that lemongrass EO and citral show potential vapor toxicity against *R. flaviceps*. The higher vapor toxicity of the lemongrass EO can be ascribed to its major constituent, citral. In addition, the lemongrass EO and citral significantly inhibited the walking and gripping abilities of *R. flaviceps* workers. Analysis results indicate that citral significantly reduced the content of chitin and changed the chemical structure in *R. flaviceps*. As such, we expected that citral has great potential for development into a new type of termite control agent. Future research should include formulation development, in order to determine whether sustained-release formulations can be designed, which may be delivered by termite workers to the nest to achieve the vapor of other individuals in the nest, as well as study the effects of temperature on such a slow-release application.

Supplementary Materials: The following supporting information can be downloaded at: https://www.mdpi.com/article/10.3390/insects13090812/s1, Table S1: Major components of *Cymbopogon citratus* EO from different origins previously reported.

Author Contributions: Conceptualization, Y.X. and C.J.; methodology, C.J., B.L. and D.Z.; software, C.J.; validation, C.J., B.L. and Z.Z.; formal analysis, C.J.; investigation, H.H.; resources, H.H.; data curation, H.H.; writing—original draft preparation, Y.X. and C.J.; writing—review and editing, Y.X., B.L., Z.Z. and D.Z.; visualization, Z.Z. and D.Z.; supervision, Y.X. and D.Z.; project administration, Y.X.; funding acquisition, Y.X. All authors have read and agreed to the published version of the manuscript.

Funding: This research received no external funding.

Institutional Review Board Statement: Not applicable.

Informed Consent Statement: Not applicable.

Data Availability Statement: The dataset utilized in this study is available upon request.

Conflicts of Interest: The authors declare no conflict of interest.

References

1. Ahmad, F.; Fouad, H.; Liang, S.Y.; Hu, Y.; Mo, J.C. Termites and Chinese agricultural system: Applications and advances in integrated termite management and chemical control. *Insect Sci.* **2021**, *28*, 2–20. [CrossRef] [PubMed]
2. Zhou, J.; Bai, Y.; Zhong, H.; Li, G. Effect of nitenpyram on the control of *Reticulitermes flaviceps*. *Int. J. Trop. Insect Sci.* **2021**, *41*, 471–477. [CrossRef]
3. Yang, X.; Han, H.; Li, B.; Zhang, D.; Zhang, Z.; Xie, Y. Fumigant toxicity and physiological effects of spearmint (*Mentha spicata*, Lamiaceae) essential oil and its major constituents against *Reticulitermes dabieshanensis*. *Ind. Crop. Prod.* **2021**, *171*, 113894. [CrossRef]

4. Benelli, G.; Pavela, R.; Petrelli, R.; Cappellacci, L.; Bartolucci, F.; Canale, A.; Maggi, F. *Origanum syriacum* subsp. Syriacum: From an ingredient of Lebanese 'manoushe' to a source of effective and eco-friendly botanical insecticides. *Ind. Crop. Prod.* **2019**, *134*, 26–32. [CrossRef]
5. Yagi, S.; Mohammed, A.B.A.; Tzanova, T.; Schohn, H.; Abdelgadir, H.; Stefanucci, A.; Mollica, A.; Zengin, G. Chemical profile, antiproliferative, antioxidant, and enzyme inhibition activities and docking studies of *Cymbopogon schoenanthus* (L.) Spreng. and *Cymbopogon nervatus* (Hochst.) Chiov. from Sudan. *J. Food Biochem.* **2019**, *44*, e13107. [CrossRef]
6. Barbosa, L.C.A.; Pereira, U.A.; Martinazzo, A.P.; Maltha, C.R.Á.; Teixeira, R.R.; Melo, E.D.C. Evaluation of the chemical composition of Brazilian commercial *Cymbopogon citratus* (DC) Stapf samples. *Molecules* **2008**, *13*, 1864–1874. [CrossRef]
7. Li, C.; Luo, Y.; Zhang, W.; Cai, Q.; Wu, X.; Tan, Z.; Chen, R.; Chen, Z.; Wang, S.; Zhang, L. A comparative study on chemical compositions and biological activities of four essential oils: *Cymbopogon citratus* (DC.) Stapf, *Cinnamomum cassia* (L.) Presl, *Salvia japonica* Thunb. and *Rosa rugosa* Thunb. *J. Ethnopharmacol.* **2021**, *280*, 114472. [CrossRef] [PubMed]
8. Zhang, X.; Zhu, H.; Wang, J.; Li, F.; Wang, J.; Ma, X.; Li, Y.; Huang, Y.; Liu, Z.; Zhang, L.; et al. Anti-microbial activity of citronella (*Cymbopogon citratus*) essential oil separation by ultrasound assisted ohmic heating hydrodistillation. *Ind. Crop. Prod.* **2022**, *176*, 114299. [CrossRef]
9. Padalia, R.C.; Verma, R.C.; Chauhan, A.; Goswami, P.; Singh, V.R.; Verma, S.K.; Singh, N.; Kurmi, A.; Darokar, M.P.; Saikia, D. p-Menthenols chemotype of *Cymbopogon distans* from India: Composition, antibacterial and antifungal activity of the essential oil against pathogens. *J. Essent. Oil Res.* **2018**, *30*, 40–46. [CrossRef]
10. Gao, S.; Liu, G.; Li, J.; Chen, J.; Li, L.; Li, Z.; Zhang, X.; Zhang, S.; Thorne, R.F.; Zhang, S. Antimicrobial activity of Lemongrass essential oil (*Cymbopogon flexuosus*) and its active component citral against dual-species biofilms of *Staphylococcus aureus* and Candida species. *Front. Cell. Infect. Microbiol.* **2020**, *10*, 603858. [CrossRef]
11. Verma, R.S.; Padalia, R.C.; Goswami, P.; Verma, S.K.; Chauhan, A.; Singh, V.R.; Darokar, M.P. Chemical composition and antibacterial activity of p-menthane chemotype of *Cymbopogon martini* (Roxb.) W. Watson (Poaceae) from India. *J. Essent. Oil Res.* **2018**, *30*, 182–188. [CrossRef]
12. Nikolić, M.; Marković, T.; Marković, D.; Calhelha, R.C.; Fernandes, Â.; Ferreira, I.C.F.R.; Stojković, D.; Ćirić, A.; Glamočlija, J.; Soković, M. Chemical composition and biological properties of *Pelargonium graveolens*, *Leptospermum petersonii* and *Cymbopogon martini* var. motia essential oils and of *Rosa centifolia* absolute. *J. Serb. Chem. Soc.* **2021**, *86*, 1291–1303. [CrossRef]
13. Kandimalla, R.; Kalita, S.; Choudhury, B.; Dash, S.; Kalita, K.; Kotoky, J. Chemical composition and anti-candidiasis mediated wound healing property of *Cymbopogon nardus* essential oil on chronic diabetic wounds. *Front. Pharmacol.* **2016**, *7*, 198. [CrossRef] [PubMed]
14. Kaur, H.; Bhardwaj, U.; Kaur, R.; Kaur, H. Chemical composition and antifungal potential of citronella (*Cymbopogon nardus*) leaves essential oil and its major compounds. *J. Essent. Oil Bear. Plants* **2021**, *24*, 571–581. [CrossRef]
15. Sawadogo, I.; Paré, A.; Kaboré, D.; Montet, D.; Durand, N.; Bouajila, J.; Zida, E.P.; Sawadogo-Lingani, H.; Nikiéma, P.A.; Nebié, R.H.C.; et al. Antifungal and antiaflatoxinogenic effects of *Cymbopogon citratus*, *Cymbopogon nardus*, and *Cymbopogon schoenanthus* essential oils alone and in combination. *J. Fungi* **2022**, *8*, 117. [CrossRef]
16. Omar, E.; Pavlović, I.; Drobac, M.; Radenković, M.; Branković, S.; Kovačević, N. Chemical composition and spasmolytic activity of *Cymbopogon nervatus* (Hochst.) Chiov. (Poaceae) essential oil. *Ind. Crop. Prod.* **2016**, *91*, 249–254. [CrossRef]
17. Pavlović, I.; Omar, E.; Drobac, M.; Radenković, M.; Branković, S.; Kovačević, N. Chemical composition and spasmolytic activity of *Cymbopogon schoenanthus* (L.) Spreng. (Poaceae) essential oil from Sudan. *Arch. Biol. Sci.* **2017**, *69*, 409–415. [CrossRef]
18. Bellik, F.Z.; Benkaci-Ali, F.; Alsafra, Z.; Eppe, G.; Tata, S.; Sabaou, N.; Zidani, R. Chemical composition, kinetic study and antimicrobial activity of essential oils from *Cymbopogon schoenanthus* L. Spreng extracted by conventional and microwave-assisted techniques using cryogenic grinding. *Ind. Crop. Prod.* **2019**, *139*, 111505. [CrossRef]
19. Verma, R.S.; Verma, S.K.; Tandon, S.; Padalia, R.C.; Darokar, M.P. Chemical composition and antimicrobial activity of Java citronella (*Cymbopogon winterianus* Jowitt ex Bor) essential oil extracted by different methods. *J. Essent. Oil Res.* **2020**, *32*, 449–455. [CrossRef]
20. Kumar, A.; Jnanesha, A.C.; Chanotiya, C.S.; Srivastava, S.; Pant, Y. Biplot investigation for essential oil yield and chemical compositions under the Deccan Plateau region of southern India in cultivars of Java citronella (*Cymbopogon winterianus* Jowitt). *Ind. Crop. Prod.* **2022**, *175*, 114249. [CrossRef]
21. Feriotto, G.; Marchetti, N.; Costa, V.; Beninati, S.; Tagliati, F.; Mischiati, C. Chemical composition of essential oils from *Thymus vulgaris*, *Cymbopogon citratus*, and *Rosmarinus officinalis*, and their effects on the HIV-1 tat protein function. *Chem. Biodivers.* **2018**, *15*, e1700436. [CrossRef]
22. Loko, Y.L.E.; Fagla, S.M.; Kassa, P.; Ahouansou, C.A.; Toffa, J.; Glinma, B.; Dougnon, V.; Koukoui, O.; Djogbenou, S.L.; Tamò, M.; et al. Bioactivity of essential oils of *Cymbopogon citratus* (DC) Stapf and *Cymbopogon nardus* (L.) W. Watson from Benin against *Dinoderus porcellus* Lesne (Coleoptera: Bostrichidae) infesting yam chips. *Int. J. Trop. Insect Sci.* **2021**, *41*, 511–524. [CrossRef]
23. Caballero-Gallardo, K.; Rodriguez-Niño, D.; Fuentes-Lopez, K.; Stashenko, E.; Olivero-Verbel, J. Chemical composition and bioactivity of essential oils from *Cymbopogon nardus* L. and *Rosmarinus officinalis* L. against *Ulomoides dermestoides* (Fairmaire, 1893) (Coleoptera: Tenebrionidae). *J. Essent. Oil Bear. Plants* **2021**, *24*, 547–560. [CrossRef]
24. Malti, C.E.W.; Haci, I.A.E.; Hassani, F.; Paoli, M.; Gibernau, M.; Tomi, F.; Casanova, J.; Bekhechi, C. Composition, chemical variability and biological activity of *Cymbopogon schoenanthus* essential oil from Central Algeria. *Chem. Biodivers.* **2020**, *17*, e2000138. [CrossRef] [PubMed]

25. Barbosa, D.R.S.; Santos, R.B.V.; Santos, F.M.P.; Junior, P.J.S.; Neto, F.M.O.; Silva, G.N.; Dutra, K.A.; Navarro, D.M.A.F. Evaluation of *Cymbopogon flexuosus* and *Alpinia zerumbet* essential oils as biopesticides against *Callosobruchus maculate*. *J. Plant Dis. Prot.* **2022**, *129*, 125–136. [CrossRef]
26. Piasecki, B.; Biernasiuk, A.; Skiba, A.; Skalicka-Woźniak, K.; Ludwiczuk, A. Composition, Anti-MRSA activity and toxicity of essential oils from *Cymbopogon* Species. *Molecules* **2021**, *26*, 7542. [CrossRef] [PubMed]
27. Ntonga, P.A.; Baldovini, N.; Mouray, E.; Mambu, L.; Belong, P.; Grellier, P. Activity of *Ocimum basilicum*, *Ocimum canum*, and *Cymbopogon citratus* essential oils against *Plasmodium falciparum* and mature-stage larvae of *Anopheles funestus* s.s. *Parasite* **2014**, *21*, 33. [CrossRef]
28. Manh, H.D.; Hue, D.T.; Hieu, N.T.T.; Tuyen, D.T.T.; Tuyet, O.T. The mosquito larvicidal activity of essential oils from *Cymbopogon* and *Eucalyptus* species in Vietnam. *Insects* **2020**, *11*, 128. [CrossRef]
29. Castillo-Morales, R.M.; Serrano, S.O.; Villamizar, A.L.R.; Mendez-Sanchez, S.C.; Duque, J.E. Impact of *Cymbopogon flexuosus* (Poaceae) essential oil and primary components on the eclosion and larval development of *Aedes aegypti*. *Sci. Rep.* **2021**, *11*, 24291. [CrossRef] [PubMed]
30. Soonwera, M.; Sittichok, S. Adulticidal activities of *Cymbopogon citratus* (Stapf.) and *Eucalyptus globulus* (Labill.) essential oils and of their synergistic combinations against *Aedes aegypti* (L.), *Aedes albopictus* (Skuse), and *Musca domestica* (L.). *Environ. Sci. Pollut. Res.* **2020**, *27*, 20201–20214. [CrossRef] [PubMed]
31. Bricarello, P.A.; Barros, G.P.; Seugling, J.; Podestá, R.; Velerinho, M.B.; Mazzarino, L. Ovicidal, larvicidal and oviposition repelling action of a nanoemulsion of citronella essential oil (*Cymbopogon winterianus*) on *Cochliomyia hominivorax* (Diptera: Calliphoridae). *J. Asia-Pac. Èntomol.* **2021**, *24*, 724–730. [CrossRef]
32. Ilahi, I.; Yousafzai, A.M.; Hao, T.U.; Ali, H.; Rahim, A.; Sajad, M.A.; Khan, A.N.; Ahmad, A.; Ullah, S.; Zaman, S.; et al. Oviposition deterrence and adult emergence inhibition activities of *Cymbopogon nardus* against *Culex quinquefasciatus* with study on non-target organisms. *Appl. Ecol. Environ. Res.* **2019**, *17*, 4915–4931. [CrossRef]
33. Agwunobi, D.O.; Pei, T.; Wang, K.; Yu, Z.; Liu, J. Effects of the essential oil from *Cymbopogon citratus* on mortality and morphology of the tick *Haemaphysalis longicornis* (Acari: Ixodidae). *Exp. Appl. Acarol.* **2020**, *81*, 37–50. [CrossRef] [PubMed]
34. Aungtikun, J.; Soonwera, M.; Sittichok, S. Insecticidal synergy of essential oils from *Cymbopogon citratus* (Stapf.), *Myristica fragrans* (Houtt.), and *Illicium verum* Hook. f. and their major active constituents. *Ind. Crop. Prod.* **2021**, *164*, 113386. [CrossRef]
35. Silva, L.C.; Perinotto, W.M.S.; Sá, A.; Souza, M.A.A.; Bitencourt, R.O.B.; Sanavria, A.; Santos, H.A.; Marie-Magdeleine, C.; Angelo, I.C. In vitro acaricidal activity of *Cymbopogon citratus*, *Cymbopogon nardus* and *Mentha arvensis* against *Rhipicephalus microplus* (Acari: Ixodidae). *Exp. Parasitol.* **2020**, *216*, 107937. [CrossRef] [PubMed]
36. Li, M.; Liu, B.; Bernigaud, C.; Fischer, K.; Guillot, J.; Fang, F. Lemongrass (*Cymbopogon citratus*) oil: A promising miticidal and ovicidal agent against *Sarcoptes scabiei*. *PLoS Negl Trop Dis* **2020**, *14*, e0008225. [CrossRef] [PubMed]
37. Rodríguez-González, Á.; Álvarez-García, S.; González-López, Ó.; Silva, F.D.; Casquero, P.A. Insecticidal properties of *Ocimum basilicum* and *Cymbopogon winterianus* against *Acanthoscelides obtectus*, insect pest of the common Bean (*Phaseolus vulgaris*, L.). *Insects* **2019**, *10*, 151. [CrossRef] [PubMed]
38. Alves, M.S.; Campos, I.M.; Brito, D.M.C.; Cardoso, C.M.; Pontes, E.G.; Souza, M.A.A. Efficacy of lemongrass essential oil and citral in controlling *Callosobruchus maculatus* (Coleoptera: Chrysomelidae) a post-harvest cowpea insect pest. *Crop Prot.* **2019**, *119*, 191–196. [CrossRef]
39. Aous, W.; Benchabane, O.; Outaleb, T.; Hazzit, M.; Mouhouche, F.; Yekkour, A.; Baaliouamer, A. Essential oils of *Cymbopogon schoenanthus* (L.) Spreng from Algerian Sahara: Chemical variability, antioxidant, antimicrobial and insecticidal properties. *J. Essent. Oil Res.* **2019**, *31*, 562–572. [CrossRef]
40. Utono, I.M.; Coote, C.; Gibson, G. Field study of the repellent activity of 'Lem-ocimum'-treated double bags against the insect pests of stored sorghum, *Tribolium castaneum* and *Rhyzopertha dominica*, in northern Nigeria. *J. Stored Prod. Res.* **2014**, *59*, 222–230. [CrossRef]
41. Moutassem, D.; Bellik, Y.; Sannef, M.E.H. Toxicity and repellent activities of Thymus pallescens and Cymbopogon citratus essential oils against *Sitophilus granaries*. *Plant Prot. Sci.* **2021**, *57*, 297–309. [CrossRef]
42. Franz, A.R.; Knaak, N.; Fiuza, L.M. Toxic effects of essential plant oils in adult *Sitophilus oryzae* (Linnaeus) (Coleoptera: Curculionidae). *Rev. Bras. Èntomol.* **2011**, *55*, 116–120. [CrossRef]
43. Devi, M.A.; Nameirakpam, B.; Devi, T.B.; Mayanglambam, S.; Singh, K.D.; Sougrakpam, S.S.; Shadia, S.; Tongbram, M.; Singh, S.D.; Sahoo, D.; et al. Chemical compositions and insecticidal efficacies of four aromatic essential oils on rice weevil *Sitophilus oryzae* L. *Int. J. Trop. Insect Sci.* **2020**, *40*, 549–559. [CrossRef]
44. Tawfeek, M.E.; Ali, H.M.; Akrami, M.; Salem, M.Z.M. Potential insecticidal activity of four essential oils against the rice weevil, *Sitophilus oryzae* (L.) (Coleoptera: Curculionidae). *BioResources* **2021**, *16*, 7767–7783. [CrossRef]
45. Hernández-Lambraño, R.; Pajaro-Castro, N.; Caballero-Gallardo, K.; Stashenko, E.; Olivero-Verbel, J. Essential oils from plants of the genus *Cymbopogon* as natural insecticides to control stored product pests. *J. Stored Prod. Res.* **2015**, *62*, 81–83. [CrossRef]
46. Araújo, A.M.N.; Oliveira, J.V.; França, S.M.; Navarro, D.N.A.F.; Barbosa, D.R.S.; Dutra, K.A. Toxicity and repellency of essential oils in the management of *Sitophilus zeamais*. *Rev. Bras. Eng. Agríc. Ambient.* **2019**, *23*, 372–377. [CrossRef]
47. Wang, X.; Hao, Q.; Chen, Y.; Jiang, S.; Yang, Q.; Li, Q. The effect of chemical composition and bioactivity of several essential oils on *Tenebrio molitor* (Coleoptera: Tenebrionidae). *J. Insect Sci.* **2015**, *15*, 116. [CrossRef] [PubMed]

48. Feroz, A. Efficacy and cytotoxic potential of deltamethrin, essential oils of *Cymbopogon citratus* and *Cinnamonum camphora* and their synergistic combinations against stored product pest, *Trogoderma granarium* (Everts). *J. Stored Prod. Res.* **2020**, *87*, 101614. [CrossRef]
49. Olivero-Verbel, J.; Nerio, L.S.; Stashenko, E.E. Bioactivity against *Tribolium castaneum* Herbst (Coleoptera: Tenebrionidae) of *Cymbopogon citratus* and *Eucalyptus citriodora* essential oils grown in Colombia. *Pest Manag. Sci.* **2010**, *66*, 664–668. [CrossRef]
50. Caballero-Gallardo, K.; Olivero-Verbel, J.; Stashenko, E.E. Repellency and toxicity of essential oils from *Cymbopogon martinii*, *Cymbopogon flexuosus* and *Lippia origanoides* cultivated in Colombia against *Tribolium castaneum*. *J. Stored Prod. Res.* **2012**, *50*, 62–65. [CrossRef]
51. Bossou, A.D.; Ahoussi, E.; Ruysbergh, E.; Adams, A.; Smagghe, G.; De Kimpe, N.; Avlessi, F.; Sohounhloue, D.C.K.; Mangelinckx, S. Characterization of volatile compounds from three *Cymbopogon* species and *Eucalyptus citriodora* from Benin and their insecticidal activities against *Tribolium castaneum*. *Ind. Crop. Prod.* **2015**, *76*, 306–317. [CrossRef]
52. Ahmad, F.; Iqbal, N.; Zaka, S.M.; Qureshi, M.K.; Saeed, Q.; Khan, K.A.; Ghramhe, H.A.; Ansari, M.J.; Jaleel, W.; Aasim, M.; et al. Comparative insecticidal activity of different plant materials from six common plant species against *Tribolium castaneum* (Herbst) (Coleoptera: Tenebrionidae). *Saudi J. Biol. Sci.* **2019**, *26*, 1804–1808. [CrossRef]
53. Devi, M.A.; Sahoo, D.; Singh, T.B.; Rajashekar, Y. Toxicity, repellency and chemical composition of essential oils from *Cymbopogon* species against red flour beetle *Tribolium castaneum* Herbst (Coleoptera: Tenebrionidae). *J. Consum. Prot. Food Saf.* **2020**, *15*, 181–191. [CrossRef]
54. Plata-Rueda, A.; Martínez, L.C.; Rolim, G.S.; Coelho, R.P.; Santos, M.H.; Tavares, W.S.; Zanuncio, J.C.; Serrão, J.E. Insecticidal and repellent activities of *Cymbopogon citratus* (Poaceae) essential oil and its terpenoids (citral and geranyl acetate) against *Ulomoides dermestoides*. *Crop Prot.* **2020**, *137*, 105299. [CrossRef]
55. Kobenan, K.C.; Bini, K.K.N.; Kouakou, M.; Kouadio, I.S.; Zengin, G.; Ochou, G.E.C.; Boka, N.R.K.; Menozzi, P.; Ochou, O.G.; Dick, A.E. Chemical composition and spectrum of insecticidal activity of the essential oils of *Ocimum gratissimum* L. and *Cymbopogon citratus* Stapf on the main insects of the cotton entomofauna in Côte d'Ivoire. *Chem. Biodivers.* **2021**, *18*, e2100497. [CrossRef]
56. Hernández-Lambraño, R.; Caballero-Gallardo, K.; Olivero-Verbel, J. Toxicity and antifeedant activity of essential oils from three aromatic plants grown in Colombia against *Euprosterna elaeasa* and *Acharia fusca* (Lepidoptera: Limacodidae). *Asian Pac. J. Trop. Biomed.* **2014**, *4*, 695–700. [CrossRef]
57. Diabate, S.; Martin, T.; Murungi, L.K.; Fiaboe, K.K.M.; Subramanian, S.; Wesonga, J.; Deletre, E. Repellent activity of *Cymbopogon citratus* and *Tagetes minuta* and their specific volatiles against *Megalurothrips sjostedti*. *J. Appl. Entomol.* **2019**, *143*, 855–866. [CrossRef]
58. Jovanović, J.; Krnjajić, S.; Ćirković, J.; Radojković, A.; Popović, T.; Branković, G.; Branković, Z. Effect of encapsulated lemongrass (*Cymbopogon citratus* L.) essential oil against potato tuber moth *Phthorimaea operculella*. *Crop Prot.* **2020**, *132*, 105109. [CrossRef]
59. Murcia-Meseguer, A.; Alves, T.J.; Budia, F.; Ortiz, A.; Medina, P. Insecticidal toxicity of thirteen commercial plant essential oils against *Spodoptera exigua* (Lepidoptera: Noctuidae). *Phytoparasitica* **2018**, *46*, 233–245. [CrossRef]
60. Labinas, M.A.; Crocomo, W.B. Effect of java grass (*Cymbopogon winteranus*) essential oil on fall armyworm *Spodoptera frugiperda* (J. E. Smith, 1979) (Lepidoptera, Noctuidae). *Acta Sci.* **2002**, *24*, 1401–1405.
61. Oliveira, E.R.; Alves, D.S.; Carvalho, G.A.; Oliveira, B.M.R.G.; Aazza, S.; Bertolucci, S.K.V. Toxicity of *Cymbopogon flexuosus* essential oil and citral for *Spodoptera frugiperda*. *Ciênc. Agrotec.* **2018**, *42*, 408–419. [CrossRef]
62. Jiang, Z.L.; Akhtar, Y.; Zhang, X.; Bradbury, R.; Isman, M.B. Insecticidal and feeding deterrent activities of essential oils in the cabbage looper, *Trichoplusia ni* (Lepidoptera: Noctuidae). *J. Appl. Entomol.* **2012**, *136*, 191–202. [CrossRef]
63. Tak, J.H.; Jovel, E.; Isman, M.B. Synergistic interactions among the major constituents of lemongrass essential oil against larvae and an ovarian cell line of the cabbage looper, *Trichoplusia ni*. *J. Pest Sci.* **2017**, *90*, 735–744. [CrossRef]
64. Ngongang, M.D.T.; Eke, P.; Sameza, M.L.; Mback, M.N.L.N.; Djiéto-Lordon, C.; Boyom, F.F. Chemical constituents of essential oils from *Thymus vulgaris* and *Cymbopogon citratus* and their insecticidal potential against the tomato borer, *Tuta absoluta* (Lepidoptera: Gelechiidae). *Int. J. Trop. Insect Sci.* **2022**, *42*, 31–43. [CrossRef]
65. Adams, R.P. *Identification of Essential Oil Components by Gas Chromatography/Mass Spectrometry*, 4th ed.; Allured Publishing Corporation: Carol Stream, IL, USA, 2007.
66. Shah, S.; Ma, M.; Ali, A.; Kaya, M.; Li, X.G.; Wu, G.; Yang, F.L. Effects of diallyl trisulfide, an active substance from garlic essential oil, on structural chemistry of chitin in *Sitotroga cerealella* (Lepidoptera: Gelechiidae). *Pestic. Biochem. Physiol.* **2021**, *172*, 104765. [CrossRef] [PubMed]
67. Boukhatem, M.N.; Ferhat, M.A.; Kameli, A.; Saidi, F.; Kebir, H.T. Lemon grass (*Cymbopogon citratus*) essential oil as a potent anti-inflammatory and antifungal drugs. *Libyan J. Med.* **2014**, *9*, 25431. [CrossRef] [PubMed]
68. Pinto, Z.T.; Sánchez, F.F.; dos Santos, A.R.; Amaral, A.C.; Ferreira, J.L.; Escalona-Arranz, J.C.; Queiroz, M.M. Chemical composition and insecticidal activity of *Cymbopogon citratus* essential oil from Cuba and Brazil against housefly. *Rev. Bras. Parasitol. Vet.* **2015**, *24*, 36–44. [CrossRef]
69. Brugger, B.P.; Martinez, L.C.; Plata-Rueda, A.; Castrol, B.M.C.; Soares, M.A.; Wilcken, C.F.; Carvalho, A.G.; Serao, J.E.; Zanuncio, J.C. Bioactivity of the *Cymbopogon citratus* (Poaceae) essential oil and its terpenoid constituents on the predatory bug, *Podisus nigrispinus* (Heteroptera: Pentatomidae). *Sci. Rep.* **2019**, *9*, 8358. [CrossRef] [PubMed]
70. Xie, Y.; Yang, Z.; Cao, D.; Rong, F.; Ding, H.; Zhang, D. Antitermitic and antifungal activities of eugenol and its congeners from the flower buds of *Syzgium aromaticum* (Clove). *Ind. Crop. Prod.* **2015**, *77*, 780–786. [CrossRef]

71. Zhu, B.C.; Henderson, G.; Chen, F.; Fei, H.; Laine, R.A. Evaluation of vetiver oil and seven insect-active essential oils against the Formosan subterranean termite. *J. Chem. Ecol.* **2001**, *27*, 1617–1625. [CrossRef] [PubMed]
72. Park, I.K.; Shin, S.C. Fumigant activity of plant essential oils and components from garlic (*Allium sativum*) and clove bud (*Eugenia caryophyllata*) oils against the Japanese termite (*Reticulitermes speratus* Kolbe). *J. Agric. Food Chem.* **2005**, *53*, 4388–4392. [CrossRef] [PubMed]
73. Gupta, A.; Sharma, S.; Naik, S.N. Biopesticidal value of selected essential oils against pathogenic fungus, termites, and nematodes. *Int. Biodeterior. Biodegrad.* **2011**, *65*, 703–707. [CrossRef]
74. Pandey, A.; Chattopadhyay, P.; Banerjee, S.; Pakshirajan, K.; Singh, L. Antitermitic activity of plant essential oils and their major constituents against termite *Odontotermes assamensis* Holmgren (Isoptera: Termitidae) of North East India. *Int. Biodeterior. Biodegrad.* **2012**, *75*, 63–67. [CrossRef]
75. Kabera, J.; Gasogo, A.; Uwamariya, A.; Ugirinshuti, V.; Nyetera, P. Insecticidal effects of essential oils of *Pelargonium raveolens* and *Cymbopogon citratus* on *Sitophilus zeamais* (Motsch). *Afr. J. Food Sci.* **2011**, *5*, 366–375. [CrossRef]
76. Xie, Y.; Wang, K.; Huang, Q.; Lei, C. Evaluation toxicity of monoterpenes to subterranean termite, *Reticulitermes chinensis* Snyder. *Ind. Crop. Prod.* **2014**, *53*, 163–166. [CrossRef]
77. Zhang, Z.; Xie, Y.; Wang, Y.; Lin, Z.; Wang, L.; Li, G. Toxicities of monoterpenes against housefly, *Musca domestica* L. (Diptera: Muscidae). *Environ. Sci. Pollut. Res.* **2017**, *24*, 24708–24713. [CrossRef] [PubMed]
78. Lee, S.; Peterson, C.J.; Coats, J.R. Fumigation toxicity of monoterpenoids to several stored product insects. *J. Stored Prod. Res.* **2003**, *39*, 77–85. [CrossRef]
79. Palacios, S.M.; Bertoni, A.; Rossi, Y.; Santander, R. Insecticidal activity of essential oils from native medicinal plants of Central Argentina against the housefly, *Musca domestica* (L.). *Parasitol. Res.* **2009**, *106*, 207–212. [CrossRef]
80. Kumar, P.; Mishra, S.; Malik, A.; Satya, S. Biocontrol potential of essential oil monoterpenes against *Musca domestica* (Diptera Muscidae). *Ecotoxicol. Environ. Saf.* **2014**, *100*, 1–6. [CrossRef]
81. Zhang, N.; Liao, Y.; Xie, L.; Zhang, Z.; Hu, W. Using essential oils from *Citrus paradisi* as a fumigant for *Solenopsis invicta* workers and evaluating the oils' effect on worker behavior. *Environ. Sci. Pollut. Res.* **2021**, *28*, 59665–59672. [CrossRef] [PubMed]
82. Fu, J.T.; Tang, L.; Li, W.S.; Wang, K.; Cheng, D.M.; Zhang, Z.X. Fumigant toxicity and repellence activity of camphor essential oil from *Cinnamonum camphora* Siebold against *Solenopsis invicta* workers (Hymenoptera: Formicidae). *J. Insect Sci.* **2015**, *15*, 129. [CrossRef]
83. Zhang, M.; Haga, A.; Sekiguchi, H.; Hirano, S. Structure of insect chitin isolated from beetle larva cuticle and silkworm (*Bombyx mori*) pupa exuvia. *Int. J. Biol. Macromol.* **2000**, *27*, 99–105. [CrossRef]
84. Waśko, A.; Bulak, P.; Polak-Berecka, M.; Nowak, K.; Polakowski, C.; Bieganowski, A. The first report of the physicochemical structure of chitin isolated from *Hermetia illucens*. *Int. J. Biol. Macromol.* **2016**, *92*, 316–320. [CrossRef]
85. Kaya, M.; Bitim, B.; Mujtaba, M.; Koyuncu, T. Surface morphology of chitin highly related with the isolated body part of butterfly (*Argynnis pandora*). *Int. J. Biol. Macromol.* **2015**, *81*, 443–449. [CrossRef] [PubMed]
86. Kaya, M.; Mujtaba, M.; Ehrlich, H.; Salaberria, A.M.; Baran, T.; Amemiya, C.T.; Galli, R.; Akyuz, L.; Sargin, I.; Labidi, J. On chemistry of γ-chitin. *Carbohydr. Polym.* **2017**, *176*, 177–186. [CrossRef] [PubMed]
87. Paulino, A.T.; Simionato, J.I.; Garcia, J.C.; Nozaki, J. Characterization of chitosan and chitin produced from silkworm crysalides. *Carbohydr. Polym.* **2006**, *64*, 98–103. [CrossRef]
88. Wang, Y.; Chang, Y.; Yu, L.; Zhang, C.; Xu, X.; Xue, Y.; Li, Z.; Xue, C. Crystalline structure and thermal property characterization of chitin from *Antarctic krill* (*Euphausia superba*). *Carbohydr. Polym.* **2013**, *92*, 90–97. [CrossRef]
89. Aranaz, I.; Mengíbar, M.; Harris, R.; Pãnos, I.; Miralles, B.; Acosta, N.; Galed, G.; Heras, Á. Functional characterization of chitin and chitosan. *Curr. Chem. Biol.* **2009**, *3*, 203–230.
90. Zia, K.M.; Bhatti, I.A.; Barikani, M.; Zuber, M.; Sheikh, M.A. XRD studies of chitin-based polyurethane elastomers. *Int. J. Biol. Macromol.* **2008**, *43*, 136–141. [CrossRef]
91. Wada, M.; Saito, Y. Lateral thermal expansion of chitin crystals. *J. Poly. Sci. Part B Poly. Phys.* **2001**, *39*, 168–174. [CrossRef]
92. Jayakumar, R.; Selvamurugan, N.; Nair, S.V.; Tokura, S.; Tamura, H. Preparative methods of phosphorylated chitin and chitosan—An overview. *Int. J. Biol. Macromol.* **2008**, *43*, 221–225. [CrossRef] [PubMed]

agronomy

MDPI

Article

Design, Synthesis, Mode of Action and Herbicidal Evaluation of Quinazolin-4(3*H*)-one Derivatives Based on Aryloxyphenoxypropionate Motif

Chaochao Wang [1], Ke Chen [2], Na Li [1], Shuyue Fu [1], Pan Li [1], Lusha Ji [1], Guoyun Liu [1], Xuekun Wang [1] and Kang Lei [1,*]

[1] School of Pharmaceutical Sciences, Liaocheng University, Liaocheng 252000, China; kelsey9709@163.com (C.W.); 17863527116@163.com (N.L.); fsyue1213@163.com (S.F.); lipan@lcu.edu.cn (P.L.); jilusha@lcu.edu.cn (L.J.); liuguoyun@lcu.edu.cn (G.L.); wangxuekun@lcu.edu.cn (X.W.)

[2] Medical School, Liaocheng University, Liaocheng 252000, China; chenke7758@163.com

* Correspondence: leikang@lcu.edu.cn

Abstract: To discover new acetyl-CoA carboxylase (ACCase) inhibiting-based herbicides, twenty-nine novel quinazolin-4(3*H*)-one derivatives were designed and synthesized based on the aryloxyphenoxypropionate motif. The bioassay results showed that most of the target compounds showed better pre-emergent herbicidal activity against monocotyledonous weeds in a greenhouse. Especially, when applied at 375 g ha^{-1} under pre-emergence conditions, compound **QPP-7** displayed excellent herbicidal activity against monocotyledonous weeds (i.e., *E. crusgalli, D. sanguinalis, P. alopecuroides, S. viridis, E. indica, A. fatua, E. dahuricu, S. alterniflora*) with inhibition rate >90%, and displayed excellent crop safety to *O. sativa, T. aestivum, G. spp*, and *A. hypogaea*. The study of structure-activity relationship (SAR) revealed that the herbicidal activity of target compounds is strongly influenced by the spatial position of R group and the bulk of R_1 group on quinazolin-4(3*H*)-one, and the (R = 6-F, R_1 = Me) pattern is confirmed as the optimal orientation. Furthermore, the molecular docking study and the good inhibitory activity of **QPP-7** against *E. crusgalli* ACCase enzyme (IC_{50} = 54.65 nM) indicated that it may be a ACCase inhibitor. Taken together, the present work demonstrated that compound **QPP-7** could serve as a potential lead structure for further developing novel ACCase inhibiting-based herbicide.

Keywords: quinazolin-4(3*H*)-one derivatives; synthesis; herbicidal activity; molecular docking; ACCase inhibitor

1. Introduction

Herbicides play an important role in weeds control, protecting crops, and increase yields in agriculture. Among the known herbicides, aryloxyphenoxypropionate (APP) are a class of herbicides that inhibit the synthesis of fatty acids and destroy the membrane structure by inhibiting the activity of acetyl-CoA carboxylase in gramineous plants to achieve herbicidal effects [1–4]. Since the first launch of diclofop-methyl in 1971, many of APP herbicides, such as haloxyfop-P-methyl, fluazifop-P-butyl, fenoxaprop-P-ethyl, quizalofop-P-ethyl, have been reached the marketplace (Figure 1A). However, an inevitable problem associated with long-term irrational use of APP herbicides is the reduced efficacy due to weed resistance [5–10]. To overcome this problem, developing APP herbicides with novel structure or improved herbicidal activity is necessary.

Biologically active natural products (NPs) are often served as the lead structures for novel agrochemical discovery in that their advantages associated with unique mode of action, easy degradation, and good environmental compatibility [11–14]. In addition, many of previous works have shown that the introduction of natural active groups is also an efficient method for agrichemical discovery, and a variety of NP-derived pesticides have been successfully developed and brought to market [15,16]. Quinazolin-4(3*H*)-ones, an important class of *N*-containing heterocyclic compounds based on a benzopyrimidone alkaloid structure, are widely distributed in plants and microorganisms (Figure 1B) [17]. Over the last few decades, natural quinazolin-4(3*H*)-ones have been found to possess a wide range of biological activities such as antifungal, [18] anticancer, [19] antiviral, [20,21] radical-scavenging, [22] antimicrobial, [23] cytotoxicity, [24] and anti-inflammatory, [25] and anti-malaria activities [26]. As such, the quinazolin-4(3*H*)-one skeleton have received considerable attention in recent years and is considered to be a privileged structure for developing drugs and pesticides [27–33]. So far, many of the quinazolin-4(3*H*)-one derivatives have been introduced into the market as drugs or pesticides, such as diproqualone (an anti-rheumatic drug), methaqualone (anti-convulsant drugs), raltitrexed (an anticancer drug), and fluquinconazole (an agricultural fungicide) have been developed to reach the market. Although there are many therapeutic drugs and pesticides based on the quinazolin-4(3*H*)-one skeleton, commercial herbicides based on the quinazolin-4(3*H*)-one skeleton are rarely reported.

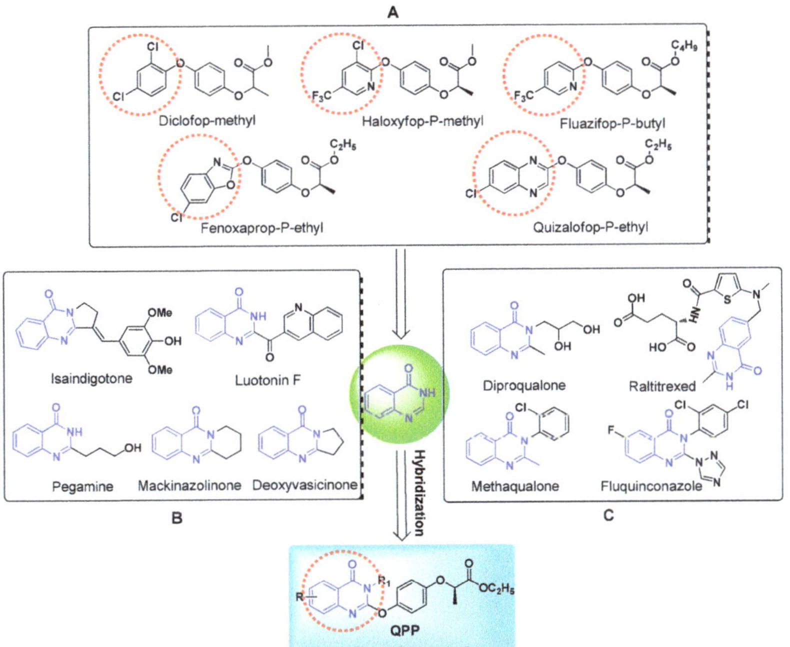

Figure 1. Design of target compound **QPP** by molecular hybridization strategy. (**A**)commercial APP herbicides; (**B**) the natural quinazolin-4(3H)-ones; (**C**) commercial drugs containing quinazolin-4(3H)-one motif.

Based on the above facts, in order to develop novel APP herbicides containing quinazolin-4(3*H*)-one skeleton with commercial potential, we intend to replace the aromatic ring part of the APP herbicides with quinazolin-4-one motif to construct quinazolinone-APP hybrids (Figure 1), which is expected to possess good herbicidal activity. Therefore, as part of our continuous efforts to develop novel structures with potential use as herbicides, [34–37] twenty-nine novel quinazolin-4-one derivatives based on the APP motif were designed, synthesized and tested for herbicidal activity and molecular mode action. To the best of our knowledge, this is the first report on the herbicidal activity of quinazolin-4(3*H*)-one derivatives with an APP motif.

2. Materials and Methods

2.1. General Information

In most cases, the reagents and solvents, purchased form Energy Chemical or Tokyo Chemical Industry, were analytical grade and used without further purification. Column chromatography purification was carried out using silica gel column chromatography (silica gel 200–300 mesh) (Qingdao Makall Group Co., Ltd., Qingdao, China). ^{1}H and ^{13}C NMR spectrum are obtained at 500 MHz and 125 MHz, respectively, using an AV-500 spectrometer (Bruker, Billerica, MA, USA) in $CDCl_3$ or DMSO-d_6 solution with tetramethylsilane (TMS) as the internal standard. The chemical shifts are reported as δ values relative to TMS. High-resolution mass spectra is conducted using an Ionspec 7.0 T spectrometer (Varian, Palo Alto, CA, USA) by the electrospray ionisation fourier transform ion cyclotron resonance (ESI-FTICR) technique. The crystal structure was determined on a Saturn 724 CCD area-detector diffractometer (Rigaku, Tokyo, Japan). High-performance liquid chromatography (HPLC) data is obtained on a SHIMADZU LC-20AT (Japan).

2.2. Chemical Synthesis Procedures

The synthetic pathway used to prepare the target compounds **QPP-1** to **QPP-29** is outlined in Scheme 1. The yields were not optimized.

Route A

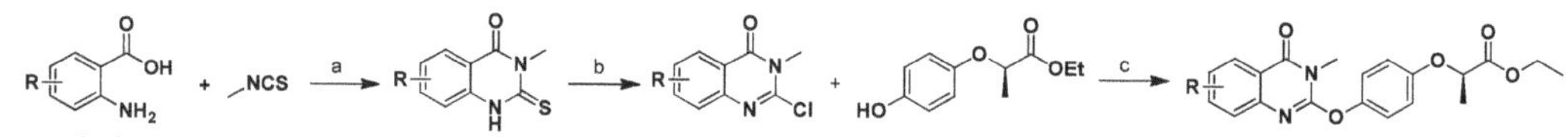

1a-1o **2a-2o** **3a-3o** **QPP-1** to **QPP-15**

1a R= H; **1b** R= 6-Me; **1c** R= 5-Me; **1d** R= 4-Me; **1e** R= 3-Me; **1f** R= 6-F; **1g** R= 5-F; **1h** R= 4-F; **1i** R= 3-F; **1j** R= 6-Cl; **1k** R= 5-Cl; **1l** R= 4-Cl; **1m** R= 3-Cl; **1n** R= 4,5-2Me; **1o** R= 4,5-2F; **2a/3a** R= H; **2b/3b** R= 5-Me; **2c/3c** R= 6-Me; **2d/3d** R= 7-Me; **2e/3e** R= 8-Me; **2f/3f** R= 5-F; **2g/3g** R= 6-F; **2h/3h** R= 7-F; **2i/3i** R= 8-F; **2j/3j** R= 5-Cl; **2k/3k** R= 6-Cl; **2l/3l** R= 7-Cl; **2m/3m** R= 8-Cl; **2n/3n** R= 6,7-2Me; **2o/3o** R= 6,7-2F;**QPP-1** R= H; **QPP-2** R= 5-Me; **QPP-3** R= 6-Me; **QPP-4** R= 7-Me; **QPP-5** R= 8-Me; **QPP-6** R= 5-F; **QPP-7** R= 6-F; **QPP-8** R= 7-F; **QPP-9** R= 8-F; **QPP-10** R= 5-Cl; **QPP-11** R= 6-Cl; **QPP-12** R= 7-Cl; **QPP-13** R= 8-Cl; **QPP-14** R= 6,7-2Me; **QPP-15** R= 6,7-2F.

Route B

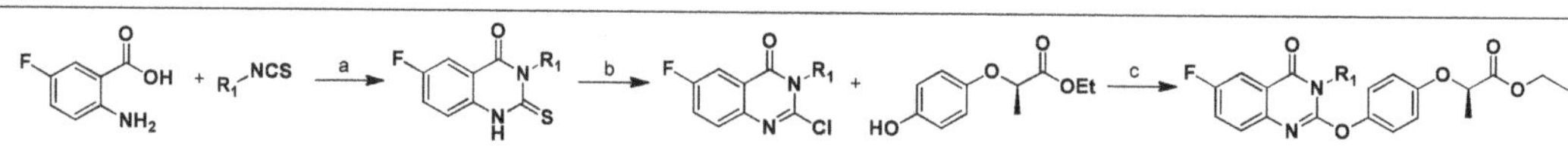

1g **4a-4n** **5a-5n** **6a-6n** **QPP-16** to **QPP-29**

4a/5a/6a R_1= Et; **4b/5b/6b** R_1= *n*-Pr; **4c/5c/6c** R_1= *n*-Butyl; **4d/5d/6d** R_1= *i*-Pr; **4e/5e/6e** R_1= Phenyl; **4f/5f/6f** R_1= 2-Methylphenyl; **4g/5g/6g** R_1= 3-Methylphenyl; **4h/5h/6h** R_1= 4-Methylphenyl; **4i/5i/6i** R_1= 2-Chlorophenyl; **4j/5j/6j** R_1= 3-Chlorophenyl; **4k/5k/6k** R_1= 4-Chlorophenyl; **4l/5l/6l** R_1= 2-Trifluoromethylphenyl; **4m/5m/6m** R_1= 3-Trifluoromethylphenyl; **4n/5n/6n** R_1= 4-Trifluoromethylphenyl; **QPP-16** R_1= Et; **QPP-17** R_1= *n*-Pr; **QPP-18** R_1= *n*-Butyl; **QPP-19** R_1= *i*-Pr; **QPP-20** R_1= Phenyl; **QPP-21** R_1= 2-Methylphenyl; **QPP-22** R_1= 3-Methylphenyl; **QPP-23** R_1= 4-Methylphenyl; **QPP-24** R_1= 2-Chlorophenyl; **QPP-25** R_1= 3-Chlorophenyl; **QPP-26** R_1= 4-Chlorophenyl; **QPP-27** R_1= 2-Trifluoromethylphenyl; **QPP-28** R_1= 3-Trifluoromethylphenyl; **QPP-29** R_1= 4-Trifluoromethylphenyl.

[a] Regeant and conditions: (a) Et_3N, EtOH, 80 °C, 3 h; (b) SO_2Cl_2, $CHCl_3$, 60 °C, 2h; (c) K_2CO_3, CH_3CN, 90 °C, 2h.

Scheme 1. Synthetic route of preparing target compounds **QPP-1** to **QPP-29**.

2.2.1. General Procedure for the Synthesis of Intermediates **2a–2o** and **5a–5n**

Intermediates **2a–2o** and **5a–5n** were prepared following a reported method [38]. To a 100 mL round-bottom flask was added anthranilic acid **1a** (2.74 g, 20.0 mmol), methyl isothiocyanate (1.61 g, 22.0 mmol), Et_3N (2.22 g, 22.0 mmol) and EtOH (30 mL). The reaction mixture was stirred at 80 °C for 3 h. After the reaction cooled to room temperature, the resulting precipitates was filtered, and the solid was washed with 20 mL EtOH/20 mL hexane, and dried to acquire the pure product **2a** as a white solid (3.56 g, yield: 92.7%).

Intermediates **2b–2o** and **5a–5n** were prepared by the similar procedure to **2a**. For data on **2a–2o** and **5a–5n**, see the supporting information.

2.2.2. General Procedure for the Synthesis of Intermediates **3a–3o** and **6a–6n**

Intermediates **3a–3o** and **6a–6n** were prepared following a reported method [38]. To a suspension of compound **2a** (1.92 g, 10.0 mmol) in $CHCl_3$ (25 mL) was added SO_2Cl_2 (1.46 g, 11.0 mmol). The reaction mixture was stirred at 60 °C for 2 h. After the completion of the reaction, the mixture was cooled to room temperature and diluted with CH_2Cl_2 (30 mL). The organic mixture was washed with brine, dried with Na_2SO_4, filtered and concentrated to be purified through chromatograph on silica gel using petroleum ether/ethyl acetate (V:V = 20:1) as eluent to give white solid **3a** (1.17 g, yield: 60.3%).

Intermediates **3b–3o** and **6a–6n** were prepared by the similar procedure to **3a**. For data on **3a–3o** and **6a–6n**, see the supporting information.

2.2.3. General Procedure for the Synthesis of Target Compounds **QPP-1** to **QPP-29**

Compound **3a** (194 mg, 1.0 mmol) was dissolved in 20 mL acetonitrile followed by addition of (R)-ethyl 2-(4-hydroxyphenoxy)propanoate (210 mg, 1.0 mmol) and potassium carbonate (207 mg, 1.5 mmol). The reaction mixture was stirred at 90 °C for 2 h. After the reaction was completed according to TLC detection, the solvent was removed under reduced pressure. The residue was purified through chromatograph on silica gel using petroleum ether/ethyl acetate (V:V = 10:1) as eluent to give target compound **QPP-1** as a white solid (317 mg, yield: 86.2%). Target compounds **QPP-2** to **QPP-29** were prepared by the similar procedure to **QPP-1**.

(R)-ethyl 2-(4-((3-methyl-4-oxo-3,4-dihydroquinazolin-2-yl)oxy)phenoxy)propanoate (**QPP-1**): white solid, yield 86.2%, m.p. 78–80 °C; ^{1}H NMR (500 MHz, $CDCl_3$) δ: 8.21 (dd, J = 8.0, 1.3 Hz, 1H), 7.62–7.58 (m, 1H), 7.38–7.30 (m, 2H), 7.20–7.12 (m, 2H), 6.98–6.88 (m, 2H), 4.76 (q, J = 6.8 Hz, 1H), 4.31–4.20 (m, 2H), 3.70 (s, 3H), 1.65 (d, J = 6.8 Hz, 3H), 1.28 (t, J = 7.1 Hz, 3H); ^{13}C NMR (125 MHz, $CDCl_3$) δ: 172.1, 163.1, 155.4, 152.6, 146.6, 145.9, 134.3, 127.1, 126.0, 124.9, 122.7, 118.9, 115.9, 73.2, 61.4, 28.8, 18.6, 14.2; HRMS, m/z calcd. for $C_{20}H_{21}N_2O_5^+$ $[M + H]^+$ 369.1445, found 369.1452.

(R)-ethyl 2-(4-((3,5-dimethyl-4-oxo-3,4-dihydroquinazolin-2-yl)oxy)phenoxy)propanoate (**QPP-2**): white solid, yield 85.1%, m.p. 101–104 °C; ^{1}H NMR (500 MHz, $CDCl_3$) δ: 7.45–7.38 (m, 1H), 7.19–7.13 (m, 3H), 7.06 (d, J = 7.3 Hz, 1H), 6.96–6.91 (m, 2H), 4.76 (q, J = 6.8 Hz, 1H), 4.30–4.18 (m, 2H), 3.63 (s, 3H), 2.85 (s, 3H), 1.65 (d, J = 6.8 Hz, 3H), 1.28 (t, J = 7.1 Hz, 3H); ^{13}C NMR (126 MHz, $CDCl_3$) δ: 172.1, 163.5, 155.3, 152.2, 148.0, 145.9, 141.2, 133.3, 127.7, 124.2, 122.7, 117.4, 115.9, 73.2, 61.4, 28.6, 22.9, 18.6, 14.2; HRMS, m/z calcd. for $C_{21}H_{23}N_2O_5^+$ $[M + H]^+$ 383.1601, found 383.1612.

(R)-ethyl 2-(4-((3,6-dimethyl-4-oxo-3,4-dihydroquinazolin-2-yl)oxy)phenoxy)propanoate (**QPP-3**): white solid, yield 83.3%, m.p. 91–94 °C; ^{1}H NMR (500 MHz, $CDCl_3$) δ: 7.99 (d, J = 0.8 Hz, 1H), 7.41 (dd, J = 8.3, 2.0 Hz, 1H), 7.26 (d, J = 8.6 Hz, 1H), 7.19–7.13 (m, 2H), 6.99–6.90 (m, 2H), 4.75 (q, J = 6.8 Hz, 1H), 4.27–4.23 (m, 2H), 3.68 (s, 3H), 2.42 (s, 3H), 1.64 (d, J = 6.8 Hz, 3H), 1.28 (t, J = 7.1 Hz, 3H); ^{13}C NMR (125 MHz, $CDCl_3$) δ: 172.1, 163.1, 155.4, 152.1, 145.9, 144.4, 135.8, 134.9, 126.5, 125.8, 122.7, 118.6, 116.7, 116.1, 115.9, 73.2, 61.4, 28.8, 21.1, 18.6, 14.2; HRMS, m/z calcd. for $C_{21}H_{23}N_2O_5^+$ $[M + H]^+$ 383.1601, found 383.1607.

(R)-ethyl 2-(4-((3,7-dimethyl-4-oxo-3,4-dihydroquinazolin-2-yl)oxy)phenoxy)propanoate (**QPP-4**): white solid, yield 81.6%, m.p. 87–90 °C; ^{1}H NMR (500 MHz, $CDCl_3$) δ: 8.09 (d, J = 8.1 Hz, 1H), 7.20–7.09 (m, 4H), 6.98–6.87 (m, 2H), 4.76 (q, J = 6.8 Hz, 1H), 4.25 (qd, J = 7.1, 1.6 Hz, 2H), 3.68 (s, 3H), 2.39 (s, 3H), 1.65 (d, J = 6.8 Hz, 3H), 1.29 (t, J = 7.1 Hz, 3H); ^{13}C NMR (125 MHz, $CDCl_3$) δ: 172.1, 162.9, 155.4, 152.7, 146.7, 145.9, 145.3, 126.9, 126.5, 125.8, 122.7, 116.5, 115.9, 73.2, 61.4, 28.7, 21.8, 18.6, 14.2; HRMS, m/z calcd. for $C_{21}H_{23}N_2O_5^+$ $[M + H]^+$ 383.1601, found 383.1612.

(R)-ethyl 2-(4-((3,8-dimethyl-4-oxo-3,4-dihydroquinazolin-2-yl)oxy)phenoxy)propanoate (**QPP-5**): white solid, yield 80.3%, m.p. 70–72 °C; ^{1}H NMR (500 MHz, $CDCl_3$) δ: 8.05 (d, J = 7.9 Hz, 1H), 7.45 (d, J = 7.2 Hz, 1H), 7.24–7.16 (m, 3H), 7.01–6.90 (m, 2H), 4.77 (q, J = 6.8 Hz, 1H), 4.30–4.14 (m, 2H), 3.69 (s, 3H), 2.24 (s, 3H), 1.65 (d, J = 6.8 Hz, 3H), 1.26 (t, J = 7.1 Hz, 3H); ^{13}C NMR (125 MHz, $CDCl_3$) δ: 172.1, 163.4, 155.2, 151.6, 146.2, 145.1, 134.8,

134.3, 124.7, 124.5, 122.7, 118.7, 115.7, 73.3, 61.4, 28.7, 18.6, 16.7, 14.1; HRMS, m/z calcd. for $C_{21}H_{23}N_2O_5^+$ [M + H]$^+$ 383.1601, found 383.1610.

(R)-ethyl 2-(4-((5-fluoro-3-methyl-4-oxo-3,4-dihydroquinazolin-2-yl)oxy)phenoxy)propanoate (**QPP-6**): white solid, yield 78.0%, m.p. 82–85 °C; ^{1}H NMR (500 MHz, $CDCl_3$) δ: 7.52–7.48 (m, 1H), 7.17–7.11 (m, 3H), 6.99–6.91 (m, 3H), 4.76 (q, *J* = 6.8 Hz, 1H), 4.32–4.18 (m, 2H), 3.66 (s, 3H), 1.65 (d, *J* = 6.8 Hz, 3H), 1.28 (t, *J* = 7.1 Hz, 3H); ^{13}C NMR (125 MHz, $CDCl_3$) δ: 172.0, 161.52 (d, *J* = 265.1 Hz), 159.9 (d, *J* = 4.0 Hz), 155.5, 153.2, 148.6, 145.7, 134.5 (d, *J* = 10.5 Hz), 122.7, 121.9 (d, *J* = 4.0 Hz), 115.9, 111.6 (d, *J* = 20.7 Hz), 73.2, 61.4, 28.5, 18.6, 14.2; HRMS, m/z calcd. for $C_{20}H_{20}FN_2O_5^+$ [M + H]$^+$ 387.1351, found 387.1355.

(R)-ethyl 2-(4-((6-fluoro-3-methyl-4-oxo-3,4-dihydroquinazolin-2-yl)oxy)phenoxy)propanoate (**QPP-7**): white solid, yield 87.5%, m.p. 88–91 °C; ^{1}H NMR (500 MHz, $CDCl_3$) δ: 7.86–7.80 (m, 1H), 7.38–7.28 (m, 2H), 7.19–7.11 (m, 2H), 6.99–6.89 (m, 2H), 4.76 (q, *J* = 6.8 Hz, 1H), 4.28–4.23 (m, 2H), 3.69 (s, 3H), 1.65 (d, *J* = 6.8 Hz, 3H), 1.28 (t, *J* = 7.1 Hz, 3H); ^{13}C NMR (125 MHz, $CDCl_3$) δ: 172.0, 162.4 (d, *J* = 3.6 Hz), 159.7 (d, *J* = 245.5 Hz), 155.5, 152.2 (d, *J* = 1.3 Hz), 145.8, 143.1 (d, *J* = 1.5 Hz), 128.1 (d, *J* = 8.1 Hz), 122.8 (d, *J* = 24.1 Hz), 122.7, 119.8 (d, *J* = 8.5 Hz), 116.7, 116.1, 115.9, 111.9 (d, *J* = 23.7 Hz), 73.2, 61.4, 28.9, 18.6, 14.2; HRMS, m/z calcd. for $C_{20}H_{20}FN_2O_5^+$ [M + H]$^+$ 387.1351, found 387.1533.

(R)-ethyl 2-(4-((7-fluoro-3-methyl-4-oxo-3,4-dihydroquinazolin-2-yl)oxy)phenoxy)propanoate (**QPP-8**): white solid, yield 88.2%, m.p. 61–64 °C; ^{1}H NMR (500 MHz, $CDCl_3$) *δ*: 8.19 (dd, *J* = 8.8, 6.2 Hz, 1H), 7.22–7.11 (m, 2H), 7.05–6.97 (m, 2H), 6.97–6.92 (m, 2H), 4.77 (q, *J* = 6.8 Hz, 1H), 4.25 (qd, *J* = 7.1, 3.2 Hz, 2H), 3.67 (s, 3H), 1.65 (d, *J* = 6.8 Hz, 3H), 1.28 (t, *J* = 7.1 Hz, 3H); ^{13}C NMR (125 MHz, $CDCl_3$) *δ*: 172.0, 166.6 (d, *J* = 253.1 Hz), 162.2, 155.5, 153.5, 148.8 (d, *J* = 13.8 Hz), 145.7, 129.7 (d, *J* = 10.9 Hz), 122.7, 115.9, 115.6, 113.6 (d, *J* = 23.4 Hz), 111.4 (d, *J* = 22.5 Hz), 73.2, 61.4, 28.8, 18.6, 14.2; HRMS, m/z calcd. for $C_{20}H_{20}FN_2O_5^+$ [M + H]$^+$ 387.1351, found 387.1365.

(R)-ethyl 2-(4-((8-fluoro-3-methyl-4-oxo-3,4-dihydroquinazolin-2-yl)oxy)phenoxy)propanoate (**QPP-9**): white solid, yield 80.0%, m.p. 67–70 °C; ^{1}H NMR (500 MHz, $CDCl_3$) δ: 7.99 (d, *J* = 8.0 Hz, 1H), 7.36–7.30 (m, 1H), 7.25 (dd, *J* = 8.0, 3.4 Hz, 1H), 7.23–7.19 (m, 2H), 7.02–6.87 (m, 2H), 4.77 (q, *J* = 6.8 Hz, 1H), 4.30–4.15 (m, 2H), 3.70 (s, 3H), 1.65 (d, *J* = 6.8 Hz, 3H), 1.27 (t, *J* = 7.1 Hz, 3H); ^{13}C NMR (125 MHz, $CDCl_3$) δ: 172.1, 162.2, 162.1, 154.9 (d, *J* = 273.1 Hz), 155.5, 152.8, 145.8, 136.1 (d, *J* = 12.3 Hz), 124.7 (d, *J* = 7.2 Hz), 122.5, 120.9 (d, *J* = 1.6 Hz), 119.8 (d, *J* = 18.5 Hz), 115.9, 73.3, 61.4, 28.9, 18.6, 14.1; HRMS, m/z calcd. for $C_{20}H_{20}FN_2O_5^+$ [M + H]$^+$ 387.1351, found 387.1361.

(R)-ethyl 2-(4-((5-chloro-3-methyl-4-oxo-3,4-dihydroquinazolin-2-yl)oxy)phenoxy)propanoate (**QPP-10**): white solid, yield 71.5%, m.p. 90–92 °C; ^{1}H NMR (500 MHz, $CDCl_3$) δ: 8.12–8.10 (m, 1H), 7.68 (dd, *J* = 7.7, 1.3 Hz, 1H), 7.31–7.26 (m, 2H), 7.25–7.20 (m, 1H), 6.97–6.91 (m, 2H), 4.78 (q, *J* = 6.8 Hz, 1H), 4.28–4.14 (m, 2H), 3.69 (s, 3H), 1.65 (d, *J* = 6.8 Hz, 3H), 1.26 (t, *J* = 7.1 Hz, 3H); ^{13}C NMR (125 MHz, $CDCl_3$) δ: 172.1, 162.6, 155.3, 152.7, 145.9, 143.5, 134.5, 130.4, 125.8, 124.9, 122.5, 120.4, 115.8, 73.2, 61.4, 28.9, 18.6, 14.1; HRMS, m/z calcd. for $C_{20}H_{20}ClN_2O_5^+$ [M + H]$^+$ 403.1055, found 403.1060.

(R)-ethyl 2-(4-((6-chloro-3-methyl-4-oxo-3,4-dihydroquinazolin-2-yl)oxy)phenoxy)propanoate (**QPP-11**): white solid, yield 91.7%, m.p. 100–102 °C; ^{1}H NMR (500 MHz, $CDCl_3$) δ: 8.16 (d, *J* = 2.4 Hz, 1H), 7.52 (dd, *J* = 8.7, 2.5 Hz, 1H), 7.28 (d, *J* = 8.7 Hz, 1H), 7.17–7.12 (m, 2H), 6.97–6.92 (m, 2H), 4.76 (q, *J* = 6.8 Hz, 1H), 4.25 (qd, *J* = 7.1, 2.6 Hz, 2H), 3.69 (s, 3H), 1.65 (d, *J* = 6.8 Hz, 3H), 1.28 (t, *J* = 7.1 Hz, 3H); ^{13}C NMR (125 MHz, $CDCl_3$) δ:172.0, 162.0, 155.5, 152.8, 145.7, 145.2, 134.6, 130.5, 127.6, 126.4, 122.7, 119.9, 115.9, 73.2, 61.4, 28.9, 18.6, 14.2; HRMS, m/z calcd. for $C_{20}H_{20}ClN_2O_5^+$ [M + H]$^+$ 403.1055, found 403.1066.

(R)-ethyl 2-(4-((7-chloro-3-methyl-4-oxo-3,4-dihydroquinazolin-2-yl)oxy)phenoxy)propanoate (**QPP-12**): white solid, yield 80.6%, m.p. 77–80 °C; ^{1}H NMR (500 MHz, $CDCl_3$) *δ*: 7.27 (t, *J* = 8.0 Hz, 1H), 7.17–7.13 (m, 1H), 7.10–7.02 (m, 3H), 6.83 (dd, *J* = 9.8, 2.9 Hz, 2H), 4.65 (q, *J* = 6.8 Hz, 1H), 4.16–4.10 (m, 2H), 3.52 (s, 3H), 1.53 (d, *J* = 6.8 Hz, 3H), 1.16 (t, *J* = 7.2 Hz, 3H); ^{13}C NMR (125 MHz, $CDCl_3$) δ:170.9, 159.8, 154.4, 151.7, 148.1, 144.6, 133.1, 132.4, 126.6, 124.2, 121.7, 114.9, 114.8, 72.1, 60.3, 27.8, 17.5, 13.1; HRMS, m/z calcd. for $C_{20}H_{20}ClN_2O_5^+$ [M + H]$^+$ 403.1055, found 403.1064.

(R)-ethyl 2-(4-((8-chloro-3-methyl-4-oxo-3,4-dihydroquinazolin-2-yl)oxy)phenoxy)propanoate (**QPP-13**): white solid, yield 85.8%, m.p. 97–98 °C; ^{1}H NMR (500 MHz, $CDCl_3$) δ: 7.27 (t, *J* = 8.0 Hz, 1H), 7.17–7.12 (m, 1H), 7.10–7.02 (m, 3H), 6.83 (dd, *J* = 9.8, 2.9 Hz, 2H), 4.65 (q, *J* = 6.8 Hz, 1H), 4.16–4.10 (m, 2H), 3.52 (s, 3H), 1.53 (d, *J* = 6.8 Hz, 3H), 1.16 (t, *J* = 7.2 Hz, 3H); ^{13}C NMR (125 MHz, $CDCl_3$) δ: 170.9, 159.8, 154.4, 151.7, 148.1, 144.6, 133.1, 132.4, 126.6, 124.2, 121.7, 114.9, 114.8, 72.1, 60.3, 27.8, 17.5, 13.1; HRMS, m/z calcd. for $C_{20}H_{20}ClN_2O_5^+$ $[M + H]^+$ 403.1055, found 403.1067.

(R)-ethyl 2-(4-((3,6,7-trimethyl-4-oxo-3,4-dihydroquinazolin-2-yl)oxy)phenoxy)propanoate (**QPP-14**): white solid, yield 89.8%, m.p. 124–125 °C; ^{1}H NMR (500 MHz, $CDCl_3$) δ: 7.93 (s, 1H), 7.19–7.09 (m, 3H), 6.94 (d, *J* = 8.6 Hz, 2H), 4.75 (q, *J* = 6.7 Hz, 1H), 4.31–4.17 (m, 2H), 3.67 (dd, *J* = 3.5, 2.5 Hz, 3H), 2.31 (d, *J* = 17.4 Hz, 3H), 1.65 (d, *J* = 6.8 Hz, 3H), 1.28 (t, *J* = 7.1 Hz, 3H); ^{13}C NMR (125 MHz, $CDCl_3$) δ: 172.1, 162.9, 155.3, 152.3, 146.0, 144.8, 144.5, 134.2, 134.2, 126.8, 126.3, 122.7, 116.6, 115.9, 73.2, 61.4, 28.7, 20.2, 19.5, 18.6, 14.2; HRMS, m/z calcd. for $C_{22}H_{25}N_2O_5^+$ $[M + H]^+$ 397.1758, found 397.1765.

(R)-ethyl 2-(4-((6,7-difluoro-3-methyl-4-oxo-3,4-dihydroquinazolin-2-yl)oxy)phenoxy)-propanoate (**QPP-15**): white solid, yield 84.6%, m.p. 76–79 °C; ^{1}H NMR (500 MHz, $CDCl_3$) δ: 7.81 (dd, *J* = 9.8, 8.9 Hz, 1H), 7.07–7.02 (m, 2H), 7.02–6.97 (m, 1H), 6.88–6.80 (m, 2H), 4.67 (q, *J* = 6.8 Hz, 1H), 4.19–4.12 (m, 2H), 3.57 (s, 3H), 1.55 (d, *J* = 6.8 Hz, 3H), 1.19 (t, *J* = 7.1 Hz, 3H); ^{13}C NMR (125 MHz, $CDCl_3$) δ: 171.9, 161.5 (d, *J* = 2.7 Hz), 155.6, 154.84 (dd, *J* = 256.25 Hz,14.5 Hz), 153.8 (dd, *J* = 246.25 Hz, 14.6 Hz), 153.2, 145.6, 144.29 (dd, *J* = 11.8, 1.6 Hz), 122.6, 115.9, 115.43 (d, *J* = 4.7 Hz), 114.26 (dd, *J* = 19.2, 1.5 Hz), 113.72 (d, *J* = 18.3 Hz), 73.1, 61.4, 28.9, 18.5, 14.1; HRMS, m/z calcd. for $C_{20}H_{19}F_2N_2O_5^+$ $[M + H]^+$ 405.1257, found 405.1266.

(R)-ethyl 2-(4-((3-ethyl-6-fluoro-4-oxo-3,4-dihydroquinazolin-2-yl)oxy)phenoxy)propanoate (**QPP-16**): white solid, yield 91.4%, m.p. 124–127 °C; ^{1}H NMR (500 MHz, $CDCl_3$) δ: 7.90–7.77 (m, 1H), 7.35–7.29 (m, 2H), 7.17–7.12 (m, 2H), 6.97–6.90 (m, 2H), 4.76 (q, *J* = 6.8 Hz, 1H), 4.33 (q, *J* = 7.1 Hz, 2H), 4.29–4.21 (m, 2H), 1.65 (d, *J* = 6.8 Hz, 3H), 1.42 (t, *J* = 7.1 Hz, 3H), 1.28 (t, *J* = 7.1 Hz, 3H); ^{13}C NMR (125 MHz, $CDCl_3$) δ: 172.0, 161.9 (d, *J* = 3.4 Hz), 159.7 (d, *J* = 245.3 Hz), 155.4, 152.1, 145.8, 143.2, 128.1 (d, *J* = 8.0 Hz), 122.8, 122.6, 120.2 (d, *J* = 8.5 Hz), 115.9, 111.9 (d, *J* = 23.5 Hz), 73.2, 61.4, 37.6, 18.6, 14.2, 13.8; HRMS, m/z calcd. for $C_{21}H_{22}FN_2O_5^+$ $[M + H]^+$ 401.1507, found 405.1514.

(R)-ethyl 2-(4-((6-fluoro-4-oxo-3-propyl-3,4-dihydroquinazolin-2-yl)oxy)phenoxy)propanoate (**QPP-17**): white solid, yield 79.7%, m.p. 122–123 °C; ^{1}H NMR (500 MHz, $CDCl_3$) δ: 7.83 (d, *J* = 8.5 Hz, 1H), 7.32 (d, *J* = 5.5 Hz, 2H), 7.14 (d, *J* = 8.9 Hz, 2H), 6.94 (d, *J* = 8.8 Hz, 2H), 4.76 (q, *J* = 6.8 Hz, 1H), 4.28–4.21 (m, 4H), 1.97–1.78 (m, 2H), 1.65 (d, *J* = 6.8 Hz, 3H), 1.29 (t, *J* = 7.1 Hz, 3H), 1.03 (t, *J* = 7.4 Hz, 3H); ^{13}C NMR (125 MHz, $CDCl_3$) δ: 172.0, 162.1 (d, *J* = 3.1 Hz), 159.7 (d, *J* = 245.5 Hz), 155.4, 152.2, 145.9, 143.2, 128.1 (d, *J* = 8.0 Hz), 122.8, 122.6, 120.1 (d, *J* = 8.4 Hz), 115.9, 111.9 (d, *J* = 23.7 Hz), 73.2, 61.4, 43.9, 21.9, 18.6, 14.2, 11.4; HRMS, m/z calcd. for $C_{22}H_{24}FN_2O_5^+$ $[M + H]^+$ 415.1664, found 415.1675.

(R)-ethyl 2-(4-((3-butyl-6-fluoro-4-oxo-3,4-dihydroquinazolin-2-yl)oxy)phenoxy)propanoate (**QPP-18**): white solid, yield 75.1%, m.p. 108–109 °C; ^{1}H NMR (500 MHz, $CDCl_3$) δ: 7.89–7.71 (m, 1H), 7.35–7.31 (m, 2H), 7.16–7.11 (m, 2H), 6.97–6.92 (m, 2H), 4.76 (q, *J* = 6.8 Hz, 1H), 4.34–4.18 (m, 4H), 1.89–1.74 (m, 2H), 1.65 (d, *J* = 6.8 Hz, 3H), 1.55–1.38 (m, 2H), 1.29 (t, *J* = 7.1 Hz, 3H), 0.99 (t, *J* = 7.4 Hz, 3H); ^{13}C NMR (125 MHz, $CDCl_3$) δ: 172.0, 162.1 (d, *J* = 3.3 Hz), 159.7 (d, *J* = 245.3 Hz), 155.4, 152.2, 145.9, 143.2, 128.1 (d, *J* = 7.8 Hz), 122.8, 122.6, 120.1 (d, *J* = 8.4 Hz), 115.9, 111.9 (d, *J* = 23.5 Hz), 73.2, 61.4, 42.3, 30.6, 20.2, 18.6, 14.2, 13.8; HRMS, m/z calcd. for $C_{23}H_{26}FN_2O_5^+$ $[M + H]^+$ 429.1820, found 429.1830.

(R)-ethyl 2-(4-((6-fluoro-3-isobutyl-4-oxo-3,4-dihydroquinazolin-2-yl)oxy)phenoxy)propanoate (**QPP-19**): white solid, yield 85.8%, m.p. 100–103 °C; ^{1}H NMR (500 MHz, $CDCl_3$) δ: 7.85–7.80 (m, 1H), 7.37–7.30 (m, 2H), 7.17–7.10 (m, 2H), 6.98–6.92 (m, 2H), 4.76 (q, *J* = 6.8 Hz, 1H), 4.31–4.20 (m, 2H), 4.10 (d, *J* = 7.5 Hz, 2H), 2.29 (dp, *J* = 14.0, 7.0 Hz, 1H), 1.65 (d, *J* = 6.8 Hz, 3H), 1.29 (t, *J* = 7.1 Hz, 3H), 1.02 (s, 3H), 1.01 (s, 3H); ^{13}C NMR (125 MHz, $CDCl_3$) δ: 172.0, 162.3 (d, *J* = 3.2 Hz), 159.7 (d, *J* = 245.4 Hz), 155.4, 152.4, 145.8, 143.2, 128.1 (d, *J* = 7.7 Hz), 122.8, 122.6 (d, *J* = 2.6 Hz), 120.1 (d, *J* = 8.4 Hz), 115.9, 112.0 (d, *J* = 23.8 Hz), 73.2, 61.4,

49.2, 29.7, 27.8, 20.2, 18.6, 14.2; HRMS, m/z calcd. for $C_{23}H_{26}FN_2O_5^+$ [M + H]$^+$ 429.1820, found 429.1830.

(R)-ethyl 2-(4-((6-fluoro-4-oxo-3-phenyl-3,4-dihydroquinazolin-2-yl)oxy)phenoxy)propanoate (**QPP-20**): white solid, yield 69.7%, m.p. 135–136 °C; ^{1}H NMR (500 MHz, $CDCl_3$) δ: 7.86 (dd, J = 8.4, 2.8 Hz, 1H), 7.54 (t, J = 7.5 Hz, 2H), 7.51–7.46 (m, 1H), 7.43–7.36 (m, 4H), 7.05 (d, J = 9.0 Hz, 2H), 6.87 (d, J = 9.0 Hz, 2H), 4.72 (q, J = 6.8 Hz, 1H), 4.27–4.17 (m, 2H), 1.62 (d, J = 6.8 Hz, 3H), 1.25 (t, J = 7.1 Hz, 3H); ^{13}C NMR (125 MHz, $CDCl_3$) δ: 172.0, 162.2 (d, J = 3.5 Hz), 159.9 (d, J = 246.2 Hz), 155.4, 151.6, 145.8, 143.3, 134.9, 129.5, 129.1, 128.3 (d, J = 7.8 Hz), 128.1, 123.1 (d, J = 24.0 Hz), 122.5, 120.5 (d, J = 8.8 Hz), 115.8, 112.3 (d, J = 23.7 Hz), 73.1, 61.4, 18.6, 14.1; HRMS, m/z calcd. for $C_{25}H_{22}FN_2O_5^+$ [M + H]$^+$ 449.1507, found 449.1516.

(R)-ethyl 2-(4-((6-fluoro-4-oxo-3-(*o*-tolyl)-3,4-dihydroquinazolin-2-yl)oxy)phenoxy)propanoate (**QPP-21**): white solid, yield 69.5%, m.p. 47–49 °C; ^{1}H NMR (500 MHz, $CDCl_3$) δ: 7.88 (dd, J = 8.4, 2.8 Hz, 1H), 7.47–7.33 (m, 5H), 7.25 (t, J = 3.6 Hz, 1H), 7.06–7.01 (m, 2H), 6.90–6.84 (m, 2H), 4.72 (q, J = 6.8 Hz, 1H), 4.22 (q, J = 7.1 Hz, 2H), 2.23 (s, 3H), 1.62 (d, J = 6.8 Hz, 3H), 1.25 (t, J = 7.1, 3H); ^{13}C NMR (125 MHz, $CDCl_3$) δ: 171.9, 161.8 (d, J = 3.5 Hz), 159.9 (d, J = 246.2 Hz), 155.4, 151.7, 145.8, 143.6, 135.5, 134.2, 131.2, 129.5, 127.3, 123.1 (d, J = 24.0 Hz), 122.5, 120.5 (d, J = 8.6 Hz), 115.8, 112.4 (d, J = 23.8 Hz), 73.1, 61.4, 18.6, 17.6, 14.2; HRMS, m/z calcd. for $C_{26}H_{24}FN_2O_5^+$ [M + H]$^+$ 463.1664, found 463.1672.

(R)-ethyl 2-(4-((6-fluoro-4-oxo-3-(*m*-tolyl)-3,4-dihydroquinazolin-2-yl)oxy)phenoxy)propanoate (**QPP-22**): white solid, yield 84.3%, m.p. 119–122 °C; ^{1}H NMR (500 MHz, $CDCl_3$) δ: 7.86 (dd, J = 8.4, 2.8 Hz, 1H), 7.39 (ddt, J = 11.8, 8.9, 5.4 Hz, 3H), 7.29 (d, J = 7.7 Hz, 1H), 7.16 (d, J = 8.5 Hz, 2H), 7.08–7.02 (m, 2H), 6.91–6.85 (m, 2H), 4.72 (q, J = 6.8 Hz, 1H), 4.27–4.13 (m, 2H), 2.43 (s, 3H), 1.62 (d, J = 6.8 Hz, 3H), 1.26 (t, J = 7.1 Hz, 3H); ^{13}C NMR (125 MHz, $CDCl_3$) δ: 172.0, 162.3 (d, J = 3.5 Hz), 159.9 (d, J = 246.0 Hz), 155.4, 151.7, 145.9, 143.4, 139.6, 134.9, 129.9, 129.3, 128.6, 128.3 (d, J = 8.0 Hz), 124.9, 123.1 (d, J = 23.9 Hz), 122.6, 120.6 (d, J = 8.8 Hz), 115.8, 112.3 (d, J = 23.7 Hz), 73.2, 61.4, 29.7, 21.4, 18.6, 14.2; HRMS, m/z calcd. for $C_{26}H_{24}FN_2O_5^+$ [M + H]$^+$ 463.1664, found 463.1669.

(R)-ethyl 2-(4-((6-fluoro-4-oxo-3-(*p*-tolyl)-3,4-dihydroquinazolin-2-yl)oxy)phenoxy)propanoate (**QPP-23**): white solid, yield 68.0%, m.p. 151–152 °C; ^{1}H NMR (500 MHz, $CDCl_3$) δ: 7.86 (dd, J = 8.4, 2.8 Hz, 1H), 7.44–7.32 (m, 4H), 7.24 (d, J = 8.3 Hz, 2H), 7.07–7.02 (m, 2H), 6.90–6.85 (m, 2H), 4.72 (q, J = 6.8 Hz, 1H), 4.25–4.19 (m, 2H), 2.43 (s, 3H), 1.62 (d, J = 6.8 Hz, 3H), 1.25 (t, J = 7.1 Hz, 3H); ^{13}C NMR (125 MHz, $CDCl_3$) δ: 172.0, 162.3 (d, J = 3.5 Hz), 159.9 (d, J = 245.7 Hz), 155.3, 151.8, 145.9, 143.3, 139.1, 132.3, 130.2, 128.3 (d, J = 8.0 Hz), 127.7, 123.1 (d, J = 24.0 Hz), 122.5, 120.6 (d, J = 8.6 Hz), 115.8, 112.4 (d, J = 23.8 Hz), 73.2, 61.4, 21.3, 18.6, 14.2; HRMS, m/z calcd. for $C_{26}H_{24}FN_2O_5^+$ [M + H]$^+$ 463.1664, found 463.1670.

(R)-ethyl 2-(4-((3-(2-chlorophenyl)-6-fluoro-4-oxo-3,4-dihydroquinazolin-2-yl)oxy)phenoxy)propanoate (**QPP-24**): white solid, yield 74.3%, m.p. 51–54 °C; ^{1}H NMR (500 MHz, $CDCl_3$) δ: 7.88 (dd, J = 8.3, 2.8 Hz, 1H), 7.64–7.57 (m, 1H), 7.50–7.36 (m, 5H), 7.10–7.05 (m, 2H), 6.93–6.84 (m, 2H), 4.72 (q, J = 6.8 Hz, 1H), 4.22 (q, J = 7.1 Hz, 2H), 1.62 (d, J = 6.8 Hz, 3H), 1.26 (t, J = 7.1 Hz, 3H); ^{13}C NMR (125 MHz, $CDCl_3$) δ: 171.9, 161.5 (d, J = 3.0 Hz), 159.9 (d, J = 246.3 Hz), 155.5, 151.2, 145.7, 143.5, 132.9, 132.4, 130.5 (d, J = 29.2 Hz), 130.1, 128.4 (d, J = 8.0 Hz), 128.0, 123.3 (d, J = 23.9 Hz), 122.6, 120.3 (d, J = 8.8 Hz), 115.8, 112.5 (d, J = 23.7 Hz), 73.1, 61.4, 18.6, 14.2; HRMS, m/z calcd. for $C_{25}H_{21}ClFN_2O_5^+$ [M + H]$^+$ 483.1118, found 483.1122.

(R)-ethyl 2-(4-((3-(3-chlorophenyl)-6-fluoro-4-oxo-3,4-dihydroquinazolin-2-yl)oxy)phenoxy)propanoate (**QPP-25**): white solid, yield 82.0%, m.p. 113–116 °C; ^{1}H NMR (500 MHz, $CDCl_3$) δ: 7.85 (dd, J = 8.3, 2.6 Hz, 1H), 7.51–7.45 (m, 2H), 7.42–7.38 (m, 3H), 7.29–7.26 (m, 1H), 7.08–7.03 (m, 2H), 6.92–6.85 (m, 2H), 4.73 (q, J = 6.8 Hz, 1H), 4.22 (qd, J = 7.1, 0.7 Hz, 2H), 1.63 (d, J = 6.8 Hz, 3H), 1.26 (t, J = 7.1 Hz, 3H); ^{13}C NMR (125 MHz, $CDCl_3$) δ: 171.9, 161.9 (d, J = 3.5 Hz), 160.0 (d, J = 246.5 Hz), 155.5, 151.1, 145.7, 143.2, 135.9, 135.1, 130.5, 129.5, 128.6, 128.4 (d, J = 8.1 Hz), 126.6, 123.3 (d, J = 24.0 Hz), 122.5, 120.4 (d, J = 8.7 Hz), 115.8, 112.4 (d, J = 23.9 Hz), 73.1, 61.4, 18.6, 14.2; HRMS, m/z calcd. for $C_{25}H_{21}ClFN_2O_5^+$ [M + H]$^+$ 483.1118, found 483.1124.

(R)-ethyl 2-(4-((3-(4-chlorophenyl)-6-fluoro-4-oxo-3,4-dihydroquinazolin-2-yl)oxy)phenoxy)propanoate (**QPP-26**): white solid, yield 78.1%, m.p. 146–148 °C; ^{1}H NMR (500 MHz, $CDCl_3$) δ: 7.85 (dd, J = 8.3, 2.5 Hz, 1H), 7.52 (d, J = 8.4 Hz, 2H), 7.45–7.36 (m, 2H), 7.31 (d, J = 8.5 Hz, 2H), 7.05 (d, J = 8.5 Hz, 2H), 6.88 (d, J = 8.3 Hz, 2H), 4.72 (q, J = 6.7 Hz, 1H), 4.22 (q, J = 7.0 Hz, 2H), 1.63 (d, J = 6.8 Hz, 3H), 1.26 (t, J = 7.1 Hz, 3H); ^{13}C NMR (125 MHz, $CDCl_3$) δ: 171.9, 162.0 (d, J = 3.1 Hz), 160.0 (d, J = 246.2 Hz), 155.5, 151.2, 145.7, 143.2, 135.2, 133.4, 128.4 (d, J = 8.0 Hz), 123.3 (d, J = 24.0 Hz), 122.4, 115.9, 112.4 (d, J = 23.9 Hz), 73.2, 61.4, 18.6, 14.2; HRMS, m/z calcd. for $C_{25}H_{21}ClFN_2O_5^+$ [M + H]$^+$ 483.1118, found 483.1123.

(R)-ethyl 2-(4-((6-fluoro-4-oxo-3-(2-(trifluoromethyl)phenyl)-3,4-dihydroquinazolin-2-yl)oxy)phenoxy)propanoate (**QPP-27**): white solid, yield 96.9%, m.p. 45–48 °C; ^{1}H NMR (500 MHz, $CDCl_3$) δ: 7.77 (d, J = 8.1 Hz, 2H), 7.66 (t, J = 7.7 Hz, 1H), 7.54 (t, J = 7.7 Hz, 1H), 7.38 (d, J = 7.9 Hz, 1H), 7.35–7.26 (m, 2H), 6.93 (d, J = 8.8 Hz, 2H), 6.78 (d, J = 8.8 Hz, 2H), 4.62 (q, J = 6.8 Hz, 1H), 4.12 (q, J = 7.1 Hz, 2H), 1.52 (d, J = 6.8 Hz, 3H), 1.16 (t, J = 7.1 Hz, 3H); ^{13}C NMR (125 MHz, $CDCl_3$) δ: 171.9, 161.9 (d, J = 3.3 Hz), 159.9 (d, J = 246.2 Hz), 155.5, 151.3, 145.6, 143.5, 133.4, 133.2 (d, J = 1.5 Hz), 130.8, 129.9, 128.5 (d, J = 7.9 Hz), 127.9 (q, J = 31.2 Hz), 127.7 (q, J = 4.2 Hz), 123.3 (d, J = 23.9 Hz), 123.1 (q, J = 216.7 Hz), 122.5, 120.1 (d, J = 8.8 Hz), 115.8, 112.4 (d, J = 23.9 Hz), 73.1, 61.4, 18.6, 14.1; HRMS, m/z calcd. for $C_{26}H_{21}F_4N_2O_5^+$ [M + H]$^+$ 517.1381, found 517.1386.

(R)-ethyl 2-(4-((6-fluoro-4-oxo-3-(3-(trifluoromethyl)phenyl)-3,4-dihydroquinazolin-2-yl)oxy)phenoxy)propanoate (**QPP-28**): white solid, yield 85.0%, m.p. 39–42 °C; ^{1}H NMR (500 MHz, $CDCl_3$) δ: 7.86 (dd, J = 8.2, 2.6 Hz, 1H), 7.76 (d, J = 7.9 Hz, 1H), 7.72–7.66 (m, 2H), 7.58 (d, J = 7.9 Hz, 1H), 7.45–7.38 (m, 2H), 7.08–7.03 (m, 2H), 6.91–6.86 (m, 2H), 4.72 (q, J = 6.8 Hz, 1H), 4.22 (q, J = 7.1 Hz, 2H), 1.63 (d, J = 6.8 Hz, 3H), 1.26 (t, J = 7.1 Hz, 3H); ^{13}C NMR (125 MHz, $CDCl_3$) δ: 171.9, 161.9 (d, J = 3.2 Hz), 160.1 (d, J = 246.6 Hz), 155.5, 150.9, 145.6, 143.2, 135.5, 132.2 (q, J = 33.1 Hz), 131.8, 130.2, 128.5 (d, J = 7.8 Hz), 126.1 (q, J = 3.2 Hz), 125.5 (q, J = 3.6 Hz), 123.5 (d, J = 23.9 Hz), 122.4, 120.3 (d, J = 8.4 Hz), 115.8, 112.4 (d, J = 23.9 Hz), 73.1, 61.4, 18.6, 14.2; HRMS, m/z calcd. for $C_{26}H_{21}F_4N_2O_5^+$ [M + H]$^+$ 517.1381, found 517.1387.

(R)-ethyl 2-(4-((6-fluoro-4-oxo-3-(4-(trifluoromethyl)phenyl)-3,4-dihydroquinazolin-2-yl)oxy)phenoxy)propanoate (**QPP-29**): white solid, yield 77.0%, m.p. 146–147 °C; ^{1}H NMR (500 MHz, $CDCl_3$) δ: 7.86 (dd, J = 8.2, 2.6 Hz, 1H), 7.82 (d, J = 8.4 Hz, 2H), 7.52 (d, J = 8.2 Hz, 2H), 7.46–7.37 (m, 2H), 7.07–7.02 (m, 2H), 6.91–6.85 (m, 2H), 4.72 (q, J = 6.8 Hz, 1H), 4.22 (qd, J = 7.1, 0.9 Hz, 2H), 1.63 (d, J = 6.8 Hz, 3H), 1.26 (t, J = 7.1 Hz, 3H); ^{13}C NMR (125 MHz, $CDCl_3$) δ: 171.9, 161.9 (d, J = 3.4 Hz), 160.1 (d, J = 246.8 Hz), 155.5, 150.9, 145.6, 143.2, 138.1, 131.4 (q, J = 32.9 Hz), 128.9, 128.5 (d, J = 7.9 Hz), 126.7 (q, J = 7.3 Hz), 123.5 (d, J = 23.9 Hz), 122.4, 120.3 (d, J = 8.3 Hz), 112.4 (d, J = 23.8 Hz), 73.1, 61.4, 18.6, 14.2; HRMS, m/z calcd. for $C_{26}H_{21}F_4N_2O_5^+$ [M + H]$^+$ 517.1381, found 517.1387.

2.3. X-ray Diffraction Analysis of Target Compound QPP-7

The single crystal of **QPP-7** was slowly cultivated from a mixture of dichloromethane and ethanol (1/1 by volume). Crystallographic data of compound **QPP-7** had been deposited with the Cambridge Crystallographic Data Centre as supplementary publications with the deposition number 2183936. The detail data can be acquired free of charge from http://www.ccdc.cam.ac.uk/ (15 July 2022).

2.4. Evaluation of Herbicidal Activity

Herbicidal activity was evaluated based on the reported methods [34]. Commercial herbicide quizalofop-P-ethyl (QZ) was selected as the positive control. The preliminary in vitro herbicidal activity of target compounds **QPP-1** to **QPP-15** was determined with *Brassica campestris* root test and *Echinochloa crusgalli* cup test at a dosage of 10 μg/mL. Further herbicidal activity of target compounds **QPP-1** to **QPP-29** against two dicotyledonous species *Brassica campestris* and *Amaranthus retroflexus*, and two monocotyledonous species *Echinochloa crusgalli* and *Digitaria sanguinalis* was tested in a greenhouse. Briefly, the target compounds were dissolved in 100 μL of *N*, *N*-dimethylformamide with the addition of a little Tween 80 and then were sprayed using a laboratory belt sprayer delivering a 750 L/ha spray volume. The dosage (activity ingredient) for each compound corresponded to 1500 g/ha. Compounds were sprayed immediately after seed planting (preemergence treatment) or after the expansion of the first true leaf (postemergence treatment). The mixture of same amount of water, *N, N*-dimethylformamide, and Tween 80 was sprayed as the control. The fresh weight of the above ground tissues was measured 14 days after treatment. The inhibition percent was used to describe the control efficiency of the compounds. The data represented the percent displaying herbicidal damage as compared to the control, where complete control of the target is 100 and no control is 0. Compounds **QPP-3**, **QPP-7**, **QPP-11** were selected to study the herbicidal activity at 750 g/ha, 375 g/ha and 187.5 g/ha, respectively. Compound **QPP-7** was selected to study the herbicidal spectrum. All of the bioassays were tested for three parallel experiments. Details of the experimental procedure is given in the Supplementary Materials section.

2.5. Crop Selectivity

Based on the reported methods, [34] the crop selectivity of target compound **QPP-7** was evaluated with three replicates per treatment. Six representative crops, namely, *Oryza sativa*, *Zea mays*, *Triticum aestivum*, *Gossypium spp*, *Glycine max* and *Arachis hypogaea*, were selected for crop selectivity studies in the greenhouse. The procedure is given in the Supplementary Materials section.

2.6. Molecular Docking Study

As we known, the APP herbicides as predrug, played major role in plants through hydrolysis into acids [39]. Therefore, the free acid of compound **QPP-7** and QZ were dock into the binding site of ACCase by using the molecular docking module (CDOCKER) in the DS software (Discovery Studio 2020, Dassault Systemes, France). The crystal structure of ACCase enzyme was extracted from crystal structure PDB code 1UYR, and H_2O and all binding ligands were deleted. The structures of small molecules were optimized with the ligand minimization protocol. The free acid of compound **QPP-7** and QZ were set near the original diclofop-methyl binding site (amino acid residues from 1596 to 2025) in ACCase, respectively. The combined spherical area were (x = 26.6004, y = 40.6731, z = 77.6454, Radius = 10). After calculations with the parameters set as default values, the generated conformations were clustered together and ranked by the lowest docking energy, and a cluster analysis was performed.

2.7. ACCase Extraction and Inhibition Activity Assay

When *E. crusgalli* was grown to the 3-leaf stage in a greenhouse, shoots were cut at the base and stored at −80 °C. ACCase was extracted and partially purified using the method described by Cocker et al. [40]. The ACCase enzyme inhibition assay was performed by Shanghai Enzyme-linked Biotechnology Co., Ltd. (Shanghai, China). The ACCase enzyme was treated with the inhibitors **QPP-7** and QZ, and the enzyme activity was measured using the method of enzyme-linked immune-sorbent assay (ELISA) according to the manufacturer's instructions (Shanghai Enzyme-linked Biotechnology Co., Ltd., Shanghai, China). The inhibitor concentrations ranged from 6.25 to 100 nM, and each experiment was repeated at least three times. The absorbance [optical density (OD) value] was determined at 450 nm and used to calculate the half maximal inhibitory concentration (IC_{50}) values.

3. Results and Discussion

3.1. Synthetic Chemistry

As depicted in Scheme 1, the target compounds **QPP-1** to **QPP-15** were prepared via a three-step synthetic route using methyl isothiocyanate and several anthranilic acids as the starting material (**Route A**); the target compounds **QPP-16** to **QPP-29** were prepared via a three-step synthetic route using 2-amino-5-fluorobenzoic acid and several isothiocyanates as the starting material (**Route B**). Briefly, anthranilic acids **1a–1o** reacted with isothiocyanates in ethanol by using triethylamine as a base to provide intermediates **2a–2o** and **5a–5n**, which was then reacted with sulfuryl chloride in trichloromethane to provide intermediates **3a–3o** and **6a–6n**. Finally, intermediates **3a–3o** and **6a–6n** reacted with commercial (R)-2-(4-hydroxyphenoxy)propionic acid in acetonitrile using potassium carbonate as a base to provide target compounds **QPP-1** to **QPP-29** in 68.0% to 96.9% yields. The structures of all the target compounds were confirmed by ^{1}H NMR, ^{13}C NMR, and HRMS.

Furthermore, to demonstrate that the reaction conditions have little effect on the chiral carbon configuration of the target compounds, the enantiomeric excesses (*ee*) values of the target compound **QPP-7** and intermediate (R)-2-(4-hydroxyphenoxy)propionic acid were tested. The enantiomeric excesses were determined by HPLC analysis over a chiral column (Daicel Chiralcel OD-H, eluted with hexane-isopropyl alcohol; monitored by UV detector). The results showed that the *ee* values of (R)-2-(4-hydroxyphenoxy)propionic acid and **QPP-7** are 100% and 95.8%, respectively. In addition, the configuration of compound **QPP-7** was confirmed using the X-ray diffraction analysis (CCDC 2183936, Figure 2).

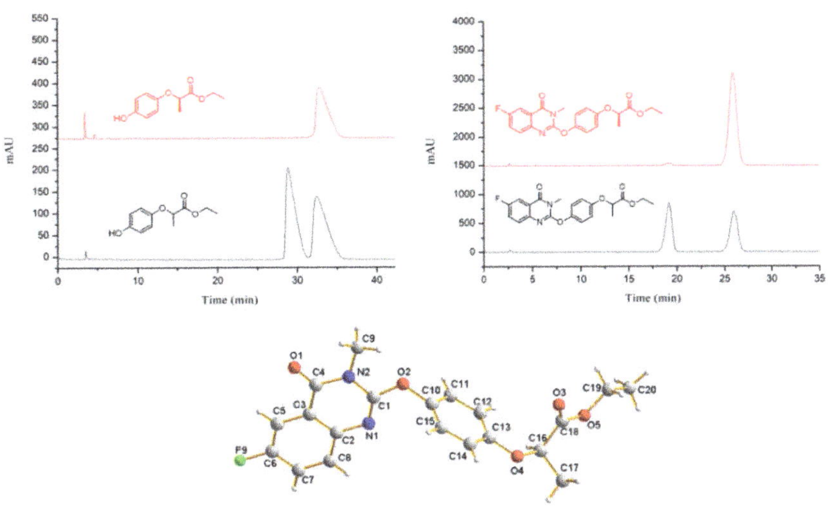

Figure 2. *ee* value testing and X-ray crystal structure of target compound **QPP-7**.

3.2. *In Vitro Herbicidal Activity of Target Compounds **QPP-1** to **QPP-15***

The in vitro herbicidal activities of the target compounds **QPP-1** to **QPP-15** were preliminarily determined by the *B. campestris* root test and *E. crusgalli* cup test at a dosage of 10 μg/mL. Commercial herbicide quizalofop-P-ethyl was selected as the positive control sample. As shown in Figure 3, some of the target compounds, such as **QPP-1** to **QPP-3**, **QPP-7** to **QPP-9**, **QPP-11**, **QPP-12**, and **QPP-15**, exhibited good herbicidal activity against the monocotyledonous plant *E. crusgalli* with >50% inhibition. Among them, compound **QPP-7** exhibited excellent herbicidal activity against *E. crusgalli* with 100% inhibition, which is equal to commercial herbicide quizalofop-P-ethyl. However, the target compounds have no inhibition against the dicotyledonous plant *B. campestris*. This finding indicated that the target compounds have stronger herbicidal activity against monocotyledonous plant.

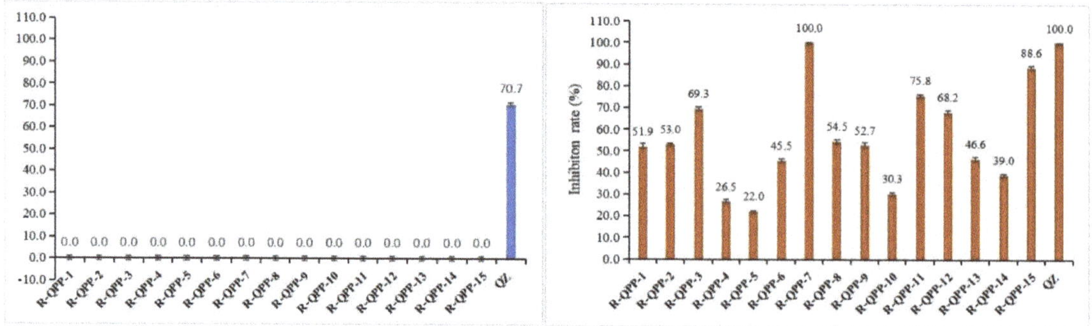

Figure 3. In vitro herbicidal activity of target compounds **QPP-1** to **QPP-15** at a dosage of 10 μg/mL; *B. campestris* root test (**Left**), *E. crusgalli* cup test (**Right**).

3.3. Herbicidal Activity of Target Compounds QPP-1 to QPP-29 in Greenhouse Tests and SAR Study

Based on the above preliminary bioassay results, the herbicidal activity of target compounds **QPP-1** to **QPP-15** was further tested on four species that were representative of monocotyledonous and dicotyledonous plants at a dosage of 1500 g ha^{-1} located in a greenhouse. As shown in Figure 4, in most cases, the target compounds displayed stronger herbicidal activity against monocotyledonous plants than dicotyledonous plants. Moreover, it was found that most of the target compounds have stronger pre-emergent herbicidal activity than post-emergent herbicidal activity. For example, compounds **QPP-1**, **QPP-3**, **QPP-7**, and **QPP-11** exhibited good herbicidal activity against all the weeds tested with sum inhibition 167.7%, 215.7%, 259.4%, and 225.8% under pre-emergence conditions, respectively, while these compounds have lower sum inhibition (75.9%, 60.0%, 208.4%, 107.2%, respectively) under post-emergence conditions.

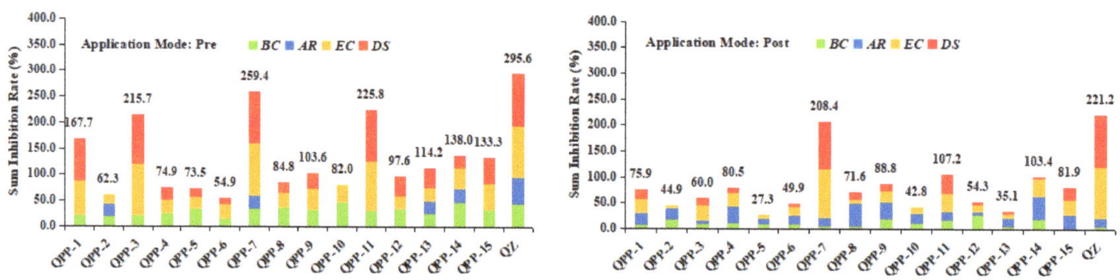

Figure 4. Effects (% inhibition) of compounds **QPP-1** to **QPP-15** on the loss of plant weight at a dosage of 1500 g ha^{-1} in greenhouse testing; pre: pre-emergence; post: post-emergence; BC: *B. campestris*; AR: *A. retroflexus*; EC: *E. crusgalli*; DS: *D. sanguinalis*.

Analyzing the herbicidal activity of **QPP-1** to **QPP-15** under pre-emergence conditions, it was found that R group on the benzene ring of quinazolin-4-one has significant influence on the herbicidal activity. Generally, when a single substituent was introduced at the 6-position on benzene ring of quinazolin-4-one, the herbicidal activity of target compounds was improved. For example, compounds **QPP-3** (R = 6-Me, sum inhibition = 215.7%), **QPP-7** (R = 6-F, sum inhibition = 259.4%), and **QPP-11** (R = 6-Cl, sum inhibition = 225.8%) exhibited stronger herbicidal activity than that of compound **QPP-1** (R = H, sum inhibition = 167.7%). Simultaneously, the herbicidal activity was enhanced with increasing the electron-withdrawing ability of the substituent at the 6-position, i.e., R = 6-F (**QPP-7**) > 6-Cl (**QPP-11**) > 6-Me (**QPP-3**). When a single substituent (regardless of whether electron-withdrawing group or electron-donating group) was introduced at the 5-position (i.e., **QPP-2**, **QPP-6**, **QPP-10**), or 7-position (i.e., **QPP-4**, **QPP-8**, **QPP-12**), or 8-position (i.e., **QPP-5**, **QPP-9**, **QPP-13**) on the benzene ring of quinazolin-4-one, the herbicidal activity of target compound was decreased sharply. In addition, when a substituent was introduced at the 7-position of **QPP-3** and **QPP-7**, the corresponding disubstituted compounds **QPP-14** and **QPP-15** showed lower herbicidal activity than that of **QPP-3** and **QPP-7**. These results suggested that the spatial position of R group on the benzene ring of quinazolin-4-one has more important influence on the herbicidal activity than that of electronic effect and introducing a single electron-withdrawing group at the 6-position on benzene ring of quinazolin-4-one would be essential for improving herbicidal activity.

As a result of the higher sum inhibition against all the weeds tested, compounds **QPP-3**, **QPP-7**, and **QPP-11** were chosen for further testing at lower doses. As shown in Table 1, these compounds displayed stronger herbicidal activity against monocotyledonous plants than dicotyledonous plants. Moreover, upon decreasing the dosage, the herbicidal activity of these compounds under post-emergence conditions decreased faster than that observed under pre-emergence conditions. These findings confirmed that the target compounds have better selective to monocotyledonous plants and exhibit stronger herbicidal activity under pre-emergence conditions than under post-emergence conditions. It was found that compounds **QPP-3**, **QPP-7**, and **QPP-11** exhibit good herbicidal activity against monocotyledonous plants under pre-emergence conditions at a dosage of 750 g ha^{-1}. Unfortunately, these compounds have lower herbicidal activity against monocotyledonous plants than that of QZ when the dosage reduced to 187.5 g ha^{-1}. Nevertheless, to our relief, compound **QPP-7** still exhibited excellent pre-emergent herbicidal activities against *E. crusgalli* and *D. sanguinalis* with inhibition 96.7% and 100% at the dosage of 375 g ha^{-1}, respectively, which are almost equal to QZ (Figure 5). This promising result indicate that compound **QPP-7** may serve as a potential lead compound for further optimization.

Table 1. Effects (inhibition/%) of compounds **QPP-3**, **QPP-7**, and **QPP-11** on loss of plant weight at lower dosage in greenhouse testing [a].

Comp.	Rate (g ha^{-1})	*B. campestris*		*A. retroflexus*		*E. crusgalli*		*D. sanguinalis*	
		Pre	Post	Pre	Post	Pre	Post	Pre	Post
QPP-3	750	8.3 ± 1.7	0	0	0	94.5 ± 1.3	24.7 ± 1.6	88.9 ± 0.2	1.9 ± 0.4
	375	0	0	0	0	60.4 ± 1.4	5.0 ± 1.0	76.7 ± 1.3	0
	187.5	0	0	0	0	45.0 ± 2.1	0	48.9 ± 1.8	0
QPP-7	750	7.8 ± 0.9	0	4.8 ± 0.2	0	100	42.1 ± 0.6	100	61.6 ± 1.9
	375	0	0	0	0	96.7 ± 1.6	24.0 ± 0.6	100	21.3 ± 1.2
	187.5	0	0	0	0	64.8 ± 3.1	0	55.6 ± 1.4	0
QPP-11	750	4.2 ± 0.3	6.6 ± 0.4	0	0	91.2 ± 1.5	15.7 ± 1.3	84.4 ± 0.8	13.2 ± 0.4
	375	0	0	0	0	75.8 ± 0.4	10.2 ± 1.0	72.2 ± 1.1	6.6 ± 0.1
	187.5	0	0	0	0	60.4 ± 0.8	0	53.3 ± 2.0	0
QZ	750	38.6 ± 2.3	7.0 ± 1.1	31.2 ± 0.7	0	100	100	100	100
	375	24.5 ± 1.7	0	20.6 ± 1.3	0	100	100	100	100
	187.5	11.1 ± 0.6	0	10.5 ± 1.7	0	97.8 ± 1.7	100	100	100

[a] Each value represents the mean ± SD of three experiments.

In order to explore the effect of R_1 group on herbicidal activity, subsequently, the target compounds **QPP-16** to **QPP-29** were synthesized by using **QPP-7** as lead compound, and their herbicidal activity was evaluated under pre-emergence conditions. Lead compound **QPP-7** and QZ were selected as the positive control samples. As shown in Figure 6, the target compounds **QPP-16** to **QPP-29** showed lower sum inhibition against all the weeds tested at a dosage of 1500 g ha^{-1} than that of lead compound **QPP-7** and QZ. Meanwhile, it was found that the decreased sum inhibition was mainly attributed to the reduced inhibitory effect of the compound on monocotyledonous plants. Analyzing the herbicidal activity of **QPP-7** and **QPP-16** to **QPP-18**, it was found that, with the extension of the carbon chain, compounds with longer carbon chain progressively lost herbicidal activity. When a branched-chain alkyl groups was introduced on the nitrogen atom, **QPP-19** showed lower herbicidal activity than that of the corresponding straight-chain compound **QPP-18**. With the alkyl group at the nitrogen atom of quinazolin-4(*H*)-one replaced by a benzene ring, the herbicidal activity of **QPP-20** did not improve. Although the introduction of substituent on the benzene ring at the nitrogen atom increased the herbicidal activity compared to **QPP-20**, the herbicidal activity of **QPP-21** to **QPP-29** was still lower than that of lead compound **QPP-7**. These results suggested that the introduction of a substituent with bulker than the methyl at the nitrogen atom did not conducive to improving the herbicidal activity.

Figure 5. Photographs illustrating the herbicidal activity of **QPP-7** under pre-emergence conditions at a dosage of 375 g ha^{-1}; From left to right: *B. campestris*, *A. retroflexus*, *E. crusgalli*, *D. sanguinalis*.

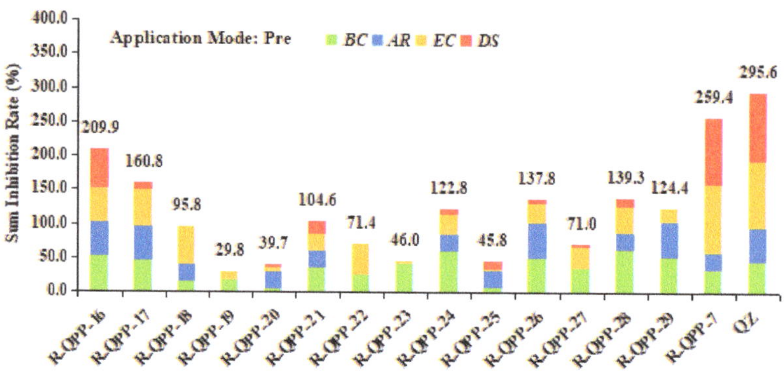

Figure 6. Effects (% inhibition) of compounds **QPP-16** to **QPP-29** on the loss of plant weight at a dosage of 1500 g ha^{-1} in greenhouse testing; pre: pre-emergence; BC: *B. campestris*; AR: *A. retroflexus*; EC: *E. crusgalli*; DS: *D. sanguinalis*.

The aforementioned results for structure-activity relationship revealed that the herbicidal activity of target compounds is strongly influenced by the spatial position of R group and the bulk of R_1 group on quinazolin-4(*H*)-one, and the (R = 6-F, R_1 = Me) pattern was confirmed as the optimal orientation.

3.4. Herbicidal Spectrum and Crop Safety of Compound *QPP-7*

To further evaluate whether compound **QPP-7** has the potential to be developed as a herbicide, its herbicidal spectrum against monocotyledonous plants and crop safety were investigated at a dosage of 375 g ha^{-1} under pre-emergence conditions. The monocotyledonous plants *Echinochloa crusgalli* (EC), *Digitaria sanguinalis* (DS), *Pennisetum alopecuroides* (PA), *Setaria viridis* (SV), *Eleusine indica* (EI), *Avena fatua* (AF), *Elymus dahuricu* (ED), *Spartina alterniflora* (SA) were chosen as the target weeds to evaluate the herbicidal spectrum of compound

QPP-7 in a greenhouse. As shown in Figure 7, **QPP-7** displays strong control with inhibition >90% against all the weeds tested, which is almost equal to QZ. This finding indicated that compound **QPP-7** has a broad herbicidal spectrum for monocotyledonous weeds control. Subsequently, six representative crops, *Oryza sativa*, *Zea mays*, *Triticum aestivum*, *Gossypium spp*, *Glycine max* and *Arachis hypogaea*, were selected for further crop selectivity study (Table 2). The results showed that *O. sativa*, *T. aestivum*, and *A. hypogaea* displayed a high tolerance toward compound **QPP-7**, while QZ was not selective for *O. sativa* and *T. aestivum* (98.7% and 59.3% injury, respectively). These promising results indicated that compound **QPP-7** has the potential to be developed as a pre-emergence herbicide lead compound for weed control in *O. sativa*, *T. aestivum*, and *A. hypogaea* Fields.

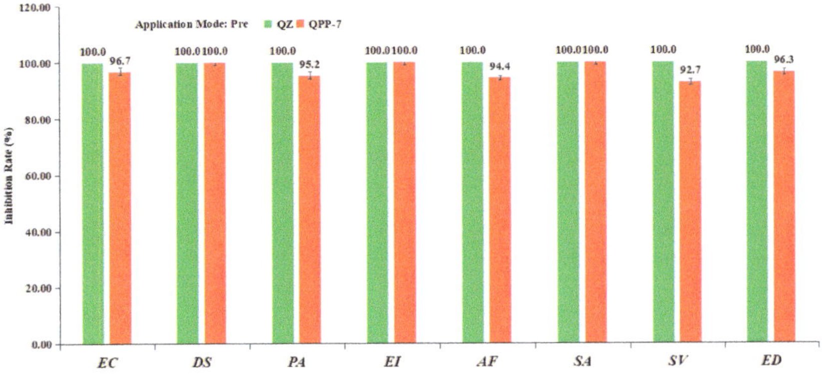

Figure 7. Herbicidal spectrum testing of compound **QPP-7** under pre-emergence conditions at a dosage of 375 g ha^{-1}; *Echinochloa crusgalli* (EC), *Digitaria sanguinalis* (DS), *Pennisetum alopecuroides* (PA), *Setaria viridis* (SV), *Eleusine indica* (EI), *Avena fatua* (AF), *Elymus dahuricu* (ED), *Spartina alterniflora* (SA).

Table 2. Pre-emergence crop selectivity of compound **QPP-7** at the dosage of 375 g ha^{-1} (Injury Inhibition) [a].

Comp.	% Injury					
	O. sativa	*Z. mays*	*T. aestivum*	*G. spp*	*G. max*	*A. hypogaea*
QPP-7	0	55.5 ± 1.4	0	10.9 ± 1.1	4.5 ± 0.4	0
QZ	98.7 ± 0.9	16.8 ± 1.7	59.3 ± 1.2	0	0	0

[a] Each value represents the mean ± SD of three experiments.

3.5. Molecular Mode of Action of the Target Compound ***QPP-7***

In order to explore the molecular mode of action of target compounds, compound **QPP-7** was selected to study the herbicidal mechanism. Since the target compounds were designed based on the APP motif, we speculated that **QPP-7** could be a ACCase inhibitor. Thus, the molecular docking simulations were first carried out. As shown in Figure 8, it was easy to find that the relevant interactions between compound **QPP-7** and the target enzyme is different to that of quizalofop-P-ethyl. In the docking complex of compound **QPP-7**, the carbonyl oxygen atom on the carboxyl group formed two non-classical hydrogen bonds with the amino acid residues of GLY1971; the carbonyl oxygen atom on the quinazolin-4(3*H*)-one formed a non-classical hydrogen bonds with the amino acid residues of GLY1997. Meanwhile, the benzene ring on the phenoxypropionic acid inserted into the active site and generated a π–π interaction with PHE1956, and the hydroxyl group generated a p–π interaction with TRP1924. In the docking complex of QZ, the hydroxyl group formed a classical hydrogen bond and two non-classical hydrogen bonds with the amino acid residues of ALA1627, GLY1626, GLY1734, respectively. Meanwhile, the benzene ring of quinoxaline formed two π–π interaction with TYR1738, and the nitrogen atom formed two non-classical hydrogen bonds with GLY1998. From the above docking results alone, it is difficult to conclude if it is **QPP-7** or QZ has a more prominent inhibitory activity against ACCase. Therefore, ACCase activity test is necessary to carry out to help us make a accurate judgment.

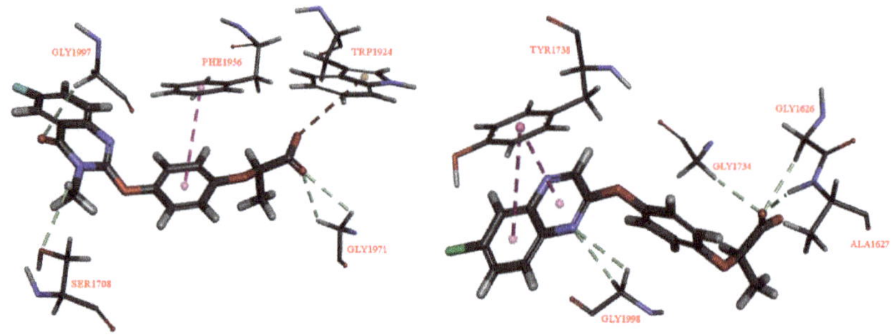

Figure 8. The docking binding mode of the free acid of **QPP-7** (**Left**) and quizalofop-P-ethyl (**Right**) to acetyl-CoA carboxylase (ACCase) (PDB code: 1UYR).

To further verify whether compound **QPP-7** is a ACCase inhibitor, the tests of *E. crusgalli* ACCase inhibition activity in vitro was performed. As shown in Table 3, compound **QPP-7** displayed good inhibitory activity against *E. crusgalli* ACCase with an IC_{50} value of 54.65 nM, which is comparable to commercial herbicide quizalofop-P-ethyl (IC_{50}= 41.19 nM). This result indicate that **QPP-7** may be an ACCase inhibitor and has an herbicidal mechanism similar to that of QZ.

Table 3. In vitro inhibitory activity of compound **QPP-7** against *E. crusgalli* ACCase [a].

Comp.	Regression Equation	IC_{50} (nM)	95% Confidence Interval	r
QPP-7	y = 0.4842x + 16.9	54.65	40.26 to 82.67	0.9751
QZ	y = 0.4449x + 24.2	41.19	27.82 to 69.67	0.9585

[a] y: ACCase inhibition rate; x: concentration of tested compounds; r: Correltion coefficient; IC_{50}: the half maximal inhibitory concentration.

It was noteworthy that although **QPP-7** has a herbicidal mechanism similar to that of QZ, however, the study of molecular docking showed that their interactions with the ACCase is different, indicating that **QPP-7** has the potential to control weeds that are resistant to APP herbicides. Furthermore, in our present work, compound **QPP-7** has the comparable ACCase inhibitory activity to that of QZ, but its herbicidal activity at a lower dosage is worse than that of QZ. The reason may be attributed to the compound **QPP-7** with natural structure fragment being easily metabolized in plants when compared to QZ.

4. Conclusions

In summary, a series of quinazolin-4(3*H*)-one derivatives based on the aryloxyphenoxypropionate motif have been designed by using molecular hybridization strategy. twenty-nine novel quinazolin-4(3*H*)-one derivatives were prepared in moderate to good yields. The bioassay results showed that compound **QPP-7** displayed good pre-emergent herbicidal activity at a dosage of 375 g ha^{-1}. The herbicidal spectrum and crop selectivity study revealed that compound **QPP-7** had a broad spectrum of monocotyledonous weed control and displayed excellent crop safety to *O. sativa*, *T. aestivum*, and *A. hypogaea*, which indicated its great potential as a herbicide lead compound. The study of structure-activity relationship showed that the spatial position of R group and the bulk of R_1 group on quinazolin-4-one have strongly influenced on the herbicidal activity of target compounds, and the (R = 6-F, R_1 = Me) pattern was confirmed as the optimal orientation. Furthermore, the inhibitory activity against *E. crusgalli* ACCase enzyme and the molecular docking simulation of the free acid of compound **QPP-7** were performed, and the results indicated that compound **QPP-7** may be a ACCase inhibitor. For developing improved herbicidal activity of APP herbicide containing quinazolin-4(3*H*)-one skeleton, further studies on the structural optimization of compound **QPP-7** are ongoing in our laboratory.

Supplementary Materials: The following supporting information can be downloaded at: https://www.mdpi.com/article/10.3390/agronomy12081840/s1, Details on the inhibition of the root growth of *B. campestris*, inhibition of the seedling growth of *E. crusgalli*, greenhouse tests, and crop selectivity, ^{1}H NMR, ^{13}C NMR and HRMS spectrum of target compounds.

Author Contributions: Conceptualization, K.L.; methodology, K.L.; validation, C.W., K.C. and N.L.; formal analysis, S.F. and P.L.; investigation, C.W., K.C. and N.L.; data curation, C.W. and K.C.; writing—original draft preparation, K.L.; writing—review and editing, K.L. and L.J.; project administration, K.L.; funding acquisition, K.L., X.W., G.L. and L.J. All authors have read and agreed to the published version of the manuscript.

Funding: This project was financially supported by the China Postdoctoral Science Foundation (No. 2020M671984); the National Natural Science Foundation of China (No. 31701827); the National Innovation and Entrepreneurship Training Program for College Students (No. 202110447022); Guangyue Young Scholar Innovation Team of Liaocheng University (No. LCUGYTD2022-04); the National Key Research and Development Program of China (No. SQ2020YFF0422322); the Natural Science Foundation of Shandong Province (No. ZR202102180037) and the Research Fund of Liaocheng University (No. 318012106).

Institutional Review Board Statement: Not applicable.

Informed Consent Statement: Not applicable.

Data Availability Statement: Not applicable.

Conflicts of Interest: The authors declare no conflict of interest.

References

1. Tong, L. Acetyl-coenzyme A carboxylase: Crucial metabolic enzyme and attractive target for drug discovery. *Cell. Mol. Life Sci.* **2005**, *62*, 1784–1803. [CrossRef] [PubMed]
2. Zagnitko, O.; Jelenska, J.; Tevzadze, G.; Haselkorn, R.; Gornicki, P. An isoleuciney leucine residue in the carboxyltransferase domain of acetyl-CoA carboxylase is critical for interaction with aryloxyphenoxypropionate and cyclohexanedione inhibitors. *Proc. Natl. Acad. Sci. USA* **2001**, *98*, 6617–6622. [CrossRef]
3. Zhang, H.L.; Yang, Z.R.; Shen, Y.; Tong, L. Crystal Structure of the Carboxyltransferase Domain of Acetyl-Coenzyme A Carboxylase. *Science* **2003**, *299*, 2064–2067. [CrossRef] [PubMed]
4. Délye, C. Weed resistance to acetyl coenzyme A carboxylase inhibitors: An update. *Weed Sci.* **2005**, *53*, 728–746. [CrossRef]
5. Leach, G.E.; Devine, M.D.; Kirkwood, R.C.; Marshall, G. Target enzyme-based resistance to acetyl-coenzyme A carboxylase inhibitors in *Eleusine indica*. *Pestic. Biochem. Physiol.* **1995**, *51*, 129–136. [CrossRef]
6. Stoltenberg, D.E.; Wiederholt, R.J. Giant foxtail (*Setaria faberi*) resistance to aryloxyphenoxypropionate and cyclohexanedione herbicides. *Weed Res.* **1995**, *43*, 527–535. [CrossRef]
7. Heap, I.M.; Morrison, I.N. Resistance to aryloxyphenoxypropionate and cyclohexanedione herbicides in green foxtail (*Setaria viridis*). *Weed Sci.* **1996**, *44*, 25–30. [CrossRef]
8. Preston, C.; Tardif, F.J.; Christoffer, J.T.; Powles, S.B. Multiple resistance to dissimilar herbicide chemistries in a biotype of *Lolium rigidum* due to enhanced activity of several herbicide degrading enzymes. *Pestic. Biochem. Physiol.* **1996**, *54*, 123–134. [CrossRef]
9. Liu, W.; Harrison, D.K.; Chalupska, D.; Gornicki, P.; O'Donnell, C.C.; Adkins, S.W.; Haselkorn, R.; Williams, R.R. Single-site mutations in the carboxyltransferase domain of plastid acetyl-CoA carboxylase confer resistance to grass-specific herbicides. *Proc. Natl. Acad. Sci. USA* **2007**, *104*, 3627–3632. [CrossRef]
10. Zhang, X.Q.; Powles, S.B. Six amino acid substitutions in the carboxyltransferase domain of the plastidic acetyl-CoA carboxylase gene are linked with resistance to herbicides in a *Lolium rigidum* population. *New Phytol.* **2006**, *172*, 636–645. [CrossRef]
11. Sparks, T.C.; Duke, S.O. Structure simplification of natural products as a lead generation approach in agrochemical discovery. *J. Agric. Food Chem.* **2021**, *69*, 8324–8346. [CrossRef] [PubMed]
12. Newman, D.J.; Cragg, G.M. Natural products as sources of new drugs over the nearly four decades from 01/1981 to 09/2019. *J. Nat. Prod.* **2020**, *83*, 770–803. [CrossRef] [PubMed]
13. Sparks, T.C.; Hahn, D.R.; Garizi, N.V. Natural products, their derivatives, mimics and synthetic equivalents: Role in agrochemical discovery. *Pest Manag. Sci.* **2017**, *73*, 700–715. [CrossRef] [PubMed]
14. Gerwick, B.C.; Sparks, T.C. Natural products for pest control: An analysis of their role, value and future. *Pest Manag. Sci.* **2014**, *70*, 1169–1185. [CrossRef]

15. Ahrens, H.; Lange, G.; Müller, T.; Rosinger, C.; Willms, L.; van Almsick, V. 4-Hydroxyphenylpyruvate dioxygenase inhibitors in combination with safeners: Solutions for modern and sustainable agriculture. *Angew. Chem. Int. Ed.* **2013**, *52*, 9388–9398. [CrossRef]
16. Marrone, P.G. Pesticidal natural products-status and future potential. *Pest Manag. Sci.* **2019**, *75*, 2325–2340. [CrossRef]
17. He, D.; Wang, M.L.; Zhao, S.Y.; Shu, Y.S.; Zeng, H.L.; Xiao, C.; Lu, C.; Liu, Y.Y. Pharmaceutical prospects of naturally occurring quinazolinone and its derivatives. *Fitoterapia* **2017**, *119*, 136–149. [CrossRef]
18. Xu, Z.; Zhang, Y.; Fu, H.; Zhong, H.; Hong, K.; Zhu, W. Antifungal quinazolinones from marine-derived *Bacillus cereus* and their preparation. *Bioorg. Med. Chem. Lett.* **2011**, *21*, 4005–4007. [CrossRef]
19. Zhao, N.; Li, Z.L.; Li, D.H.; Sun, Y.T.; Shan, D.T.; Bai, J.; Pei, Y.H.; Jing, Y.K.; Hua, H.M. Quinolone and indole alkaloids from the fruits of *Euodia rutaecarpa* and their cytotoxicity against two human cancer cell lines. *Phytochemistry* **2015**, *109*, 133–139. [CrossRef]
20. Ma, C.; Li, Y.; Niu, S.; Zhang, H.; Liu, X.; Che, Y. *N*-Hydroxypyridones, phenylhydrazones, and a quinazolinone from *Isaria farinosa*. *J. Nat. Prod.* **2011**, *74*, 32–37. [CrossRef] [PubMed]
21. Peng, J.; Lin, T.; Wang, W.; Xin, Z.; Zhu, T.; Gu, Q.; Li, D. Antiviral alkaloids produced by the mangrove-derived fungus *Cladosporium* sp. PJX-41. *J. Nat. Prod.* **2013**, *76*, 1133–1140. [CrossRef] [PubMed]
22. Cui, C.M.; Li, X.M.; Li, C.S.; Sun, H.F.; Gao, S.S.; Wang, B.G. Benzodiazepine alkaloids from marine-derived endophytic fungus *Aspergillus ochraceus*. *Helv. Chim. Acta* **2009**, *92*, 1366–1370. [CrossRef]
23. Zhuang, Y.; Teng, X.; Wang, Y.; Liu, P.; Li, G.; Zhu, W. New quinazolinone alkaloids within rare amino acid residue from coralassociated fungus, *Aspergillus versicolor* LCJ-5-4. *Org. Lett.* **2011**, *13*, 1130–1133. [CrossRef] [PubMed]
24. Cagir, A.; Jones, S.H.; Gao, R.; Eisenhauer, B.M.; Hecht, S.M. Luotonin A. A naturally occurring human DNA topoisomerase I poison. *J. Am. Chem. Soc.* **2003**, *125*, 13628–13629. [CrossRef] [PubMed]
25. Molina, P.; Tárraga, A.; Gonzalez-Tejero, A.; Rioja, I.; Ubeda, A.; Terencio, M.C.; Alcaraz, M.J. Inhibition of leukocyte functions by the alkaloid isaindigotone from *Isatis indigotica* and some new synthetic derivatives. *J. Nat. Prod.* **2001**, *64*, 1297–1300. [CrossRef] [PubMed]
26. Sen, D.; Chatterjee, T.K. Pharmacophore modeling and 3D quantitative structureactivity relationship analysis of febrifugine analogues as potent antimalarial agent. *J. Adv. Pharm. Technol. Res.* **2013**, *4*, 50–60. [CrossRef]
27. Kshirsagar, U.A. Recent developments in the chemistry of quinazolinone alkaloids. *Org. Biomol. Chem.* **2015**, *13*, 9336–9352. [CrossRef]
28. Gatadi, S.; Lakshmi, T.V.; Nanduri, S. 4(3*H*)-Quinazolinone derivatives: Promising antibacterial drug leads. *Eur. J. Med. Chem.* **2019**, *170*, 157–172. [CrossRef]
29. Amrutkar, R.D.; Amrutkar, S.V.; Ranawat, M.S. Quinazolin-4-one: A varsatile molecule. *Curr. Bioact. Compd.* **2020**, *16*, 370–382. [CrossRef]
30. Xing, Z.M.; Wu, W.H.; Miao, Y.X.; Tang, Y.Q.; Zhou, Y.K.; Zheng, L.F.; Fu, Y.; Song, Z.B.; Peng, Y.Y. Recent advances in quinazolinones as an emerging molecular platform for luminescent materials and bioimaging. *Org. Chem. Front.* **2021**, *8*, 1867–1889. [CrossRef]
31. Fan, Z.J.; Shi, J.; Luo, N.; Ding, M.H.; Bao, X.P. Synthesis, crystal structure, and agricultural antimicrobial evaluation of novel quinazoline thioether derivatives incorporating the 1,2,4-triazolo[4,3-a]pyridine moiety. *J. Agric. Food Chem.* **2019**, *67*, 11598–11606. [CrossRef]
32. Peng, J.W.; Yin, X.D.; Li, H.; Ma, K.Y.; Zhang, Z.J.; Zhou, R.; Wang, Y.L.; Hu, G.F.; Liu, Y.Q. Design, synthesis, and structure-activity relationship of quinazolinone derivatives as potential fungicides. *J. Agric. Food Chem.* **2021**, *69*, 4604–4614. [CrossRef] [PubMed]
33. Wang, X.; Li, P.; Li, Z.N.; Yin, J.; He, M.; Xue, W.; Chen, Z.W.; Song, B.A. Synthesis and bioactivity evaluation of novel arylimines containing a 3-aminoethyl-2-[(*p*-trifluoromethoxy)anilino]-4(3*H*)-quinazolinone moiety. *J. Agric. Food Chem.* **2013**, *61*, 9575–9582. [CrossRef]
34. Lei, K.; Li, P.; Yang, X.F.; Wang, S.B.; Wang, X.K.; Hua, X.W.; Sun, B.; Ji, L.S.; Xu, X.H. Design and synthesis of novel 4-hydroxyl-3-(2-phenoxyacetyl)-pyran-2-one derivatives for use as herbicides and evaluation of their mode of action. *J. Agric. Food Chem.* **2019**, *67*, 10489–10497. [CrossRef] [PubMed]
35. Lei, K.; Liu, Y.; Wang, S.B.; Sun, B.; Hua, X.W.; Xu, X.H. Synthesis and herbicidal activity of 3-Acetyl-4-hydroxy-2,1-benzothiazine derivatives. *Chem. Res. Chin. Univ.* **2019**, *35*, 609–615. [CrossRef]
36. Lei, K.; Li, P.; Zhou, X.Y.; Wang, S.B.; Wang, X.K.; Ji, L.S.; Liu, R.M.; Xu, X.H. Design, synthesis and herbicidal activity of 5-acylbarbituric acid derivatives and study of molecular mode of action. *Chin. J. Org. Chem.* **2020**, *40*, 2788–2797. [CrossRef]
37. Lei, K.; Hua, X.W.; Tao, Y.-Y.; Liu, Y.; Liu, N.; Ma, Y.; Li, Y.H.; Xu, X.H.; Kong, C.H. Discovery of (2-benzoylethen-1-ol)-containing 1,2-benzothiazine derivatives as novel 4-hydroxyphenylpyruvate dioxygenase (HPPD) inhibiting-based herbicide lead compounds. *Bioorg. Med. Chem.* **2016**, *24*, 92–103. [CrossRef] [PubMed]
38. Yan, G.; Zekarias, B.L.; Li, X.Y.; Jaffett, V.A.; Guzei, I.A.; Golden, J.E. Divergent 2-chloroquinazolin-4(3*H*)-one rearrangement: Twistedcyclic guanidine formation or ring-fused *N*-acylguanidines via a domino process. *Chem. Eur. J.* **2020**, *26*, 2486–2492. [CrossRef] [PubMed]

39. Zhang, H.; Tweel, B.; Tong, L. Molecular basis for the inhibition of the carboxyltransferase domain of acetyl-coenzyme-A carboxylase by haloxyfop and diclofop. *Proc. Natl. Acad. Sci. USA* **2004**, *101*, 5910–5915. [CrossRef]
40. Cocker, K.M.; Moss, S.R.; Coleman, J.O.D. Multiple mechanisms of resistance to fenoxaprop-P-ethyl in United Kingdom and other European populations of herbicideresistant *Alopecurus myosuroides* (black-grass). *Pestic. Biochem. Physiol.* **1999**, *65*, 169–180. [CrossRef]

MDPI AG
Grosspeteranlage 5
4052 Basel
Switzerland
Tel.: +41 61 683 77 34

MDPI Books Editorial Office
E-mail: books@mdpi.com
www.mdpi.com/books

www.ingramcontent.com/pod-product-compliance
Lightning Source LLC
Chambersburg PA
CBHW061619210926
51850CB00009B/2378
* 9 7 8 3 7 2 5 8 7 5 3 4 4 *